A System Architecture Approach to the Brain: From Neurons to Consciousness

A System Architecture Approach to the Brain: From Neurons to Consciousness

L. Andrew Coward

Nova Biomedical Books
New York

For permission to use material from this book please contact us:
Telephone 631-231-7269; Fax 631-231-8175
Web Site: http://www.novapublishers.com

LIBRARY OF CONGRESS CATALOGING-IN-PUBLICATION DATA:

A system architecture approach to the brain : from neurons to consciousness / L. Andrew Coward (editor).

p. cm.

Includes bibliographical references and index.

ISBN 1-59454-433-6

1. Brain--Computer simulation. 2. Cognition. 3. System analysis. 4. Computer architecture. I. Coward, L. Andrew.

QP376.S948 2005

612.8'2--dc22 2005011107

Published by Nova Science Publishers, Inc. ✣ New York

Contents

Preface

This book is the integrated presentation of a large body of work on understanding the operation of biological brains as systems. The work has been carried out by the author over the last 22 years, and leads to a claim that it is relatively straightforward to understand how human cognition results from and is supported by physiological processes in the brain.

This claim has roots in the technology for designing and manufacturing electronic systems which manage extremely complex telecommunications networks with high reliability, in real time and with no human intervention. Such systems perform very large numbers of interacting control features. Practical considerations including limits to information recording and processing resources and the need to add to and change the features severely constrain the architectures of such systems.

Although there is little direct resemblance between such systems and biological brains, the ways in which these practical considerations force system architectures within some specific bounds leads to an understanding of how different but analogous practical considerations constrain the architectures of brains within different bounds called the Recommendation Architecture. These architectural bounds make it possible to relate cognitive phenomena to physiological processes.

In the late 1960s the author began work for a Canadian telecommunications equipment manufacturer called Northern Electric. The company has gone through various name changes since then, but when I left Nortel Networks thirty years later in 1999 the organizational culture had a strong continuity with the company I joined at the beginning of my career. The company has a long history as a supplier of equipment to the telephone industry, but until the late 1950s had done relatively little original design. For historical reasons, Northern Electric was able to manufacture and sell equipment designs acquired from the much larger US company AT&T. However, because of antitrust developments in the US, this source of design was cut off beginning in the late 1950s, and Northern Electric was forced to create its own indigenous capabilities. The company had a strong culture of self dependence. One way in which this culture was expressed in the late 1960s was that everything required for its systems, including even screws and metal frames, was manufactured internally with only raw materials like sheet metal and wire being purchased externally. Once the source of design in the US began to be cut off, immense energy went into creating an internal design capability.

At that time, electronic technology was at the beginning of its rapid growth. Northern Electric had begun to manufacture transistors in the mid 1950s, and by the late 1960s was a significant player in silicon memory devices. Perhaps because the company was forced to take a new look at design, Northern Electric and its then parent company Bell Canada were among the first to recognize that the burgeoning software and silicon technologies would have a profound effect on the telecommunications network. In the 1960s, connecting two telephones required creation of a physical path using mechanical switches called relays. Control over setting up the appropriate path also used relays. Relays were organized into systems called central office switches (COs) which provided telephone service to customers in a geographic area or organized the connections between such areas. Slightly simpler systems called private branch exchange switches (PBXs) provided telephone service within large office buildings.

Such switches were immensely complex, but provided telephone service with an extremely high degree of reliability. Applying the new electronic technology to such systems was a considerable risk to the company. Furthermore, in the 1970s and early 1980s the hardware memory and processing resources were much more limited than today, and software design technology much more primitive. Curiously, a major factor in the market acceptance of such switches once they became available was the cost of real estate in city centres, because switches needed to be located in prime downtown office space, and electronic technology had the potential for requiring much less such space.

After an initial period in semiconductor research, and with some background in software from university, the author was assigned the responsibility for investigating how the new semiconductor and software technologies would affect the integrity of the telephone network. The particular problem was how to ensure that the quality and reliability provided by older technologies could be maintained. In order to provide an equivalent quality of service, the out-of-service time for one of the new electronic central office switches was required to be less than 2 hours in forty years. Note that out-of-service time could be caused by hardware failure or software bug, but in addition any down time needed for upgrading hardware or software or for changing the services provided to users had to be included in this 3 minutes per year. The architecture for the system therefore had to be able to support such upgrades and changes "on the fly" without service interruption.

The central office switches delivered by Nortel Networks to the telephone network beginning in the early 1980s had over 20 million lines of software code and over 4 billion transistors, but the average out-of-service time achieved once the system designs had matured was less than 30 seconds per year. The design of such systems required the coordinated efforts of thousands of engineers over periods of many years. The Nortel Networks culture of self reliance resulted in the switch hardware using internally designed custom integrated circuits, and software written in internally developed languages. The hardware and software design environments and design information management systems were also created internally. In practice, digital switches could probably not have been designed any other way at that point in technology evolution.

Within this large design community, the author had a range of assignments which in retrospect provided a particularly good set of vantage points for appreciating the system design process and the ways in which practical considerations strongly constrained system architecture. Over a 30 year period, my role included responsibilities in semiconductor technology, integrated circuit design, software development, software language and development environment development, hardware design, and system architecture and reliability.

In early 1980s, I began to wonder whether the approaches to system design and management with which I had become familiar could have any application to understanding the brain. The work began very much as a hobby in the background to my job responsibilities in telecommunications technology, but in 1984 the architecture which became the basis for all the subsequent work came together in my imagination in a café in Ottawa one evening in October. A couple of years later, my evenings and weekends were devoted to writing a book describing this architecture which was published by Praeger. However, during the working day I was responsible for providing technical support to PBX technology groups in their use complex custom designed integrated circuits, and was very pleased that I was judged to have exceeded the objectives for this "day job" almost completely unrelated to the book.

Having said that, there is no question that the intellectual environment in the technology organization of Nortel was probably the best in the world for thinking about extremely complex systems, because the company had to solve the problems with designing, manufacturing and supporting such systems on a day to day basis. I am therefore indebted to the many people in that technology organization.

After the book was published in 1990, the project was largely put on hold, although in the next few years I did implement an electronic version of the architecture. Then in 1997 I wrote a couple of conference papers and in 1998 had the opportunity to take early retirement from Nortel to focus on my brain architecture research. Just before leaving I applied for a patent on the electronic version of the architecture which was eventually approved as US patent 6363420 "Heuristically designing and managing a network".

I am indebted to Tom Gedeon for his interest and discussion of ideas, and especially for providing an institutional base for my research first at Murdoch University and then at the Australian National University.

Scott Jordan (then at Xavier University in Chicago) encouraged and edited for publication my first work on the application of my architectural ideas to understanding the origins of a range of phenomena often labeled consciousness. Gunther Knoblich (at the Planck Institute for Psychology in Munich) made an excellent suggestion that I should create descriptions of psychological phenomena at an "intermediate design" level of description within my architecture. Dave Gibson (then at Nortel) made considerable efficiency enhancements to my first electronic implementation of a system within the Recommendation Architecture bounds. Nick Swindale at the University of British Columbia asked some good questions about neuron algorithms and pointed me towards some key work on primate brain physiology. I had some excellent discussions with Jack Martin at Simon Fraser University on a number of topics including the development of self awareness in children.

The perspective created by experience within a project involving thousands of designer man-years for design of a complex real time system with minimal use of "off-the shelf" components is not widely familiar outside a narrow range of high tech design communities. This perspective is different from that gained through design of relatively "simple" systems involving less than 10 man-years effort to integrate standard computing components which is probably the practical upper limit for an academic project. The human brain must learn to perform very large numbers of different cognitive and control features with a limited pool of physiological resources, and it is plausible that the architectural constraints exerted by this requirement can be better appreciated from the complex system design perspective. This perspective is therefore the appropriate starting point for understanding the brain. That is the thesis of this book.

L. Andrew Coward
Department of Computer Science
Australian National University
Canberra
January 2005

Glossary

Certain terms are widely used in and critical to understanding of the recommendation architecture, but their meaning is somewhat different from general usage. These terms are defined in the book when they first appear, but are collected here for easy reference.

A motor action, combination of motor actions, or sequence of different combinations of motor actions. Also includes indirect activations of within , and actions on as a result of detection of conditions.

The degree to which an input into a influences the probability of that component activating is the behavioural weight of the input.

A subsystem of the which defines, permanently records and detects repetitions of .

A subsystem of the which receives inputs indicating the detections of groups of ( ) detected by , interprets the detection of a as a range of behavioural recommendations, adds the recommendations of all detected to select the strongest recommendations, and uses consequence feedback to adjust the weights which recommended the selected .

a set of system resources customized to manage the selection decision for one system or behaviour type. When a component is activated, the corresponding behaviour is performed by the system. Components are separations within .

a small set of information elements from the available to the system, each with a corresponding value. A condition occurs when all (or a significant subset) of the information set defining the condition has the values specified for the condition.

Conditions are similar if a high proportion of the information defining them is drawn from a limited information set in which information sources and values are specified and each source often has its corresponding value at times when other members of the set have their corresponding values.

the complexity of a condition is the total number of that contribute to the condition, either directly or via intermediate conditions. Conditions can be activated indirectly on the basis of simultaneous past activity, and both conditions occurring in current sensory inputs and indirectly activated conditions may become incorporated in a higher complexity condition.

A system which controls physical equipment.

The most elementary unit of information handling resource in a system.

An information process performed by a in support of system features.

A feature is a consistent way in which a system responds to a set of similar environmental circumstances. The environmental circumstances and corresponding responses are similar from the point of view of an external observer, and are a useful way for such an observer to understand the system, but may not reflect similarities in the way the system detects in its environment and generates on a more detailed level.

all the information available to a control system, derived from the past and present states of the control system itself, of the physical system being controlled, and of the environment in which the physical system operates.

The ensemble of available at one point in time.

An organization of to optimize use of system resources. Detailed perform groups of very similar . Sets of detailed with a moderate degree of similarity in performed across the set form intermediate with a moderate sharing of resources. Sets of intermediate with a lower degree of similarity across the set form higher level with a lower sharing of resources, and so on.

a set of system resources which have been customized to perform a group of . The are similar in the sense that they can all be performed efficiently by the module resources. Modules are separations within .

An operationally complex system is one in which the number of control is large relative to the available information recording and processing resources.

the interpretation by one of the detection of a by another . This interpretation is that the currently appropriate system is limited to a specific subset of the influenced by the recipient . If the confidence in this interpretation is 100%, the meaning is unambiguous, otherwise it is ambiguous.

A which controls a complex combination of physical equipment in real time with no external intervention.

a set of which are similar and programmed on one physical substrate within . The substrate is activated if one or more of its are detected and generates an output to . Activation of a portfolio therefore indicates both the detection of and a range of behavioural recommendations.

A raw sensory input is the output from a sensor detecting something about the environment. The activity of one individual input in general carries no information discriminating between different behaviourally relevant states of the environment. Only combinations of inputs can carry such information.

A set of architectural bounds within which any operationally complex system which learns will tend to be constrained. A major characteristic of the recommendation architecture bounds is separation between a modular hierarchy called which defines and detects , and a subsystem which interprets detections as behavioural recommendations.

Specialized clustering systems detect indicating the consequences of a . These may be defined a priori or heuristically to some degree. Detection of a reward modulates the of recently active inputs into recently activated behavioural .

An input results in sequential detection of conditions at higher and higher levels of complexity. If the state is followed by a behaviour and the behaviour is followed by a reward, and if conditions are defined incorporating information derived from conditions detected within the input state and conditions correlating with the performance of the behaviour, such conditions occurred just before the reward generation behaviour and could acquire recommendation strength on to that behaviour. Such recommendation strength is a second order reward.

A result of preprocessing of to create combinations which carry some information which can discriminate between different behaviourally relevant states of the environment.

an external system which provides feedback to a brain or other learning system on the appropriateness or otherwise of a behaviour, but cannot access the learning system internally. Thus feedback can only arrive as sensory input to the learning system and cannot bypass sensory processing to directly change internal states or force specific behaviours.

a description of how system information storage and processing resources are organized so that can be performed efficiently.

An information process performed in support of system features. The simplest level of system operation is a .

Physical resources for recording, processing and communicating information within the system.

see supervisor.

: a description of how work in a way which can easily be understood by an outside observer.

Chapter I

Introduction

This book makes a major claim with respect to understanding the human brain and other mammal brains. The claim is that an information architecture, called the recommendation architecture, is the appropriate theoretical framework for relating cognitive phenomena to neurophysiological structures and processes. When this recommendation architecture is utilized, it becomes relatively straightforward to understand cognition at the highest levels in terms of physiology. This understanding includes phenomena often labeled consciousness. Furthermore, it is possible to understand how complex cognition can be bootstrapped from experience with minimal and plausible a priori (genetic) guidance.

The recommendation architecture represents a radically different approach to understanding the brain from other approaches such as artificial neural networks, dynamical systems, or computer modeling of cognitive processes etc. The reason for taking this radically different approach is that the other approaches fail to recognize that system architectures are severely constrained when faced with the requirement to perform very large numbers of interacting features with limited information recording and processing resources. Architectural constraints become even more severe when there is an ongoing need to change some features without undesirable side effects on the many other features. The recommendation architecture is the form into which a learning system is forced by these constraints. If a cognitive model fails to recognize these constraints, it may be able to model one or a few features, or many features if there are no limits to available resources, but will break down when requirements similar to those placed upon biological brains are imposed.

The recommendation architecture constrains the ways in which information can be recorded and processed at the device level; how devices must be organized into modules and components; the types of functions which these modules and components can perform; and the type of processes which can be performed by the system as a whole. This book will describe how the human brain strongly resembles a system subject to the constraints of the recommendation architecture.

For example, there are certain major physiological structures which appear in all mammal brains. These include the cortex, the hippocampus, the thalamus, the basal ganglia and the cerebellum. A system constrained within the recommendation architecture bounds exhibits separations into structures with functions which strongly resemble the observed functions of these physiological structures. The recommendation architecture requires that the structure corresponding with the cortex is made up of devices organized into layers, columns and areas as observed in the actual cortex.

A system within the recommendation architecture bounds requires a process which strongly resembles REM sleep to configure resources appropriately to handle future experiences. The memory mechanisms available to a system with the recommendation architecture strongly resemble the semantic, episodic, procedural and working memory types observed in human beings. For a higher cognitive process such as arithmetic processing, the difficulties humans encounter in performing such processes are typical of those expected for a system with the recommendation architecture.

A cognitive system architecture or cognitive model can be defined which is consistent with the recommendation architecture. This system architecture specifies how system information recording and processing resources are organized to support cognition. The architectural form is driven by the need to make effective use of limited such resources. For example, different groups of similar system operations are performed by different sets of resources. These sets are called modules, and the resources of a module are organized to perform the group of similar operations as efficiently as possible. Modules are therefore key to making the most effective use of system resources. However, because similar operations may be required by many different cognitive features, such modules do not correspond with cognitive categories, features or processes as they would be described by an outside observer.

This lack of correspondence means that descriptions of how the processes performed by different modules support a particular cognitive feature or process can be very complex. An analogous situation is encountered in the most complex electronic systems, where a user manual is provided to assist in understanding the features of the system, but it can be very difficult to relate easy to understand "user manual" type descriptions of features to the system architecture. Descriptions of how a particular feature is supported by the processes of the system architecture can be very complex. Many widely known cognitive models are of the "user manual" type and give little insight into how physiological resources are organized to support the processes they describe.

The system mechanisms supporting memory, cognition and consciousness in the recommendation architecture are therefore more difficult to follow than mechanisms proposed by other models, but unlike the other models can be related to underlying physiological mechanisms and processes.

A major advantage of recommendation architecture cognitive model is that the processes of the architecture do not require a priori knowledge of cognitive categories. Rather, behaviours appropriate to the presence of such categories can be learned using only information derived from sensory inputs. The primary way of organizing sensory information is in terms of similarities of different types between sensory input states at different times. In a general sense, conditions which actually occur within input states are randomly selected. These conditions are organized into groups of similar conditions called portfolios. Each portfolio acquires recommendation strengths in favour of a wide range of different behaviours, initially by random assignment. The most strongly recommended behaviour is implemented at each point in time. Recommendation strengths are adjusted by consequence feedback following a selected behaviour. Learning speed and efficiency can be greatly enhanced by specific resource biases which reduce but do not eliminate the random element, including resource biases which result in a tendency towards imitation. Such biases can plausibly be specified genetically.

There has been controversy over whether a scientific theory of human consciousness is possible even in principle. The nature of the understanding which can be achieved by a scientific theory is discussed in chapter 2, and the characteristics of a scientific theory of higher cognition described. In particular, scientific theories make it possible to describe the same phenomenon consistently on a number of different levels of detail. Thus a chemical process can be described in terms of chemicals, in terms of atoms and molecules, or in terms of quantum mechanics. Each level of description has advantages, and there is a rigorous consistency between descriptions on different levels. In the case of higher cognition, an analogous theory must make it possible to describe a full range of cognitive processes in a single brain in physiological terms, perhaps through one or more intermediate levels. This book presents the thesis that the presence of the recommendation architecture forms in the human brain is the key to creating such a scientific theory.

Theoretical arguments are made in chapter 3 that any system which performs a complex combination of control features experiences severe constraints on its architectural form. These constraints result from the need for the system to perform its features using information recording and processing resources which are not unlimited, and from the simultaneous need to support changes to its features without severe side effects on other features. It is the combination of these two needs which results in the architectural constraints, and the results of such constraints are confirmed by experience with extremely complex electronic control systems.

Limits to information handling resources force the organization of those resources into modular hierarchies. The need for change results in modules having properties which make them very useful for understanding system features in terms of device operations. Although modules are defined to achieve economy in the use of system resources and therefore do not correspond with cognitive categories or features, or types of cognitive behaviour, the modular hierarchy is the basis for creating a scientific theory of cognition of the type described in chapter 2.

If changes to features are under external intellectual control, the needs constrain the system into the form generally known as the von Neumann architecture, ubiquitous in electronic systems. If changes to features must be defined heuristically (i.e. learned) the needs constrain the system into the recommendation architecture form. Chapter 4 describes the detailed forms of a system with the recommendation architecture, and Chapter 8 describes the striking resemblances between the mammal brain and these recommendation architecture forms.

The information handling capabilities of a system with the recommendation architecture have been confirmed by electronic implementations, as described in chapters 5 and 6. Comparisons with other approaches to modeling the brain are made in chapter 7.

In chapter 9, a detailed cognitive model of the human brain within the recommendation architecture constraints is described. Chapters 10 and 11 demonstrate that the model can provide an account for a wide range of memory phenomena including episodic, semantic, procedural and working memory and priming in terms of physiological structures. Chapter 12 describes how higher cognitive processes can be supported by the cognitive model, illustrated by the example of arithmetic processing.

Chapter 13 discusses how the model can provide an account for a wide range of phenomena labeled "conscious", including phenomena often claimed to be "hard problems" beyond the reach of science. Finally, chapter 14 discusses how the model could be applied to understanding a range of other problems including the evolution and development of human cognition, the physiological basis for speech processing, and the development of self awareness in children.

There are a number of concepts which are important to understanding the recommendation architecture approach. In some cases these concepts have some similarities with those used in other theories, but with some critical differences. Such concepts are defined when first required in later chapters, but to make it easier to remain aware of the differences from other approaches the definitions are collected together in the glossary.

The rest of this introduction provides an overview of some of the key ideas developed in later chapters.

Scientific Theories

The recommendation architecture is claimed as the basis for a scientific theory of cognitive phenomena. As will be discussed in chapter 2, scientific understanding means that causal relationships observed in phenomena at high level can be precisely mapped into causal relationships at a more detailed level. In the physical sciences, this means for example that causal relationships observed in chemistry can be mapped to equivalent causal relationships between atoms, and causal relationships between atoms can be mapped into quantum mechanics. For understanding the brain, it would mean that causal relationships observed in psychology could be precisely mapped into causal relationships between higher physiological structures such as the cortex, hippocampus, thalamus etc. These relationships between higher physiological structures could then be mapped into causal relationships between more detailed physiological structures such as cortex columns and areas, and nuclei in the thalamus. The causal relationships between these more detailed structures could be mapped into causal relationships between neurons. The claim of this book is that the recommendation architecture makes this type of understanding possible for the brain.

Complex Operational Systems and Architectural Constraints

The recommendation architecture approach has its roots in the technology for design of electronic systems which control complex combinations of physical equipment in real time with no human intervention. Examples are the systems which control aircraft flight simulators or telecommunications networks. Any system which controls physical structures is a control system. A control system controlling a complex combination of physical structures, enabling those physical structures to operate effectively in some complex environment, is an operational system. An electronic operational system may control a telecommunications network. A brain is a biological operational system which controls a body.

The architectures of electronic operational systems are strongly constrained into specific forms by a number of practical considerations. These practical considerations include the need to perform large numbers of different features with limited information handling resources in such a way that it is possible to modify and add to features.

Although these electronic systems have little direct resemblance to biological brains, such brains have some analogous practical limitations. Natural selection will tend to favour brain architectures which can perform the largest number of different features with a given set of resources, and brains need to learn with limited interference between prior and later learning.

Conditions and Architectural Constraints

In very general terms, an operational system must receive inputs which provide information about the status of its environment, of the physical structures controlled, and of the operational system itself. The operational system must detect conditions within this input information space which correlate strongly with circumstances in which different behaviours are appropriate. One condition is defined by a set of inputs and a state for each of those inputs. The condition occurs if each member of its input set is in the state specified for it by the condition. The "design" problem for an operational system is to define useful conditions and their associations with appropriate behaviours.

The architectural constraints arise because for a real system the number of possible conditions is enormous. Lack of access to unlimited resources for recording and detecting conditions therefore dictates the need for economies in the selection and detection of conditions. In particular, similar conditions will need to be organized into groups which can be detected by shared resources, and any one group will influence many different behaviours. However, if one group of conditions influences many different behaviours, a problem arises if there is a need for changes or additions to system features. A change to conditions needed to support a new feature can easily have undesirable side effects on all the other features dependent on detection of the original conditions. Separate condition detection for every feature would of course avoid this problem, but at a very high resource cost. Finding a compromise between the conflicting pressures of resource limitations and modifiability of features is the primary force constraining architectural form.

The architectural compromise required is radically different depending upon whether features are defined under external intellectual control (as for electronic systems) or learned from experience (as in the human brain).

Modular Hierarchies

One effect of resource constraints which applies independent of how features are defined is the organization of system resources into a modular hierarchy. A module is defined as a group of system resources specialized for the detection of a set of similar conditions. In this context "similar" means similarity in an information sense: a significant proportion of the inputs contributing to the conditions are the same and in the same state for different conditions in the set. Outputs from a module indicate the detection of conditions within its set but do not provide complete identification of which individual conditions have been detected.

With this definition of information similarity it is possible to customize the information recording and processing resources of the module for the detection of the similar group, resulting in significant resource economies. A module on the most detailed level detects a set of very similar conditions. A group of detailed modules with somewhat less similarity between conditions across the group can share a somewhat smaller proportion of their resources across their condition sets and form an intermediate module. Groups of intermediate modules form higher level modules defined by an even smaller but still significant degree of resource sharing across even less similar condition sets and so on.

The resultant modular hierarchy is thus defined on the basis of resource sharing. However, an information similarity between two conditions does not necessarily imply that the same system behaviours are always appropriate when the two different conditions are present. Similar conditions may be present when different behaviours are appropriate, and additional condition detections are required to discriminate effectively. As a result, modules do not in general correspond exactly on any level with system features as perceived by an external observer.

This lack of correspondence is the reason for the large mismatch between user manuals and system architectures in electronic systems. A user manual describes the features of a system in such a way that a user can easily understand. A system architecture describes how the system resources are organized in order to deliver those features within the constraints of the resources available. Tracing the operation of a feature through the system architecture can be a very complex process.

Applied to the brain, the implication is that physiological structures will correspond with modules performing groups of similar system functions, and will not in general correspond with the cognitive features and processes observed by psychology. For example, as discussed in chapter 7, columns in the cortex are modules which define and detect groups of conditions which are similar in an information sense but in general do not correlate exactly with cognitive features or categories.

Information Exchange between Modules

The modular architecture organization has a specific problem related to modifiability. This problem is that if changes to modules are made to implement a change or addition to some system features, there is a high probability that the changes will result in undesirable side effects upon other features also dependent upon the changed modules. This problem could be resolved if every feature had its own set of modules, but this approach will generally be ruled out by resource constraints. The necessary compromise is that modules will be defined in such a way that one module supports as few features as possible, and the operation of one feature utilizes as few modules as possible. This compromise is equivalent to minimizing the information exchanges (i.e. minimizing the sharing of condition detections) between modules as far as is consistent with available resources.

An information exchange between two modules has two complementary meanings. One is an information meaning that a condition has been detected by the source module. The other is an operational meaning that the currently appropriate system behavior is within a specific subset of the behaviours influenced by the recipient module. This subset can of course be empty, corresponding with the meaning that none of the behaviours influenced by the recipient module are currently appropriate.

There are two possible types of operational meanings, and these two types of meaning result in two qualitatively different system architectures. One meaning is that the currently appropriate system behavior is within a specific subset of behaviours with 100% confidence. The other is that the currently appropriate system behavior is probably within the specific subset. In the first case, information exchanges are operationally unambiguous and can be interpreted as commands. In the second case such exchanges are operationally ambiguous and can only be interpreted as recommendations. In both cases, because the same condition detection by one module may be received by many different modules, one condition detection has many different operational meanings.

The Von Neumann Architecture

If information exchanges are unambiguous, the need to support such a level of meaning forces the system architecture into the familiar memory, processing separation often referred to as the von Neumann architecture. The organization of system programming into sequences of instructions or commands also reflects the use of unambiguous information meanings. Such unambiguous operational interpretations of conditions tends to minimize the volume of conditions which must be detected. However, because all information exchanges are unambiguous, it is extremely difficult to make changes to features heuristically. The difficulty arises because for features to be defined heuristically there must be some degree of random experimentation followed by consequence feedback. However, if random experimentation aimed at improving one behaviour changes an unambiguous piece of information, it changes commands directed at many different behaviours. Consequence feedback in response to the targeted behaviour might change the information again, again changing commands directed at many different behaviours. Such undesirable side effects would have to be corrected by consequence feedback following each of the affected behaviours, but the corrections would themselves introduce further side effects. This process is very unlikely to converge on a consistent set of appropriate behaviours. Learning in a von Neumann system performing a complex combination of interacting features has never been successfully implemented.

The Recommendation Architecture

If information exchanges are partially ambiguous, such exchanges can only be interpreted as operational recommendations. Any implemented behaviour must generally be supported by a relatively large number of consistent recommendations. The number of conditions detected will therefore need to be much larger than for a von Neumann system. However, because any implemented behaviour is supported by a large number of recommendations, changes to some of the conditions corresponding with those recommendations will have less effect on the behaviour accepted, and learning by random experimentation followed by consequence feedback becomes possible.

In a system utilizing partially ambiguous information, modules must detect groups of similar conditions, and module outputs must be interpreted as behavioural recommendations. A separate subsystem must determine which is the most strongly recommended behaviour. There is therefore a primary architectural separation between a modular hierarchy which defines and detects conditions in the system input space and a subsystem which uses consequence feedback to assign recommendation weights to conditions, and determines the appropriate behaviour at each point in time. The modular hierarchy is called clustering because it clusters conditions into similar groups. The subsystem which selects behaviour is called competition because it manages the competition between alternative behavioural recommendations. Any one output from clustering may be assigned recommendation weights in favour of many different behaviours, and competition determines the most strongly recommended behaviour across all the currently detected conditions.

The separation between clustering and competition is the highest architectural separation of the recommendation architecture, qualitatively different from the memory, processing separation in the von Neumann architecture, but analogous with it in the sense that both separation types are driven by the requirement to maintain adequate contexts for information exchanges.

Changes to Conditions in the Recommendation Architecture

A relatively simple learning system could have all its conditions specified a priori (for example by genetic information), and only the behavioural interpretations (i.e. recommendation weights) of the conditions would change. The insect brain corresponds roughly with this situation. For more sophisticated learning, some heuristic selection of conditions will be required. Selection of one condition from the immense number of possible conditions will inevitably require some random element.

Once a condition has been defined, it may immediately acquire many different behavioural meanings. Any subsequent changes to the condition to improve its appropriateness for one behaviour will therefore introduce undesirable side effects on the other behaviours. As the number of behaviours dependent on the typical condition increases, the less change to conditions can be tolerated. For a system with limited resources that controls a complex combination of behaviours, the limiting case is that once a condition has been defined it cannot be changed.

The recommendation architecture model for condition recording in the mammal brain is therefore that the cortex (corresponding with the clustering modular hierarchy) constantly records conditions throughout the life of the animal, but with some tightly managed exceptions does not change or delete a condition once it has been recorded. For the human brain, the implication is that on average ~ 1000 conditions per second are permanently recorded throughout life, although very large variations from the average will occur depending upon the degree of novelty in the perceived environment. This permanent recording is the basis for instantaneously created permanent memory.

Conditions are recorded on pyramidal neurons in the cortex, a condition corresponding with the set of inputs to one dendrite.

Change Management

Permanent recording means that in order to avoid excessive numbers of conditions, considerable effort must be devoted to determining when and where to record additional conditions. Much of the structure of the cortex is required to support this change management function. The primary high level criterion for initiating condition recording is the requirement to generate an adequate range of behavioural recommendations in response to every situation. This is equivalent to requiring a minimum degree of condition detections in response to every input state, and conditions are recorded until this minimum is reached.

One factor limiting the selection of conditions is that only conditions which actually occur in system experience can be recorded. A second factor is that provisional conditions are defined in advance, and a condition can only be recorded if a corresponding provisional condition already exists. The process for advance definition of provisional conditions makes the probable behavioural usefulness of recorded conditions as high as possible. Provisional conditions are defined as groups of inputs which have frequently been active at the same time in recent system experience. Recent experience is thus used as the best available estimate for future experience. In the mammal brain, REM sleep is a key element in this estimation process.

Provisional conditions within a module are similar to the conditions already recorded in the module. New conditions are therefore recorded in the modules which have already recorded the most similar conditions.

Conditions on Different Levels of Complexity

Conditions are defined on different levels of complexity, where the complexity of a condition is defined as the number of system inputs which contribute to the condition, either directly or via intermediate conditions.

Because there is a random element to condition selection, and conditions are not changed after being recorded, they do not correlate exactly with cognitive features or categories of objects. However, it is important that conditions have the capability to discriminate between such features and categories. In other words, a condition must tend to occur more often when a particular subset of behaviourally relevant categories are present than when others are present. Such a condition will acquire recommendation weights in favour of the behaviours appropriate to the particular subset.

Conditions on some levels of complexity will provide better discrimination between some categories of objects. Conditions are therefore detected on a number of different levels of complexity to gain discrimination between different types of behaviourally relevant objects. Conditions on higher levels of complexity incorporate information from several different objects perceived sequentially, and can discriminate between different groups of objects. Conditions at yet higher levels of complexity can discriminate between different groups of groups of objects.

The layering of pyramidal neurons in the cortex reflects the requirement to detect conditions on different levels of complexity.

Because conditions incorporating information from several different objects must be detected, it is necessary for the detections of independent populations of conditions derived from different objects be supported simultaneously within the same physical set of neurons. The separation of the different populations is maintained by a different phase of frequency modulation being applied to the output signals of neurons in the different populations. This frequency modulation is observed as a 40 Hz signal in the electroencephalogram (EEG) of the brain. Limitations to the number of different objects which can be considered at the same time (i.e. working memory limitations) are a consequence of a limit to the number of different phases of frequency modulation which can be supported.

Portfolios and Recommendations

In the recommendation architecture, a primary module is the column detecting similar conditions on a number of different levels of complexity. Many of these levels within a column are used to determine when and where new conditions will be recorded, and a small subset of the conditions are communicated to competition where they are interpreted as behavioural recommendations. The set of conditions communicated to competition is known as the portfolio of the column.

An array of columns can be viewed as dividing up an input space into different groups (or portfolios) of similar conditions. These groups are defined heuristically, and there is therefore a random element in the definition process. One portfolio detects the occurrence of any one of a group of similar conditions in an input state. Portfolios are added and expanded until every input state contains conditions recorded within a number of portfolios. Each portfolio will have recommendation strengths in favour of a wide range of behaviours, and competition determines the most strongly recommended behaviour across all active portfolios in each situation.

One portfolio will not correspond with a cognitive category such as a dog. However, there will be a set of portfolios which tend to detect conditions relatively frequently when dogs are perceived. Such portfolios will acquire recommendation strengths in favour of appropriate behaviours in response to the presence of a dog. Another set of portfolios tends to detect conditions relatively frequently when cats are perceived. Such portfolios will acquire recommendation strengths in favour of appropriate behaviours in response to the presence of a cat. Because there are some similarities between dogs and cats (they both have heads, four legs, tails etc.) there will be some overlap between the "dog" and the "cat" portfolio sets. In other words, some conditions which sometimes occur within perceptions of dogs will also sometimes occur within perceptions of cats. Portfolios in the overlap will have recommendation strengths in favour of appropriate behaviours in response to the presence of both dogs and cats. However, provided the portfolios have been adequately defined, the predominant recommendation strength in response to perception of a dog will be in favour of appropriate responses to dogs and so on.

The key requirement is therefore that although portfolios do not correlate exactly with behaviourally relevant categories they must be able to discriminate between such categories. The group of portfolios which frequently detects conditions in response to one category cannot be exactly the same as the group which frequently detects conditions in response to a behaviourally different category.

Because conditions can be added but not changed, portfolios can expand but not change or contract. However, there are mechanisms by which new portfolios can be added if existing portfolios do not provide adequate discrimination between behaviourally different categories. Such additions would be triggered if on a number of occasions the activation of a specific portfolio or group of portfolios was followed by one particular behaviour, but the behaviour in this situation sometimes resulted in positive consequences, sometimes negative. The implication is that the portfolio group does not discriminate adequately between similar but behaviourally different circumstances, and new portfolios in the input space of the group need to be added.

In the recommendation architecture cognitive model, the columns observed in the mammal cortex correspond with portfolios.

Hierarchy of Behavioural Components

Because consequence feedback is used within competition, information exchanges within competition cannot acquire complex behavioural meanings. Competition is therefore organized into a hierarchy of components in which each component corresponds with a specific behaviour or type of behaviour. An input to a component from clustering has the operational meaning of a recommendation in favour of or against the behaviour corresponding with the component. An input to a component from a component corresponding with a different type of behaviour is a recommendation against the behaviour corresponding with the component. An input from a component corresponding with a general type of behaviour to a component corresponding with a specific behaviour of the general type is interpreted as a recommendation in favour of the specific behaviour.

In order to minimize information flow and processing within competition and clustering, there is a major separation between a subsystem selecting the input domain which will be searched for behaviourally relevant conditions (i.e. an attention function), a subsystem selecting a general type of behaviour, and a subsystem selecting a specific behaviour within an already selected general type. This separation corresponds with the observed separation between the thalamus, basal ganglia and cerebellum in the mammal brain.

Activation of Portfolios

A portfolio is activated and produces outputs to competition if one or more of its programmed conditions is present within current sensory inputs. However, a portfolio can also be activated if it has often been active in the past at the same time as currently active portfolios, if it has often recorded conditions in the past at the same time as currently active portfolios, or if it has been recently active. These indirect activations expand the range of potentially relevant behavioural recommendations beyond the set available within current sensory inputs.

Such indirect activations are behaviours which must be adequately recommended by currently active portfolios before they can be implemented. In other words, a portfolio has recommendation strength in favour of activation of another portfolio if the two portfolios have been active at related times in the past. These recommendation strengths are modulated by consequence feedback following a sequence of behaviours which included the indirect activation.

Because new conditions are defined as sets of currently active simpler conditions, a new condition may incorporate both sub-conditions which have been directly activated by sensory inputs and indirectly activated sub-conditions. The definition of conditions in terms of current sensory inputs can therefore become very complex.

A population of indirectly activated portfolios can approximate to the population which would be activated in response to actual perception of some object. As a result, mental images of objects which are not present is possible. However, portfolios on levels of condition complexity close to sensory inputs are not in general activated indirectly, and such mental images are therefore not visual hallucinations.

Memory and Indirect Portfolio Activation

Activation on the basis of frequent past activation at the same time makes it possible to activate mental images of associated objects and therefore the behavioural recommendation strengths associated with those objects. Semantic memory is supported by this mechanism.

Activation on the basis of simultaneous past condition recording makes it possible to indirectly activate populations of portfolios approximating to the populations active during specific past experiences. These indirectly activated populations have recommendation strengths in favour of behaviours appropriate during those past experiences, including speech behaviours. Episodic memory including descriptions of past experiences is therefore supported by this mechanism.

Note that it is the permanent condition recording mechanism at the device or neuron level which makes indirect portfolio activation meaningful in an information sense and therefore results in semantic and episodic memory.

The learning of skills, or procedural memory, involves changes to recommendation weights and no record of past weights is preserved.

These different mechanisms for semantic, episodic and procedural memory provide an account for the dissociations observed between the different memory types, including the different cognitive deficits resulting from physical damage.

Indirect Activation and Consciousness

Indirect activation can result in a perceived object or situation generating a mental image of some associated other object or situation, and that object or situation generating a mental image of yet another other object or situation. The internal stream of mental images regarded as a key phenomenon of consciousness can therefore be described in terms of portfolio activations, cortex columns and ultimately in terms of neuron activations.

The often discussed phenomenon that a person can respond to an object behaviourally, but can also become "conscious" of an object in an experience which is distinctive but difficult to describe in detail can also be understood in terms of portfolio activations. For example, a person can be sufficiently aware of a tree to avoid walking into it, but can also be "aware" of the tree in a qualitatively different fashion. In the recommendation architecture model, the first situation corresponds with the activation only of portfolios containing conditions actually present in sensory input currently derived from the tree. In the second case, a "halo" of portfolios often active in the past or which often recorded conditions in the past at the same time as the direct "tree" portfolios are indirectly activated. This halo contains portfolios which would be directly activated in response to many different objects and circumstances associated in some way with trees.

The much higher level of portfolio activation accounts for the distinctiveness of the experience. Portfolios are defined heuristically within the experience of the individual, and the specific portfolios indirectly activated will be very sensitive to that individual experience. There may be many different groups of portfolios within the activated population in which all the portfolios in the group would also be activated in response to direct perceptions of a different type of object. However, in general none of the groups will initially be large enough to generate accepted behavioural recommendations appropriate to their corresponding object (such as speaking its name). It is therefore not possible to generate a verbal description of the details of this consciousness experience. However, one group may have enough recommendation strength to evolve the indirect population towards a population which has such recommendation strength for one object, which would result in a mental image of the object and the ability to generate the name of the one object.

Becoming conscious of an object can therefore be understood as acceptance of recommendations to activate portfolios often active in the past at the same time as portfolios directly activated by the perceived object. The resultant population contains large numbers of portfolios which would be directly activated by perception of objects which have often been perceived at the same time in the past as the current object, but does not have enough recommendation strength to name any of the other objects. However, the indirect population can evolve towards a mental image of one or more of these objects.

Feelings and Emotions

Subject to limitations on available condition recording resources, it is possible to create independent portfolio hierarchies for different behaviour types. Portfolios can be optimized to some degree within such a hierarchy for the different behaviour type. A compromise is required between the behavioural benefits and resource costs of many different such behaviour types. This compromise will generally mean that independent hierarchies can only be defined for a limited number of different general behaviour types such as aggressive, avoidance, food seeking, and sexual behaviours.

Such a specialization has the additional benefit that modulation of the condition detection sensitivity of different such hierarchies modulates the relative probability of different behaviour types. Neurochemicals released by indicators correlating with a particular need for a behaviour type can therefore modulate the probability of that behaviour.

In the recommendation architecture model, emotions and feelings like anger, fear, hunger and sexual arousal correspond with release of neurohormones which increase the probability of strong recommendations of the corresponding behaviour type.

Electronic Implementations

The key processes required by the recommendation architecture cognitive model have been implemented in electronic form and their capability to support cognitive processing confirmed.

Firstly, the ability of the condition recording algorithm to support heuristic definition of portfolios able to discriminate effectively between different cognitive objects and groups of objects has been confirmed. Secondly, the ability of the portfolios to manage when and where additional conditions will be recorded has been demonstrated, along with the ability of the REM sleep mechanism to manage the effective definition of provisional conditions. Thirdly, the ability of the indirect activation mechanisms to generate mental images of behaviourally relevant objects has been confirmed. Fourthly, the ability of a competition subsystem to associate portfolios with appropriate behaviours on the basis of consequence feedback alone has been demonstrated, and the improvement resulting when consequence feedback is supplemented by an imitation capability confirmed. Together, these results confirm that a system with the recommendation architecture has the ability to bootstrap cognitive like processes from experience with minimal a priori guidance.

Finally, the ability to use the proposed frequency modulation mechanism to support independent populations of portfolios corresponding with different cognitive objects in the same physical set of neuron like devices has been demonstrated.

Experimental Predictions

A number of physiological and psychological predictions can be made on the basis that the recommendation architecture is the information architecture of the mammal brain. Predictions are made of way in which pyramidal neurons change their responses following combinations of inputs from specific sources elsewhere in the cortex. Predictions are also made of the way in which dream sleep deprivation will affect behaviour.

Approach

The approach of this book has some analogies with the approach for designing very complex electronic systems. In this electronic system approach, there is an understanding of the general architectural constraints, and an understanding of the capabilities of the underlying technologies, including what general capabilities can be designed into a module. An effective design at an intermediate design level can then be constructed. This design separates system operations into modules and defines the performance of and interactions between modules in such a way that the information processes needed for system features are supported. The specified performance of and interactions between modules is kept within the available general capabilities. The flow of module operations which result in different features can be described and tested, and it can be reliably assumed that the module capabilities and interactions can in due course be implemented in the available technology.

For the brain, a cognitive architecture is described within the recommendation architecture constraints. The capabilities of all the detailed processes employed in the model have been confirmed by electronic implementations of the recommendation architecture. "Intermediate design" descriptions for cognitive phenomena like semantic, episodic and procedural memory or arithmetic processing can then be created provided they only employ the confirmed detailed processes. The performance of these intermediate design descriptions can then be compared with psychological experiments.

References

The recommendation architecture and its application to understanding the brain have been explored in a number of prior publications. This book develops the ideas previously published into a detailed and consistent exposition. Previous publications on the recommendation architecture are listed below, but reference to these previous publications is made in the rest of the book only when the reference is essential to support a point not discussed fully within the book itself.

Coward, L. A. (1990). , New York: Praeger.

Coward, L. A. (1997). The Pattern Extraction Hierarchy Architecture: a Connectionist Alternative to the von Neumann Architecture, in Mira., J., Morenzo-Diaz, R., and

Cabestanz, J. (eds.)
, 634-43, Berlin: Springer.
Coward, L. A. (1997). Unguided Categorization, Direct and Symbolic Representation, and Evolution of Cognition in a Modified Connectionist Theory, in Riegler, A. and Peschl, M. (eds.), , 139-146, Vienna: Austrian Society of Cognitive Science.
Coward, L. A. (1999). A physiologically based approach to consciousness, , 17 (3), 271-290.
Coward, L. A. (1999). A physiologically based theory of consciousness, in Jordan, S. (ed.), , 113-178, Maryland: UPA.
Gedeon, T., Coward, L. A., and Bailing, Z. (1999). Results of Simulations of a System with the Recommendation Architecture, , 78-84.
Coward, L. A. (1999). The Recommendation Architecture: Relating Cognition to Physiology, in Riegler, A. and Peschl, M. (eds.) , 91 - 105, Plenum Press.
Coward, L. A. (2000). A Functional Architecture Approach to Neural Systems. , 9, 69 - 120.
Coward, L. A. (2000). Modeling Cognitive Processes with the Recommendation Architecture. , 47-67. University of Twente: Enschede.
Coward, L. A., Gedeon, T. and Kenworthy, W. (2001). Application of the Recommendation Architecture to Telecommunications Network Management, 11(4), 323-327
Coward, L. A. (2001). The Recommendation Architecture: lessons from the design of large scale electronic systems for cognitive science. 2(2), 111-156.
Coward, L. A. and Sun, R. (2002). Explaining Consciousness at Multiple Levels, in Shohov, S. P. (ed.), 14, 51 - 86, Nova Scientific Press.
Coward, L. A. (2003). General Constraints on the Architectures of Functionally Complex Learning Systems: Implications for Understanding Human Memory. In B. Kokinov and W. Hirst, (Eds.). , NBU Series in Cognitive Science.
Coward, L. A. (2004). Simulation of a Proposed Binding Model. , BICS 2004, University of Stirling, Scotland.
Coward, L. A., Gedeon, T. D. and Ratanayake, U. (2004). Algorithms for Learning Complex Combinations of Operations. .
Coward, L. A. (2004). The Recommendation Architecture Model for Human Cognition. , University of Stirling, Scotland.
Coward, L. A. and Salingaros, N. A. (2004). An Information Architecture Approach to Understanding Cities. 30(2), 107 - 118.

Coward, L. A. and Sun, R. (2004). Some Criteria for an Effective Scientific Theory of Consciousness and Examples of Preliminary Attempts at Such a Theory. 13(2), 268 - 301.

Coward, L. A., Gedeon, T. D. and Ratanayake, U. (2004). Managing Interference between Prior and Later learning. Calcutta.

Sun, R., Coward, L. A. and Zenzen, M. J. (2005). On Levels of Cognitive Modeling. .

Coward, L. A. (2005). Accounting for Episodic, Semantic and Procedural Memory in the Recommendation Architecture Cognitive Model. .

Chapter II

Scientific Theories of Subjective Human Experience

Is it possible to achieve a scientific understanding of human behaviour and subjective experience in terms of neurons and other detailed physiological structures? Would such a scientific understanding explain human consciousness? The response to these questions begins with another question: what is the nature of "scientific understanding" and "explanation"?

The philosophy of science is the investigation of the nature of scientific understanding and the means to achieve it. Over the last 500 years of the growth of modern science, the concepts of the philosophy of science have changed as physics in particular has become more and more abstract and mathematical. For example, "explanation" is commonly interpreted as identifying causes. This is relatively straightforward when used in the context of an example such as the heat of the sun explaining rain and snow. Such heat causes evaporation of surface water, which results in precipitation when the water vapour rises to higher altitudes where the temperature is lower. However, the interest in a more specific definition of "explanation" increased as many of the scientific theories of the twentieth century came to depend upon objects like atoms, fields of force, and genes etc. which are not directly observable by human senses. Two views have developed in the philosophy of science. One is that explanatory power in a scientific theory provides insight into the causal structure of reality, the other is that theories with such power simply organize experience more effectively.

An early concept in the philosophy of science was that of natural laws, and the scientific method was viewed as a way of uncovering these laws. However, as in the case of explanatory power, natural laws are viewed in two significantly different ways. One is that they are principles which the natural world is constrained to obey. The other is that they are simply descriptions of the way the world is.

The influential nineteenth century physicist Ernst Mach argued that scientific theories are only valuable as descriptions of observed phenomena, and that viewing them as descriptions of underlying reality could only mislead. From this viewpoint scientific theories are merely provisional mathematical models which facilitate mental organization and prediction of phenomena. This viewpoint was developed in the twentieth century into a philosophical theory known as logical positivism. In logical positivism a scientific theory is a system derived by formal logic from a set of axioms. The theory acquires an empirical interpretation through a set of rules of correspondence which establish correlations between observable objects and processes and the abstract concepts and relationships of the theory. A statement within a theory is only meaningful if it can be proved true or false, at least in principle, by observation.

Although something similar to logical positivism would probably be the way most scientists would verbalize their view of scientific theory, the intuitive emotional appeal of the idea that science could determine the "principles" of the universe remains, and appears to be a significant part of the motivation of many scientists. For example, part of the attraction of the theory of physics known as string theory [Greene 1999], which models all physical particles and forces as internal states of and interactions between many instances of one single minute and elementary element of "string", is the idea that so many aspects of physical reality can be derived from "first principles".

The approach to determining natural laws, or the "scientific method" has also been viewed in many different ways. One of the earliest approaches was inductionism, associated with the name of Roger Bacon. In this approach, the first step is to collect numerous observations, as far as possible without any theoretical preconceptions. The next step is to infer theories from the observations by inductively[1] generalizing the data into hypothesized physical laws. The third step is to make more observations in order to modify or reject the hypothesis. The problem perceived with this approach is that modern physics includes concepts like forces, fields, and subatomic particles which could not be inductively derived solely from the data they explain. However, the emphasis in the neurosciences on physiological experiment and suspicion of high level theories indicates a remaining influence for this approach.

Another approach, associated with the name of Isaac Newton, is called hypothico-deductionism. In this approach the starting point is a hypothesis or provisional theory. The theory is used to deduce what will be observed in the empirical world, and its correctness demonstrated or refuted by actual observation. A problem with this approach is that more than one theory may be consistent with the empirical data. Tycho Brahe and Nicholas Copernicus had competing models for the solar system, but every observation predicted by one theory was also predicted by the other. A second problem is that it is always possible that new observations could topple the most established theory.

The first problem is addressed to some degree by the "law of parsimony" often known as Occam's Razor after William of Occam. This principle indicates that when all other aspects

---

[1] The process of induction is one of first organizing a set of statements into some sequence, then proving that if one statement in the sequence is true then the next element in the sequence is also true, then finally proving that the first element in the sequence is true. For example, in mathematics it is used to prove that the sum of the series 1 + 2 + 3 + ... + (n - 1) + n is equal to 1/2(n2 + n). If it is assumed that the sum of the series for one value v of n is equal to the formula value of 1/2(v2 + v), it is easy to show that adding one more term (v + 1) to 1/2(v2 + v) is equal to 1/2((n + 1)2 + (n + 1)), or the formula value for the next value (v + 1). Hence it the formula is correct for one value v it is correct for the next value (v + 1) and therefore correct for the next value (v + 2) and so on for all values of n greater that v. Since when n equals 1 the sum of the series is 1 and the value of the formula is also 1, the formula is true for all n greater than 1.

are equal, the simplest theory is preferred. For example, the lower the number of ad hoc hypotheses (or axioms in logical positivist terms) a scientific theory has, the better. The second problem was reinterpreted into a principle by twentieth century philosophy of science, essentially following Mach, that a scientific theory is only regarded as valid as long as there are no observations which contradict the predictions of the theory.

This reinterpretation is often associated with the name of Karl Popper, who argued that there was no one "scientific method". Science in his view is a form of problem solving, and the only criterion for a useful theory is that it makes predictions which can in principle be falsified by observations. Theories are accepted or rejected on the basis of the results of such observations. However, there are problems with this view of the scientific method also, because a prediction in general depends not just on the theory to be tested but on all the other theories used to model the system being observed and on the theories used to model the observational instruments.

In the mid twentieth century Thomas Kuhn developed a view which addresses this multiple theory dependence problem. In his view, science proceeds in two different ways. Normal science is predicated on the assumption that the body of theories used within the scientific community is consistent and correct, and observations are collected and fitted within this current scientific "paradigm". However, new and unsuspected phenomena are sometimes uncovered, and given sufficient mismatch with the current paradigm a crisis may develop leading to a new paradigm, or new body of consistent theories. The shift from the classical to a quantum mechanical basis for physics is viewed as a prime example of such a paradigm shift.

In the actual practice of the physical sciences, different sciences developed to address different areas of human experience, and a major element in the last 200 years has been the attempt to integrate these sciences, with a "theory of everything" as an ultimate goal [Greene 1999]. Investigation of motion led to Newtonian mechanics and in the nineteenth century to Hamiltonian dynamics. Investigation of the nature and motions of astronomical objects, perhaps originally motivated by the need to define when best to plant crops, were integrated with the observations that objects fall towards the ground, and with experience of ocean tides, into Newtonian gravity. Knowledge derived from experience of materials in cooking, metal working, pottery making etc. led to the science of chemistry. Before the nineteenth century, investigations of light, electricity, and magnetism were viewed as part of physics but had little integration with each other or with other topics in the physical sciences.

In the nineteenth century the independent phenomena of light, electricity, and magnetism were integrated into a single Maxwellian electrodynamic theory. Then in the late nineteenth and early twentieth centuries, special relativity integrated electrodynamic theory with the theory of motion, general relativity became the most fundamental theory of gravity, and atomic theory became the most fundamental theory of chemistry. Through the twentieth century the development of quantum mechanics led to an integration between theories of motion, materials, light, electricity and magnetism, and a partial integration with astronomy (i.e. theories of the nature of astronomical objects, and in particular the nuclear sources of the energy generating the light and heat of the sun and other stars). In the late twentieth century, string theory was offered as an integration between all these phenomena and gravity.

An important factor in physics becoming the paradigm for an effective science has been its capability to provide causal descriptions of wider and wider areas of experience with fewer and fewer ad hoc objects and causal relationships. However, this success has been independent of any resolution in the fundamental debate in the philosophy of science over whether scientific theory reflects ultimate reality or merely organizes experience effectively. The immense quantitative predictive power of quantum mechanics has been taken as a confirmation of the importance of prediction in science, and the long development of classical physics followed by the dramatic shift from Newtonian to quantum mechanical theories in the first half of the twentieth century is often taken as evidence in favour of Kuhn's distinction between normal and paradigm shift science.

However, there is an aspect to the practice of the physical sciences which is generally missing from the philosophy of science. This aspect is the existence of hierarchies of description which make it possible to describe causal relationships within a phenomenon on many different levels of detail. All the different levels are employed in order to create scientific understanding of the phenomenon. To illustrate these hierarchies, consider materials science.

In chemical engineering, a good causal understanding of processes in terms of the results of reactions between different materials under different conditions exists. In macroscopic chemistry there are concepts like chemicals and chemical reactions. Although the number of different chemicals is very large, it is much less than the conceivable number of different possible materials (i.e. mixtures of chemicals in different proportions) which could be encountered in human experience. However, the same high level chemical engineering processes can also be understood in terms of chemicals and their properties. At the macroscopic chemical level it is possible to establish laws which define what will be observed. For example, a range of different chemicals can be placed in the category "acid", others in the category "alkali" and yet others in the category "salt". A general law is that any acid plus any alkali will produce a salt plus water. Atomic theory, developed via the nineteenth century concept of chemical elements, uses less than 100 different types of atom, and by means of intermediate objects like hydroxyl ions (OH^-) and hydrogen ions (H^+) can generate the higher level chemical laws. Quantum mechanics, using a yet smaller set of objects (electron, proton, and neutron) can generate the concepts of atoms, hydroxyl and hydrogen ions and the causal relationships between them. String theory includes an attempt to derive all quantum mechanical phenomena from different states of vibration of a single string object, and even to derive parameters which are ad hoc in all other theories, such as the number of observed dimensions to space.

A causal description of exactly the same chemical process can in principle be generated on any of the different levels. The differences between the descriptions on different levels is that there are fewer ad hoc objects and causal relationships at the deeper levels. All of the more numerous objects and causal relationships at higher levels can be constructed as (sometimes very complex, and on occasion probabilistic) combinations of deeper level objects and causal relationships. However, as a result, the complete description of exactly the same phenomenon will be very much larger at the more detailed levels. Take for example the cooking of a meal. Even at the macroscopic chemistry level a full causal description would require reference to all of the hundreds or thousands of different chemicals present in various combinations in the ingredients. At the atomic level a full causal description would require reference to each of the more than 10^{24} different atoms. The quantum mechanical level would require reference to an even larger number of electrons and nuclei, with each of these objects described by probability distributions rather than by point positions and velocities.

Much of the work of science takes place at higher levels. Scientists, like all human beings, can only handle a limited volume of information at one time, and must use a higher level theory for thinking about a broader phenomenon, then focus in on a small aspect of that phenomenon if necessary to apply the more detailed theory. For example, a scientists attempting to determine the reasons for a problem in a chemical manufacturing plant will mainly think at the macroscopic chemical level, with occasional shifts to atomic or even quantum mechanical levels. The ability to shift freely between levels, or to understand the mapping between the levels, is a critical part of scientific capability.

In the ideal case the predictions made on every level would be consistent with observations. For example, the predictions made at macroscopic chemical, atomic and quantum mechanical levels are all consistent with the observation that adding sulphuric acid to caustic soda results in water and a salt called sodium sulphate[2]. However, the value of a more detailed theory is that it requires many fewer ad hoc assumptions, and sometimes leads to correct predictions of phenomena which cannot be derived from the higher level theory. In such circumstances an attempt may be made to modify the objects and causal relationships at higher level, but more often rules are established defining when higher level descriptions must be superceded by deeper level descriptions.

This use of theories on multiple levels is the rule even for research at the boundaries of quantum mechanics. As pointed out by Greene [1999] "In deriving theories, physicists often start working in a purely classical language that ignores quantum probabilities, wave functions and so forth ... subsequently overlaying quantum concepts upon a classical framework. This approach is not particularly surprising, since it directly mirrors our experience. At first blush, the universe appears to be governed by laws rooted in classical concepts such as a particle having a definite position and velocity at any given moment in time. It is only after detailed microscopic scrutiny that we realize that we must modify such familiar classical ideas". It is interesting that research into quantum mechanics thus begins at a level of description which is known to be incorrect. There is even some discomfort with this situation, but the conditions under which the higher level is incompatible with the more correct quantum mechanical predictions are well understood, and no useable alternative high level theory exists [Greene 1999].

There are in fact a very large number of different levels of description, often called models, in the physical sciences. In other words, different sets of objects and causal relationships are constructed for many different phenomenological domains. Domains are defined in such a way that interactions outside the domain can generally be neglected, or modeled in a relatively simple fashion. All models must be consistent with common deeper levels of description (and ultimately quantum mechanics), or at least to be a close model for observations with a clear understanding of when a shift to a deeper level is required. For example, at a high level the motion of the planet Earth within the solar system is modeled by

[2] At least under normal temperature and pressure conditions.

Newtonian gravity acting on a point mass. At a somewhat more detailed level the physical form of the Earth is modeled using concepts like core, mantle, and continental plates floating, moving and colliding on top of the mantle. At a yet more detailed level the origins of materials are analyzed in terms of volcanic, sedimentary and erosion processes. At the level of atomic theory, crystallography deals with the possible types of mineral crystals, and depends upon quantum mechanics when it makes use of x-ray diffraction to determine crystalline form.

The appropriate objects and causal relationships on any one level are strongly influenced by the parameters which define the phenomenological domain to be studied. Some of the most important parameters are the total amounts of matter and energy interacting within the domain and the timescales over which phenomena occur. Thus within small domains on the surface of the Earth, atoms and chemicals are useful objects. In larger domains, mineral deposits are useful, and for yet larger domains, continental plates. However, when large amounts of matter are combined with large amounts of energy at high density, as in the cores of stars, nuclei become the useful objects.

The objects on a higher level are thus defined in such a way that most of the interactions on deeper levels occur within the higher level objects, and only a small residual interaction occurs between objects. The higher level model can then explicitly include only these residual interactions and the volume of information is kept within human capabilities.

The picture of science which emerges from this discussion is of an accumulation of many different models defined on many different levels of detail. Each model has a set of objects and causal relationships between those objects, and effectively models one phenomenological domain, in the sense that predictions of the model correspond with observations of the domain. Interactions within the domain are very much more significant to prediction than those between the domain and anything outside the domain. A domain is generally defined by having particular orders of magnitude for the amounts of mass and energy, the mass and energy densities, and the timescale for phenomena within the domain. A primary driving force for the different models is to limit the volume of information which must be taken into account to achieve effective modeling to a level which can be handled by a human being.

The objects and causal relationships at higher levels can be defined as combinations of more detailed objects and causal relationships. In the ideal case, the causal relationships between objects at high levels can be defined in simple terms with 100% accuracy without reference to the internal structure of those objects as defined at more detailed levels. For example, the internal structure of atoms at the quantum mechanical level can largely be ignored in macroscopic chemistry. However, in practice this ideal is not achieved completely, and the simpler causal relationships at high level sometimes generate predictions which are less consistent with observation than those generated at a more detailed level. Each model must therefore have rules which indicate the conditions under which a more detailed model must supercede the higher level model, or in other words when the generally negligible effects of the internal structure of higher level objects must be taken into account. It must be possible to shift fairly easily between models on adjacent levels.

It is of philosophical interest that it is possible to create such a hierarchy of models in the real universe, each model within information content limitations imposed by the capacity of the human brain. However, for all models except perhaps the deepest it is clear that the models are ways to organize phenomena rather than rules which the universe must follow, since in general the higher level models are not fully consistent with observation.

The great expansion in experimental neuroscience in the twentieth century led to increased interest in the philosophy of neuroscience. This interest has developed in two major directions. One direction has been criticism of the everyday understanding of human behaviour known as folk psychology. The other has been a set of arguments that some aspects of human experience may be particularly difficult to understand in scientific terms, the so-called "hard problem" of consciousness.

Folk psychology is the body of explanatory approaches to human understanding which is widely used in everyday life. These explanatory approaches are often illustrated by homilies. For example, if asked why David is opposing the suggestion made by Mark, I might reply that when Mark first made the suggestion it came across as a criticism of David. My explanation is understood because of shared understandings of the ways human beings react to each other. This body of shared understandings is called folk psychology.

The eliminative materialism approach to the philosophy of neuroscience takes the position that folk psychology is a false and misleading account of the causes of human behaviour, and that the future effective neuroscience will replace rather than evolve out of folk psychology. The analogy is sometimes made with the early chemical theory that combustion was the release of an ethereal substance called phlogiston, an explanation completely discarded by later chemistry in favour of combustion being a reaction with atmospheric oxygen. Paul Churchland [1989] for example makes the argument that concepts like activity vectors and vector transformations may be the appropriate objects for construction of a rigorous neuroscience, and that vector mathematics is unfamiliar and alien to common sense. Dennett [1978] made the less sweeping suggestion that there is an apparent conflict between neuroscience and commonsense views of pain. A somewhat softened view has been offered by Bickle [1998], who argued that there could be different degrees to which a rigorous neuroscience might supplant folk psychology.

To summarize the different positions on the usefulness of folk psychology, suppose there is some mental state M in folk psychology. The question is whether M always has a corresponding neural state N. In the extreme eliminative materialist position, there is no such thing as a meaningful mental state M. The softened position would be that some folk psychology mental states M correspond to some degree with neural states N. Expressed this way it is clear that this issue is the same as the accuracy of the hierarchies of description in the physical sciences discussed earlier. Scientific understanding for human beings must be within the limits imposed by human information handling capabilities. There will therefore be a need in the neurosciences also for hierarchies of description, with the probability that rules will be required indicating when a less precise higher level description must be supplanted by a more precise description on a more detailed level.

The highest level could well correspond with something like folk psychology, and some form of vector mathematics might be a more detailed level. However, just as in research at the borders of quantum mechanics, it is probable that higher level descriptions with known limits to accuracy would remain a significant part of the scientific hierarchy.

The "hard problem" in the understanding of human experience has two aspects. One is the understanding of sensory perceptions, the other is the understanding of subjective experience as a whole.

The discussion of sensory perceptions has centred around the qualitative nature of introspective experiences of pain or colour. These introspective experiences have been labeled qualia. The subjectively perceived complexity of experiences of colour, the difficulty of verbalizing the complexity, and the differences between experiences of the same colour by the same individual at different times have been linked with the long standing philosophical debate over whether colours are objective properties or dependent on the specific mind perceiving the colour to conclude that no neuroscience explanation of qualia is possible.

A discussion of the explainability of subjective experience as a whole was initiated by Nagel [1974]. He asked the question "What is it like to be a bat ?" and argued that scientific knowledge cannot supply a complete answer. Chalmers [1996] attempted to clarify the argument by imagining a universe populated by beings with the same brain processes as human beings but without consciousness, and asked how a difference could be detected. This conceptual argument has been criticized by Churchland and Churchland [1997] on the basis that it is not in fact plausible to imagine creatures with the same brain processes but lacking consciousness.

To a significant degree, the directions taken by the philosophy of neuroscience over the "hard problem" appear to relate to earlier debates over whether physical laws are in some sense "absolute truth" or simply effective ways to organize observations. The physical sciences are strictly concerned with explaining what can be observed. The question of whether a universe could exist in which pseudopeople (or "zombies" behaved in exactly the same way as real people but had no "consciousness" implies a belief in a "consciousness" which cannot be observed in any way. In the physical sciences this would be analogous with postulating a force which had no effect upon any observations. Such a force would be eliminated from consideration by Occam's principle.

Qualia seem to be a fall back position in which it is argued that perceived complexity, uniqueness, and difficulty in generating adequate verbal descriptions of introspective experiences indicates that there could be something unmeasurable about such experiences. However, a neuroscience on a more detailed level might well be able to provide an account for the perceived complexity, uniqueness, and verbalization difficulty, even though a detailed account of exact content might be impractical.

Such a problem is not unknown in the physical sciences, where such problems are regarded as practical difficulties and not "in principle" barriers. For example, it has been argued that every snowflake is different, based originally on thousands of photographs of individual snowflakes taken by Wilson Bentley [1931]. Although snowflakes are generally flat in one dimension and have six fold symmetry in the other two dimensions, the detailed form of a snowflake is very sensitive to the exact values of a range of parameters during its formation. Even for one snowflake, determining all parameter values to sufficient accuracy and then calculating the exact form is impractical. This does not mean that the physical sciences cannot "in principle" explain the form of a snowflake, only that the explanation on a level of detail required to account for an exact form would require immense effort and reveal little. In general an explanation of the reasons for flatness, six fold symmetry, and extreme variability in terms of molecular properties is regarded as adequate. An explanation of the complexity, uniqueness, and verbalization difficulty of some sensory experiences will in fact be offered in chapter 13.

Dennett [2003] has pointed out some similarities between philosophical attitudes to consciousness and general attitudes to magic. Few scientists would argue against the existence of magic in the sense of tricks performed by conjurers for which a rational explanation of what is observed is not easy to find. Such explanations can be found once the trick is revealed. However, if asked the question "Do you believe in magic ?", many of the same scientists would say no, because the word "magic" in such a context implies the impossibility of rational explanation. Dennett suggests that some philosophical attitudes to consciousness resemble this attitude to magic in the sense that existence of a rational explanation implies that it is not "really" consciousness.

Summary

By analogy with the physical sciences, a full neuroscience would be made up of causal descriptions of phenomena on a number of different levels of detail. On each level there would be objects and interactions between objects.

At the highest level there will be relatively large numbers of different types of object, and complex interactions between them. The objects and causal descriptions on this level will resemble those of folk psychology, probably including sensory perceptions, mental images, and motivations. At more detailed levels there will be fewer objects with simpler patterns of interaction. Higher level objects will be constructable from more detailed objects in such a way that most of the interactions between detailed objects in the course of high level processes are within higher level objects and only a small proportion is between such objects. Only this small proportion needs to be included in the higher level descriptions.

Objects on one intermediate level will probably include brain structures like the cortex, hippocampus, thalamus etc. Objects on another yet more detailed level will probably include physiological structures like cortex columns and subcortical nuclei. Objects on an even deeper level will include neurons, and a yet deeper level synapses and neurochemicals. There will in general be bodies of rules to indicate when a higher level description needs to be supplanted by a more detailed description for adequate consistency with observation.

The need for such a hierarchy of descriptions is that direct understanding of high level psychological phenomena in terms of the behaviour of 10^{10} neurons is probably beyond the information handling capacity of a human brain. Direct understanding in terms of quantum mechanics is even less feasible. It is of course possible that some deeper level model like quantum mechanics may need to supplant higher level models under some specific circumstances. Penrose [1994] argues that such supplanting by quantum phenomena at the cell level is essential to understanding cognition. However, the ability to essentially ignore quantum mechanical phenomena in macroscopic chemistry, and the generally extreme conditions under which quantum mechanics generates different predictions from higher level models suggests that this argument is implausible.

A critical question is therefore whether it is possible to define a hierarchy of descriptions in the phenomenology of the human brain which is within the information handling capacity of that brain. It will be argued in chapter 3 that natural selection pressures resulting from the complexity of the operations performed by the brain have forced brain architecture into a form in which the appropriate objects on many levels of detail exist naturally.

References

Bentley, W (1931). Snow Crystals. McGraw-Hill {Dover Publications Reprint}

Bickle, J. (1998). Psychoneural Reduction: The New Wave. Cambridge, MA: MIT Press.

Chalmers, D. (1996). The Conscious Mind. Oxford: Oxford University Press.

Churchland, P. (1989). A Neurocomputational Perspective. Cambridge, MA: MIT Press.

Churchland, P. and Churchland, P. (1997). Recent Work on Consciousness: Philosophical, Empirical, and Theoretical. Seminars in Neurology 17, 101 - 108.

Dennett, D. (1978). Why You Can't Make a Computer That Feels Pain. Synthese 38, 415 - 456.

Greene, G. (1999). The Elegant Universe. Norton.

Nagel, T. (1974). What is it like to be a bat ? Philosophical Review 83, 435 - 450.

Penrose, R. (1994). Shadows of the Mind. New York: Oxford University Press.

Chapter III

Operationally Complex Systems

Information Systems and Operational Systems

Since the middle of the twentieth century, electronic systems have been designed to perform increasingly complex combinations of tasks. Many of these systems fall within the domain of information technology such as database management, text editing, and graphics and image processing systems. However, some of the most complex systems are in the domain of direct control of physical equipment. The equipment controlled can be nuclear power stations, chemical manufacturing plants, oil refineries, or telecommunications networks. The electronic systems control these facilities in real time with little or no human intervention. These operationally complex systems generally experience constraints on their architectural form which are significantly more severe than the constraints experienced by information technology systems. A human brain controls very complex physical "equipment" (i.e. a body) in real time with little or no external intervention. Analogous constraints on architectural form therefore exist, although the resultant form has minimal resemblance to any current commercial electronic system.

An operational system (or control system) is one which acts upon its environment and itself in order to achieve objectives. A very simple example of an operational system is a thermostat with the objective of keeping the internal temperature of a refrigerator within a specified range. The thermostat receives an electrical signal which provides information about the temperature within the refrigerator, and turns cooling on if the signal indicates a temperature above the range and off if below the range. In addition, the thermostat must respond to its input signal within a reasonable period of time.

A slightly more complex operational system would be a refrigerator thermostat with the additional objective of minimizing energy consumption as far as possible. Additional behaviours would include turning cooling on at a range of different levels where higher levels cooled faster but were less efficient in using energy. An additional sensor input signal could be an indication of the temperature external to the refrigerator. The decision on current behaviour would depend upon an interaction within the control system between information on past and current temperatures inside and outside the refrigerator and the two objectives.

For example, rates of temperature change inside and outside the refrigerator might be estimated.

Operationally Complex Systems

An operationally complex system is one in which the number of possible behaviors is very large, there are many potentially conflicting objectives, considerable interaction is required within a large body of information derived from sensory inputs to determine appropriate behaviour at each point in time, and the time available after occurrence of an input condition in which to determine appropriate behaviour is short.

To illustrate these factors, consider two types of operationally complex electronic systems, both of which have been in use for many years. One is the flight simulator [Chastek and Brownsword 1996], the other is the even more complex central office telecommunications switch [Nortel Networks 2000]. A current flight simulator may have one or two million lines of software code, and simulates the environment experienced by the pilot of an aircraft such as a military jet, including changes to that environment caused by the actions of the pilot. A central office switch with twenty or thirty million lines of code is physically connected with 100 thousand telephones and computers, and provides their users with a wide range of telecommunications services on demand.

The behaviours controlled by a flight simulator include changing the images on monitors simulating the view through cockpit windows, changing the settings on cockpit instruments, and moving the emulated cockpit platform in three dimensions as the pilot performs tasks including takeoff, landing, maneuvering in flight including combat situations, midair refueling, and deploying weapons systems. The appropriate system behaviour is determined by an interaction between: firstly the past and current settings of the simulated aircraft components such as engine, control surfaces, aircraft component failures, radar and weapons; secondly the simulated environment such as wind, turbulence, the ground or aircraft carrier, and other aircraft; and thirdly the actions of the pilot. The system must generate sufficiently accurate behaviour fast enough that the simulated response is sufficiently close to real response for pilot training to be effective. For example, slow system responses could mean that training was dangerously counterproductive. In addition, timing mismatches between visual and movement changes can result in physiological reactions known as simulator sickness, even when the mismatches are too slight to be consciously perceived.

The basic tasks of a central office switch include establishing and taking down connections between telephones and computers using a limited pool of connectivity resources; collecting billing information; collecting traffic information to determine when and where upgrades are required; making changes to services; and identifying problems. However, these basic tasks are subject to a number of considerations which make the system much more complex than might appear superficially.

The first consideration is that there are thousands of features which must be taken into account in the course of establishing a connection. Features like call forwarding, emergency calls, and toll free calls to institutions rather than to specific telephones all require decisions on destination, routing, and billing based on information in addition to the actual number dialed. Features like conferencing and call monitoring for law enforcement require connectivity to be established between more than two points. Features like fraud prevention must intervene before suspect calls are established. Features like prioritization of calls in national emergencies must permit some calls to have priority in taking connectivity resources from the common pool. Data and quality of service features must make it possible to select from different types of end-to-end connectivity. Many of these features interact in the sense that the activity of one feature can change the way another feature operates. An example would be when one telephone has the service of only accepting calls from a specific set of other telephones, and one of those other telephones invokes a service of forwarding all its calls to the telephone which restricts its callers.

The second consideration is a requirement that no single switch be totally out of service for more than two hours in a forty year period, including any outage time needed for system upgrades and other maintenance. For instance, in 1999 Nortel Networks switches were out of service for an average of 18 seconds per switch per year including failure and maintenance time. A key reason for this requirement is that one switch provides service to a significant geographical area. Failure of just one telephone leaves the option in an emergency of finding another working telephone nearby, but failure of the switch removes service from all the telephones in the area. One implication of this requirement is that the system must perform constant self diagnostics to test the performance of all hardware and software systems. Duplicates of critical subsystems must exist, with diagnostics which are capable of determining which subsystem is defective and moving all operations on to the other subsystem with no loss of service or interruption to existing connections. Any changes to individual user services such as location changes or additional features, and any upgrades to system hardware or software, must occur without service interruption.

The third consideration is that dial tone must be provided within a second of any request for service, and the connection established within a few seconds of dialing, even if hundreds of other new calls are in progress and thousands of connections already exist. Furthermore, the time available for information to pass between two points is very limited. The delay in transmitting a sound element within a conversation must be less than a few milliseconds, and the delay for real time video less than a few tens of milliseconds, even if many other functions are using the same physical connection. The quality of the received voice or image must be high, with minimal noise or distortion.

At any given instant in time, multiple behaviours could be indicated, including establishing a connection, breaking down a connection and releasing its resources for other calls, collecting billing information, performing a diagnostic task, recording the current volume of user requests for service, or making a change to the system. If processor or other resource limitations make it impossible to perform all the behaviours immediately, an interaction between internally specified objectives will determine the relative priority of different behaviours. For example, regular diagnostics may be delayed when user service demands are high, but if the delay becomes prolonged, the risk of serious undetected failures becomes significant and user services will be delayed if necessary to allow diagnostics to proceed. The combination of over 100 thousand users, thousands of interacting features, and tight reliability and real time constraints make central office switches some of the most complex electronic systems in existence.

The human brain performs an operationally complex combination of tasks. It must learn to select an appropriate response to a vast range of input conditions from within its wide range of available behaviours. These available behaviours include body movements (including changes to facial expression); verbal behaviours; and attention behaviours. Some attention behaviours are body movements (e.g. to look at different objects), others are internal actions to retain information active (e.g. the numbers when dialing a telephone call) or to activate information not currently active (e.g. mental images). There is often limited time within which appropriate behaviour must be selected.

For example, suppose that two people, Rick and Elaine, encounter each other, and consider the operational role of Rick's brain. A simple categorization problem would be to identify Elaine and speak her name. Even this problem is not as simple as it might appear, because variations in Elaine's appearance and the lighting conditions may make the raw visual input quite different from previous occasions. Furthermore, the raw visual input to Rick's brain is constantly changing as he and Elaine move, as Rick directs his attention to different parts of Elaine's face and body, and to other objects in the immediate environment. The higher level activated information in Rick's brain may change as he considers mental images of various memories. Elaine's appearance (e.g. expression) may be changing in response to Rick. To make things even more complex, the real behavioural problem is not a simple categorization. For example, depending on the location of the encounter, the appropriate response might be "Hi Elaine, what are you doing here?" Depending on her detailed appearance it might be "Hi Elaine, are you feeling okay?". This latter response might not be appropriate depending on the identities of other people present, but if the appearance indicated that Elaine was very unwell would be appropriate independent of who else was there. In addition, the appropriate response might include a hug, a kiss, or a handshake. The appropriate response also depends upon Rick's internal objectives: enhancing his business or personal relationship with Elaine; enhancing his relationships with other people present; and/or completing the task he was currently performing without being drawn into a conversation. The relative weight of these internal objectives might be affected by changes to Elaine's expression and by the responses of co-workers who happened to be present. Internal behaviours might include searching through recent memories for anything relevant to Elaine. Finally, the appropriate response must be delivered within a couple of seconds, and followed up by a sequence of appropriate and consistent other behaviours.

The human brain thus meets the definition of a operationally complex system: the number of possible behaviours is large; there are a number of potentially conflicting objectives; considerable interaction is required within a large body of information derived from sensory inputs in order to determine appropriate behaviour at each point in time; and the time available in which to determine behaviour is short.

Organization of System Resources

An operational system does not calculate mathematical functions in any useful sense. Rather, it detects conditions within the information space available to it and associates different combinations of those conditions with different behaviours. The information space available to the system is made up of inputs which provide information about the state of the equipment being controlled, the state of the environment in which the equipment operates, and the internal state of the system itself.

In principle, an operational system could be implemented as an immense look-up table, with each operationally relevant state of the total information space listed separately with its corresponding behaviour. If sequences of such states sometimes determined behaviour, all such sequences would also have to be listed in the table. Such an implementation would require impractical levels of information recording and processing resources. In practice therefore, an operational system must detect a much smaller population of conditions within different subsets of its information space, and associate many different combinations of this limited set of conditions with different behaviours. Identifying an appropriate but limited set of such conditions is a primary design problem for such systems. However, even such a comparatively limited set will in general be extremely large for an operationally complex system.

In order to conserve resources, it will therefore also be necessary to collect similar conditions into groups. The conditions in one group can be detected by a set of system resources optimized for detecting conditions of the group type. In this context, two conditions are similar if a significant proportion of the information defining them is the same. It will also be necessary to detect complex conditions by first detecting relatively simple conditions which occur in many complex conditions, then to detect the more complex conditions using the detections of the simpler conditions. A set of system resources optimized for detecting a group of similar conditions is called a module. Because similar conditions may be relevant to many different behaviours or features, modules will not correspond with such behaviours or features.

Note also that this definition of module is not the same as the concept of identical or almost identical units convenient for the construction of a system. There may be some general similarities between some modules for construction related reasons as discussed below, provided that such similarities are consistent with the primary role of modules. However, the primary role of modules is making the best use of resources in the performance of system operations, and there will be significant heterogeneity amongst even similar modules in order to achieve this primary purpose.

A system must also have a portfolio of primitive actions which it can perform on itself and its environment. For example, the brain has a portfolio of individual muscle movements. However, resource limitations will also make it impractical to associate conditions with individual actions. It will be necessary to define behaviours as frequently used sequences and combinations of actions, and types of behaviour as groups of similar sequences. A component contains a set of system resources optimized for driving a group of similar behaviours, and will be invoked by the detection of appropriate combinations of conditions.

An interaction between two modules occurs when a condition detected by one module is incorporated into a condition detected by another module. One condition detected by one module may be incorporated into different conditions detected by many other modules, either directly or via condition detections by intermediate modules. An interaction between two components occurs when both are invoked by currently detected conditions, and it is necessary to determine whether both can be invoked at the same time and if not, which of the two is most strongly invoked by currently detected conditions. All these interactions can also be viewed as information exchanges.

The User Manual and the System Architecture

The user manual for a system describes how the features of the system operate and interact from the point of view of an external observer. A feature is a group of similar behaviours invoked by similar circumstances, but in this context "similarity" is defined for the convenience of the external observer.

The system architecture describes how the operations of the system are separated into modules and components, the interactions between these modules and components, and how these modules, components and interactions result in the behaviour of the system. The separation of the system into modules and components is strongly influenced by resource constraints as discussed in the previous section. System modules and components will therefore not correspond with user manual type features. Rather, a module or component will contribute to the performance of many features and a feature will be dependent upon many modules and components. Hence the relationship between system architecture and user manual will be very complex. How the system operates is most easily understood from the user manual, but how features result from system operations can only be understood from the system architecture.

The implications for the brain are that the system architecture at the physiological level may have a very complex relationship with descriptions of cognitive phenomena as measured by psychology.

Practical Considerations which Constrain Architecture

Practical considerations which must be taken into account to some degree for any system become critical influences on the architecture of operationally complex systems. The primary such consideration is limitation to the available information recording and processing resources. As discussed earlier, one effect of this limitation is organization of resources into components and modules. However, there are some additional considerations which also have an effect on architecture.

One such consideration is the need to maintain adequate meaning for all information moving within the system. A second consideration which has important implications for information meaning is the need to modify some features without excessive undesirable side effects on other features. A third, related consideration is the need to diagnose and repair the system without too much difficulty. A fourth is the need to construct many copies of the system from some kind of design specification by a process which is not too complex and error prone, and which does not require a specification containing an excessive volume of information.

Information moving within the system must have an adequate level of meaning. Firstly, temporal relationships between information being processed by different parts of the system must be maintained: the results obtained from the processing of a group of system inputs which arrived at the same time must not be confused with the results obtained from processing inputs which arrived at a later time. Secondly, because the results of processing by one module may be communicated to many other modules, and each module may use the same result for a different purpose, any changes to modules must be such that an adequate meaning for its results is preserved for all modules receiving those results.

Module changes may be required to add and modify features, or to recover from physical damage. The need to perform such changes efficiently means that it must be possible to identify a set of module changes which can implement such modifications or recoveries with minimal undesirable side effects on other features. The identification process can only use information which is readily available to the agent performing the change. This agent may be an external designer, but when changes are learned the agent is the system itself.

Finally, the need to construct the system by a process which can only use a limited volume of design information means that systems tend to be made up of a small set of standard building block types. The system then contains many building blocks of each type. Each block within a type is fairly similar to the standard for the type. The design of a building block type can then be described just once, and individual building blocks only require description of their differences from the standard. Since the system must consist of modules, a compromise is also required between making modules as similar as possible to minimize design information and customizing each module to make the most effective use of system resources for its assigned system role.

Although the brain is not designed under external intellectual control, analogous constraints to many of the above exist. A brain able to perform a higher number of features with the same physical information handling resources will have significant natural selection advantages. The brain must not confuse information derived from different objects perceived at different times. Information moving within the brain from any one source to different locations must have an adequate level of meaning to all recipients. The brain must be able to recover to some degree from physical damage. It must be possible to add features and modify existing features without excessive side effects on other features. The genetic information space available to specify the construction of the brain is not unlimited, and the construction process must be such that it can be specified with limited genetic information and operate either without errors or with recovery from errors.

However, although these constraints are analogous with some of those on electronic systems, a critical difference is that the behaviour of an electronic system is specified under external intellectual control, but the brain must learn a significant proportion of its behaviour heuristically. The result of this difference is that the architectural form into which the brain is constrained is qualitatively different from that into which electronic systems are constrained.

Impact of the Practical Considerations on System Architecture

The various considerations place constraints on the possible form of a system architecture which become more and more severe as operational complexity increases relative to the resources available. Some of the constraints are the same for any operationally complex system. Others differ depending on whether system features are defined under external intellectual control or heuristically through experience (i.e. learned).

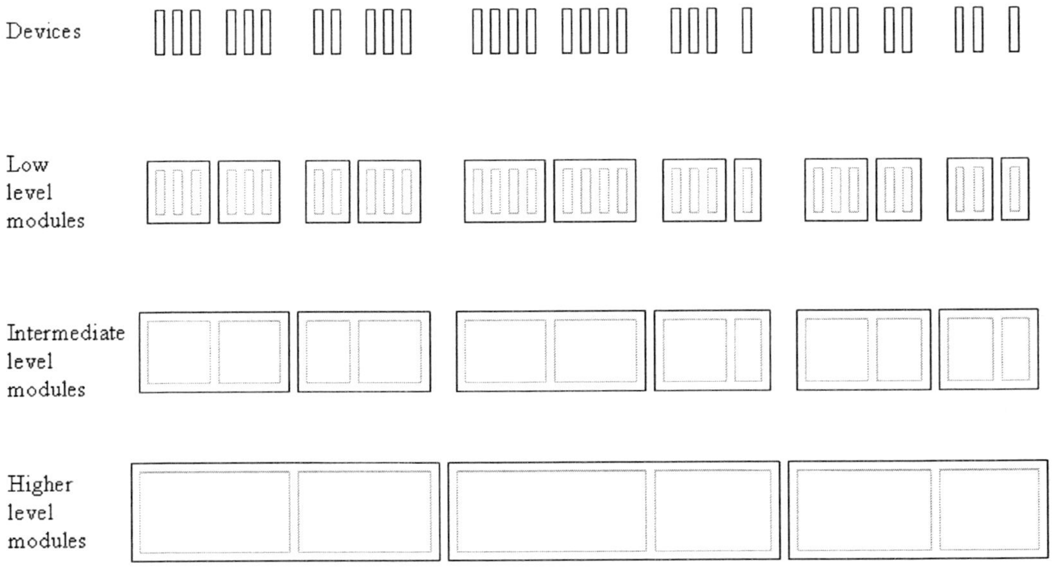

Figure 3.1 Organization of system operations into a modular hierarchy. The organization of the hierarchy is on the basis of condition information similarity with the objective of achieving information recording and processing resource economies. Devices are the most detailed "modules" and detect groups of very similar conditions. Low level modules are made up of groups of devices and detect all the conditions detected by their constituent devices. There is a strong but somewhat lower degree of similarity within this larger group of conditions and a significant degree of resource sharing in the detection of different conditions is possible. An intermediate level module is made up of a group of low level modules. The degree of condition similarity across the set detected by all the constituent low level modules is lower, and a lower but still important degree of resource sharing is possible. Higher level modules are made up of groups of intermediate level modules with a yet lower but still significant degree of condition similarity and consequent resource sharing.

Hierarchy of Modules

As discussed earlier, limitations to resources result in similar conditions being organized into groups which can be detected by a module containing resources optimized for the group type. To achieve further resource economies, intermediate level modules made up of groups of more detailed modules are defined. The set of conditions detected across such a group are more different than the set detected by one detailed module, but are sufficiently similar that some of the detection process can be performed by optimized resources shared across the group. This resource sharing defines the intermediate module. Yet higher level modules can be defined by resource sharing across a group of intermediate modules and so on. The resultant modular hierarchy is illustrated in figure 3.1.

Hierarchy of Components

The level of connectivity required to connect relevant conditions to every primitive action driver would be excessive. Frequently used combinations and sequences of actions are therefore defined as behaviours with one corresponding control component. This component receives inputs from various modules indicating the detection of conditions, and generates outputs which drive the appropriate sequence of actions. Groups of similar behaviours invoked under similar conditions are collected into higher level components where most of the condition detecting inputs can be directed to the higher level component.

Information Exchange within a Modular Hierarchy

Condition similarity in an information sense does not guarantee similarity of behavioural meaning. The conditions detected by one module will therefore be relevant to many behavioural features, and the operation of any one feature will depend upon conditions detected by many modules. Using modular hierarchies to reduce resource requirements therefore introduces a problem. If the operation of the system must be changed in some desirable manner without excessive proliferation of the conditions which must be detected, in general some previously defined groups of conditions must be modified. Such modifications will tend to introduce undesirable side effects into other operations which employ the original unmodified groups. It will therefore be necessary to find a compromise between use of resources and ease of modification. To make operational change practicable, modules must be defined in such a way that overall information exchange between modules is minimized as far as possible. In other words, conditions detected by one module must be used as little as possible by other modules, as far as is consistent with the need to economize on resources. The modular hierarchy is therefore a compromise between resource economy and ease of modification. Experience with the design of operationally complex systems indicates that such modular hierarchies are indeed essential to make operational changes possible [Kamel 1987].

Information Exchange within a Component Hierarchy

The primary role of information exchange within a component hierarchy is to coordinate different behaviours and resolve conflicts between them. Because components correspond with behaviours, the management of information meanings is simpler than within a modular hierarchy. An input to a component can only mean that the corresponding behaviour is encouraged or discouraged, and an output that the corresponding behaviour is encouraged and/or any other behaviour is discouraged.

Examples from Design of Electronic Systems

As an illustration of the compromises needed in defining the various hierarchies, consider the flight simulator example. Different modules could be organized to correspond approximately with different tasks performed by the pilot, such as takeoff, level flight, turning etc. With this approach, most of the information required at any point in time is within one module and the processing time required is minimized. It is therefore easier to achieve the necessary real time response. However, information on different parts of the aircraft must be maintained in each module and any changes to aircraft design will require changes to all modules. If modules are organized to correspond more closely with different parts of the aircraft, functional changes to the simulator to reflect aircraft technological advances are easier. Simulator architectures have therefore shifted from task modules to aircraft component modules as available processing speeds have increased and aircraft technological change has accelerated [Bass et al 1998].

In the case of the much more complex central office switch, call processing, diagnostics, billing and traffic measurement are different major customer features. However, the operational complexity of these tasks and the interactions between them is so great that if high level modules corresponded with these features, the interactions between modules and the demands for computing resources would be excessive. For example, billing requires information on the features invoked in the course of call processing, and call processing requires immediate information on problems identified by diagnostics.

As a result, there is very little correspondence on any level between modules and identifiable customer features. At the highest level, there are no modules corresponding with the major feature domains like call processing, diagnostics or billing. There are no modules on any level which correspond exactly with features like conferencing or call processing. Rather, at a detailed level the management of a connection between two users, from originating a connection through answering the call to disconnection of both users, is managed by repeated invocation of software modules performing functions with names like queue handler, task driver, event router, function processor, or arbitrator, and differences in the features invoked are only reflected in the information provided to these modules.

Modules in the Brain

Limitations to information handling resources are certainly present in biology, and will tend to result in a modular hierarchy in the brain. However, modules are not simply arbitrary parts from which the system is constructed, but must have specific properties tailored so that the brain achieves resource economies. The modular hierarchy must divide up the population of operations performed by the brain into major modules in such a way that similar operations are grouped together but the information exchange between them is minimized as far as possible. Major modules must be divided into submodules on the same basis, and so on down to the most detailed system operations. One general result of this process is that a module on any level will tend to have much more information exchange internally (i.e. between its submodules) than with other modules. Because, in the brain, information flows along physical connection pathways, this leads to the expectation of a hierarchy of anatomical modules distinguished by much more internal connectivity than external.

To illustrate the resource/modifiability problem for the brain, consider how visual input could be used to control a number of types of behaviour with respect to a river. These types of behaviour could include white water canoeing in, swimming in, fording, fishing in, drinking from, and verbally describing the river. For each of these types there are different visual characteristics which identify the optimum locations and ways to perform different behaviours within the type. These visual characteristics are all observations of the river which correlate partially with operationally significant conditions like depth of water, rate of flow, presence of visible or hidden rocks, evenness of the river bed, condition of the river banks etc. Such characteristics might include areas of white water, areas of smooth water, visible rocks, or whirlpools. The shapes and internal variations of different instances of these characteristics could be relevant to determining behaviour and therefore be part of the characteristic.

However, the shape and appearance of an area of relatively smooth water most appropriate for indicating a place to stop the canoe and plan the next part of the route through the rapids may not be exactly the same as the shape and appearance most appropriate for indicating the probable location of a fish, and the shape and appearance of a smooth area in which drinking water would contain the least sediment may be different again. Furthermore, even within one behaviour type, the shape and appearance of the relatively smooth area most appropriate for stopping the canoe may be different from the relatively smooth area indicating the best point of entry into a rapid. These high level characteristics might be constructed from more detailed characteristics like ripples, waves, relative colorations and shades, boundaries of different types etc. However, again the more detailed characteristics best for constructing the most appropriate higher level characteristics may not be the same for all behaviours.

One solution would be to define characteristics independently on all levels for each behaviour for which optimum characteristics were different in any way. This solution will in general be very resource intensive for an operationally complex system, and in most cases a compromise must be found. For example, a characteristic "patch of smooth water" might be used for multiple behaviours but for each behaviour there could be additional characteristics which could distinguish between operationally different patches.

Consider now the problem of learning an additional behaviour, such as learning to find a place to ford a river after white water canoeing, fishing, and drinking have been learned. Suppose that this learning must occur with minimum consumption of resources but also with minimum interference with already learned behaviours. Minimizing the use of resources would be achieved by minimizing the number of additional characteristics, in other words by modifying existing characteristics for the new behaviour. Minimized interference would be achieved by creating a new characteristic whenever there was a difference between the optimum characteristic for the new behaviour and any existing characteristic. Some compromise must be found between these two extremes, in which some already defined characteristics are slightly modified in a way which does not severely affect existing behaviours, and only a small number of additional characteristics are defined when the side effects of modification would be extensive and/or hard to correct. The effect of this process is that individual characteristics will not correspond with conditions which are behaviourally useful for just one behaviour. Rather, they are conditions which in different combinations are effective for managing multiple behaviours.

Cognitive science has made many attempts to define modules at relatively high psychological levels, with the hope of being able to identify such modules with physiological structures. "Module" in this context has generally been understood (following Fodor) as a unit which can only access a limited subset of the information available to the system, which performs a set of tasks following algorithms not visible to other modules, and which has a dedicated neural structure. Such a definition has some similarities with the minimized information exchange requirement. Cognitive modules of many different types have been proposed, including peripheral modules, domain specific modules, and conceptual modules. Proposed peripheral modules include early vision, face recognition, and language. Such modules take information from a particular domain and only perform a specific range of functions. Proposed domain specific modules include driving a car or flying an airplane [Hirschfeld and Gelman 1994]. In such modules highly specific knowledge and skill is well developed in a particular domain but does not translate easily into other domains. Conceptual modules are innate modules containing intuitive knowledge of broad domains, such as folk-psychology and folk-physics, with limited information flow to other parts of the brain [Leslie 1994]. However, various reasons including the difficulty of associating any such modules with plausible physiological structures has led to growing skepticism of the existence of such modules in the brain.

Because of the high operational complexity of the functions performed by the brain, the modules which would be expected to develop on the basis of practical considerations would not correspond with cognitive functions such as those described in the previous paragraph. However, the modules would be expected to correspond with physiological structures.

Module and Component Physical Similarities

Because systems must be constructed using a volume of design information which is as limited as possible, modules tend to resemble each other. Such a resemblance means that the same design information can be used with minor overlays to construct many different components and modules. For example, in electronic systems transistors are formed into integrated circuits and integrated circuits assembled on printed circuit boards, but the numbers of different types of transistors, integrated circuits, and printed circuit assemblies are kept as small as possible. Software is generally written in a high level language which can be translated (compiled and assembled) to assembly and machine code, but the portfolios of machine code, assembly code and high level language instructions are limited as much as possible.

A brain is constructed by a process which uses genetic information, and limiting the volume and complexity of this information could be expected to reduce the risk of copying errors and errors in the construction process. Brains can therefore be expected to be constructed from basic sets of modules and components on many levels of detail, which can be combined in various ways to achieve all required system functions. Note, however, that although the physical structure may be generally similar, every module and component will have a different operational role and will be physically different at a detailed level.

Condition Synchronicity

An operational system receives a constant sequence of input information. Frequently, a behaviourally relevant condition occurs within information that becomes available to the system at the same point in time. Other behaviourally relevant conditions may be made up of groups of simpler information conditions which are detected and became available to the system at different times separated by specific time intervals. Whether a condition is made up of direct system inputs or of subconditions made up directly or indirectly of system inputs, all of the information making up the condition must be derived from system inputs at the same point in time or at times with the appropriate temporal relationships.

For example, in the aircraft simulator example, a change to the orientation of the aircraft initiated by the pilot at one point in time could result in changes to many different aircraft components and to the cockpit displays and platform orientation. The module corresponding with each component could take a different time to determine its new status, and it is critical that all status changes occur synchronously.

A module will take a certain amount of time after the arrival of its inputs to detect the presence of any of its conditions and generate outputs indicating those detections. Outputs will take a certain amount of time to travel from the module generating them to the modules utilizing them. In the absence of correction, these delay times will tend to result in condition detections within information which does not have the appropriate temporal relationships.

The system architecture must therefore have the capability to ensure that conditions are detected within information with the appropriate temporal relationships. There are two possible approaches to this synchronicity management: global and local. The two approaches are illustrated in figure 3.2.

In the global approach, all the relevant inputs at the same point in time are recorded in a reference memory. Modules detect their conditions using the input information in this reference memory, and record the presence or absence of conditions they detect in this input set in the memory. Such recorded conditions can then be used by other modules. Modules must detect conditions in a sequence which ensures that the presence of a condition is not tested until all its component conditions have been tested.

In the local approach, modules are arranged in layers, and one layer can only draw condition defining information from the immediately preceding layer, or from system inputs in the case of the first layer. If all the modules in one layer take the same amount of time to detect their conditions and transmission time to any module in the next layer is the same, then all conditions will be detected within a set of system inputs at the same point in time.

The global approach requires additional resources and complexity to support explicit management of synchronicity. The local approach avoids the need for this explicit synchronicity management process, but in general will be less accurate.

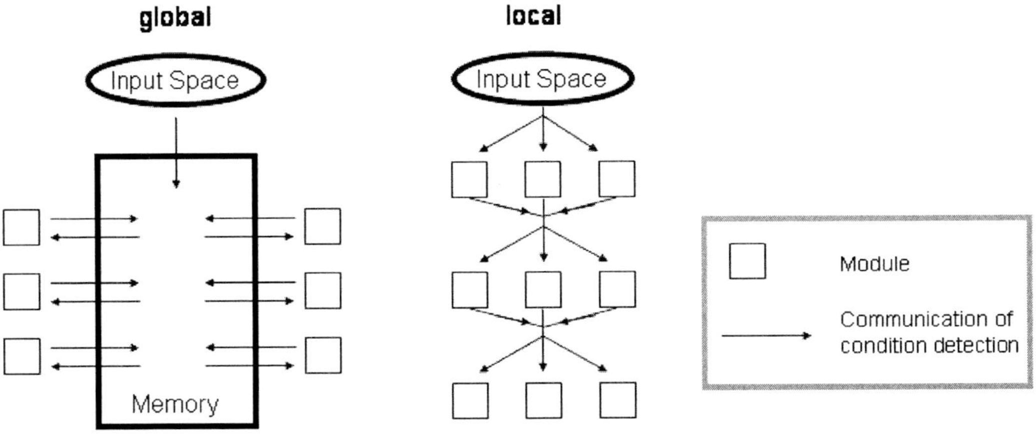

Figure 3.2. The global and local approaches to ensuring appropriate temporal relationships within the information making up conditions. In the global approach, an input state containing the input information at one point in time is recorded in a reference memory. Conditions are detected using information from that memory. The presence or absence of conditions in the input state is also recorded in the memory for use in the detection of more complex conditions. All conditions are therefore detected in a synchronous input state. In the local approach, simple conditions are detected within the information from the input state in a first layer of modules. Outputs from these modules indicating detection of their conditions are provided to a second layer which detects more complex conditions and so on. The layering and predominant connectivity only from one layer to the next layer means that all the conditions detected in one layer are within a synchronous input state provided module processing times and communication times are consistent across the layer. The local approach is therefore less complex but more prone to errors.

In commercial operational systems, global management of synchronicity is invariably used to ensure that the order in which module operations occur minimizes the probability of errors. For example, in actual flight the orientation and motion of an aircraft affects the operation of a weapon system and the operation of a weapon system affects the orientation and motion of the aircraft. The timing of the arrival of updated information between modules influencing these different components therefore affects the overall accuracy of the simulation. In a central office switch, call processing uses connectivity resources from a common pool, and both call processing and diagnostics can add and remove resources from the pool. If the knowledge of what remains in the pool is not updated in an appropriate sequence, a call in progress may plan to include a resource which is no longer in the pool, and have to recalculate its end-to-end connectivity path when it discovers that the resource is not actually available.

Modification, Diagnosis and Repair Processes

Although modules do not correspond with identifiable features, the modular hierarchy nevertheless makes it possible to add or modify features while limiting side effects and to diagnose and fix problems. To understand how the module hierarchy provides these capabilities in an electronic system, the first step is to recognize that a side effect of the way modules are defined is that the module hierarchy can operate as a hierarchy of descriptions for operational processes. If a module participates several times in the course of some process, a description which names the module rather than providing a full description of every internal operation will be simpler. Because modules are defined so that most interactions (i.e. information exchanges) are within modules and only a small proportion are with other modules, such a description can just include reference to the small external proportion. Such descriptions are more likely to be within the information handling capacity of the human brain. Although a particular feature will not in general be totally within one module, the criterion for module definition will tend to limit any one feature primarily to a relatively small set of modules.

An operational process can therefore be described at high level by making explicit reference to the names of each module which participates in the process and to the relatively small level of interaction between the modules. A new feature or a problem can be understood at that level, and the necessary changes or corrections to each module at high level understood. Consideration can then be limited to each participating high level module in turn by considering only the parts of the operational process in which each such module is active. These parts of the process are considered at the level of the submodules of each such module and the changes or corrections to each participating submodule which will create the desired module changes determined. Each submodule can then be considered separately and so on down to the level of software instructions and transistors at which changes can be implemented. In practice this change process also involves shifts back to higher levels, to determine whether a proposed change at detailed levels actually achieves the desired change at higher level, and whether it has undesirable side effects when different operational processes involving the same modules are considered at higher level. The modular hierarchy thus makes it possible to determine an appropriate set of detailed implementable changes to achieve a required high level feature or corrective change by a process which is always within the information capacity limitations of a human being.

If features must be modified heuristically, the existence of the modular hierarchy as defined is still an advantage. If circumstances are experienced which indicate the need for a change in behaviour, the set of modules most often active in those circumstances can be identified as the most appropriate targets for change. The overall minimization of information exchange will tend to limit the size of this set.

The existence of a modular hierarchy can also make recovery from damage more effective. For example, suppose that one module has been damaged. An external designer could of course identify the damaged module. A system which must use only its own resources to recover from damage is less likely to have such a direct capability. Brains need (and exhibit) the capability to recover from some of the deficits created by a stroke. It is essential that the actions selected heuristically to achieve recovery from damage have a high probability of achieving the desired change and low probability of undesirable side effects. This high probability must be achieved using only whatever information is actually available to the system. Such information could include identification that some modules were no longer receiving inputs (because these inputs came from the now damaged module). Resources for a new module to replace those inputs could be assigned, with outputs to the modules no longer receiving inputs. If the system has the capability to identify which modules have often been active in the past at the same time (see below for reasons why such a capability will in general be required), then the new module could be given provisional inputs from modules which were frequently active in the past at the same time as the modules no longer receiving inputs. These input biases will improve the probability that as the new module learns, it will replace the operations of the damaged module. Modularization and minimization of information exchange thus improves the effectiveness of damage recovery.

Context for Information Exchange within a Modular Hierarchy

As discussed earlier, the detection of a condition by one module is sometimes communicated to other modules. Such information exchanges also have an operational meaning to modules receiving them. To a recipient module, a specific condition detection means that the currently appropriate system behaviour is within a specific subset of the set of all behaviours influenced by the module.

A specific condition detection may have a different context (i.e. a different operational meaning) in each of the modules which receive it. For example, the detection of a specific condition by one module in a flight simulator might be used to indicate the need for changes to cockpit monitor images, cockpit displays, and platform motion.

Changes to a module can include both changes to its input population and changes to the conditions it detects within that input population. Such changes will affect outputs, and changes for one operational purpose may affect any module which makes use of outputs from the changed modules, directly or via intermediate modules. There is no guarantee that these secondary changes are desirable. This issue is conceptually illustrated in figure 3.3.

For example, a change to the definition of a condition detected by one module in a flight simulator made to reflect a change to the instrument displaying such conditions might have undesirable side effects on the cockpit platform motion which depended upon the original condition definition. The operational change problem can therefore be viewed as a problem of maintaining meaningful contexts for all information exchange. The seriousness of the context maintenance problem is illustrated by the experience that achieving adequate contexts for shared information is a major (and sometimes insurmountable) problem in integrating independently developed software systems, even for information technology applications [Garlan et al].

There are two different confidence levels which could be supported for operational meanings. One is unambiguous, in which a condition detection limits appropriate behaviour to the specific subset with 100% confidence. In this case a condition detection can be interpreted as a system command or instruction.

The alternative is meaningful but partially ambiguous meanings. In this case a condition detection indicates that the appropriate behaviour is probably within a specific subset. Such information exchanges can only be interpreted as operational recommendations, with the implication that a higher level of condition detection resources will be required to generate high integrity behaviours.

Although a system supporting only ambiguous contexts will require more condition detecting resources for a given set of system features, it has the advantage that heuristic changes to features are feasible in such a system and generally impractical in a system using unambiguous exchanges. To understand the practical issues around heuristic change, consider again the set of intercommunicating modules in figure 3.3.

In that figure, on average each of the eleven modules receives information from 2 modules and provides information to two modules.

Suppose that information exchanges are unambiguous. In this case the detection of a condition can be interpreted as an operational command. If a change is made to the conditions detected by the black module M in order to implement a feature change, that change will potentially affect the conditions detected by the three gray modules in figure 3.3 which receive information directly from M. Because these modules also influence features not requiring change, and the result of a change to a condition will be operational commands under the changed condition and not the original, changes may be required to the gray modules to correct for undesirable side effects of the change to M. However, any changes to the gray modules may affect the four striped modules which receive information from the gray modules and so on.

In a system in which features are defined under external intellectual control, the effects of changes to one module must be traced through all affected modules and side effects corrected explicitly. This is a difficult problem to solve and accounts for the difficulty of modification of such systems. However, if features are defined heuristically, the system must experiment with behaviours, receive consequence feedback, and adjust accordingly.

An experimental change to a feature will in general require an experimental change to conditions. Any experimental change to a condition will result in a command changes not only affecting the targeted feature but also any other features also influenced by the condition. The change will therefore introduce uncontrolled side effects on these features. However, consequence feedback is received only in response to the targeted feature and will not provide useful information on the other features until such features are invoked.

Thus any experimental change to one feature will result in multiple undesirable side effects on other features. Each of these side effects will eventually require an experimental change followed by consequence feedback to correct. Because all conditions are unambiguous commands, this process is very unlikely to converge on a high integrity set of features. In practice, for a large number of features dependent on the same group of shared resource modules, it is more likely that system behaviour will steadily diverge from desirable behaviour. This is similar to the catastrophic forgetting problem encountered in artificial neural networks, in which later learning completely overwrites and destroys earlier learning [French].

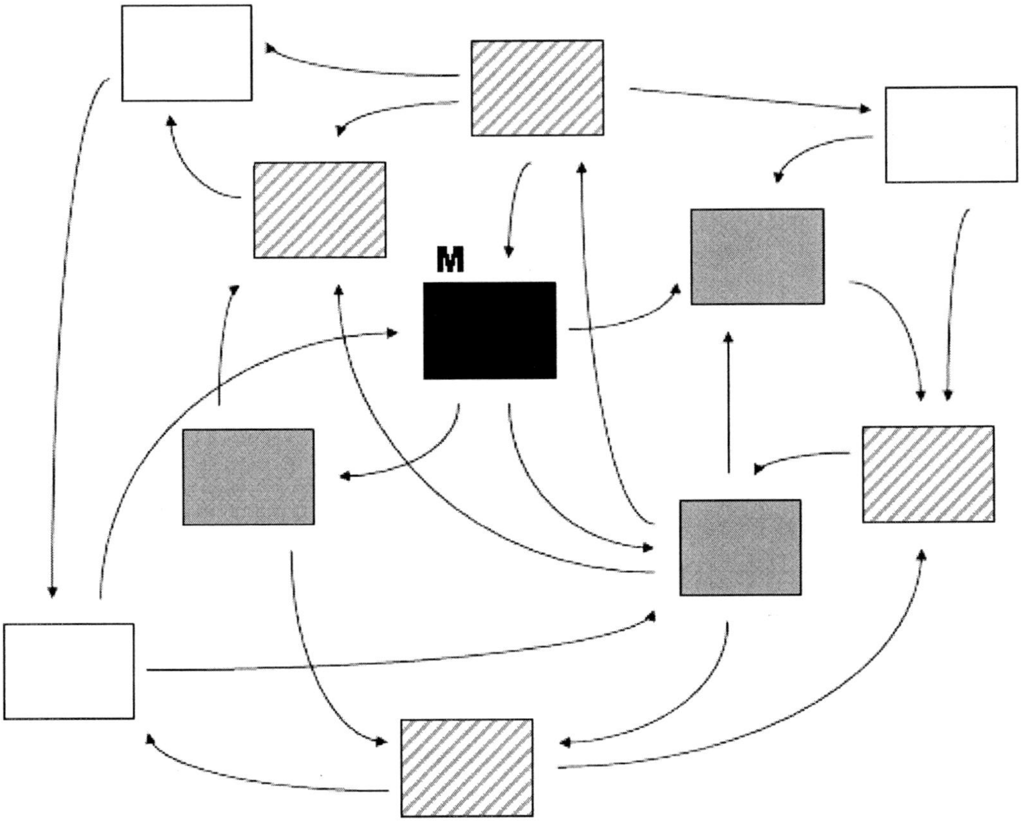

Figure 3.3 Spread of impact of information changes from one changed module. The arrows indicate information communications. In the illustration the black shaded module M is changed in some way to achieve some feature change. Module M provides condition detections to the three gray shaded modules. The features influenced by these modules may therefore be affected by the change to M. The gray shades modules provide condition detections to the striped modules which may therefore be affected by the change to M. The white modules receive inputs from the striped modules and may also be affected. There could be an exponentially increasing wave of side effects developing from the change to M.

However, if information exchange contexts are partially ambiguous, information exchanges are operationally only recommendations. Any accepted behaviour for one feature will in practice be supported by a large number of recommending condition detections. "Small" changes to a few of these conditions to support changes to other features will not necessarily affect the performance of the feature, and any effects on performance could potentially be corrected the next time the feature is invoked. Thus provided condition changes are "small", the system could converge on a set of high integrity features. Clearly the definition of "small" is critical.

Local synchronicity management is unlikely to be adequate if an unambiguous level of information meaning must be sustained. However, if information exchanges are partially ambiguous, some level of synchronicity errors resulting from local synchronicity management can be tolerated, and the resource cost and complexity of global management can be avoided.

The von Neuman Architecture

Some of the reasons for the ubiquitous appearance of the memory, processing von Neumann architecture in commercial electronic systems are now apparent. Unambiguous information exchanges are used because much less information handling resources are used. The command structure of software (e.g. if: [x = a] do: [...]) reflects use of unambiguous contexts. The use of unambiguous information mandates the memory, processing separation in order to maintain synchronicity. Finally, learning in a von Neumann system performing an operationally complex task has never been demonstrated in practice.

The Recommendation Architecture

If information which is operationally ambiguous is exchanged within a module hierarchy, module outputs indicating the presence of information conditions can only be interpreted as behavioural recommendations. Multiple recommendations will be generated in response to an overall input condition in most cases. A subsystem separate from the modular hierarchy is therefore required which can select one behaviour. To make such selections this subsystem must have access to information in addition to the overall input state, such as the consequences of behaviours under similar conditions in the past.

The separation between a modular hierarchy and a component hierarchy which interprets module outputs as alternative behavioural recommendations and selects one behaviour is one of the primary architectural bounds of the recommendation architecture. This separation is illustrated in figure 3.4. The hierarchy is called because it clusters conditions into modules detecting portfolios of similar conditions, and the selection subsystem is called because it manages the competition between alternative behavioural options.

Information on the most appropriate associations between information conditions detected within clustering and behaviours must either be derived from the experience of other systems (i.e. via design or genetic experience, or the experience of an external instructor) or be derived from the experience of the system itself (i.e. from feedback of the consequences of behaviours under various conditions in the past).

In a very simple version of the recommendation architecture, all information conditions and behaviour selection algorithms could be specified a priori. In a more sophisticated version, conditions and algorithms could be almost entirely specified a priori but slight changes could be made to conditions by adding inputs which frequently occurred at the same time as most of the condition and deleting inputs which rarely occurred. Consequence feedback could tune the behaviour selection algorithms. Learning from experience would be present but very limited and the architectural constraints deriving from the need to learn without side effects would therefore be less severe. Coward [1990] has argued that this is the case for the reptile brain. A yet more sophisticated version would allow conditions to be defined heuristically and algorithms to be evolved by consequence feedback and external instruction, which as discussed below appears to correspond with the mammal brain. Different degrees of a priori guidance in the selection of starting points for conditions and algorithms could occur.

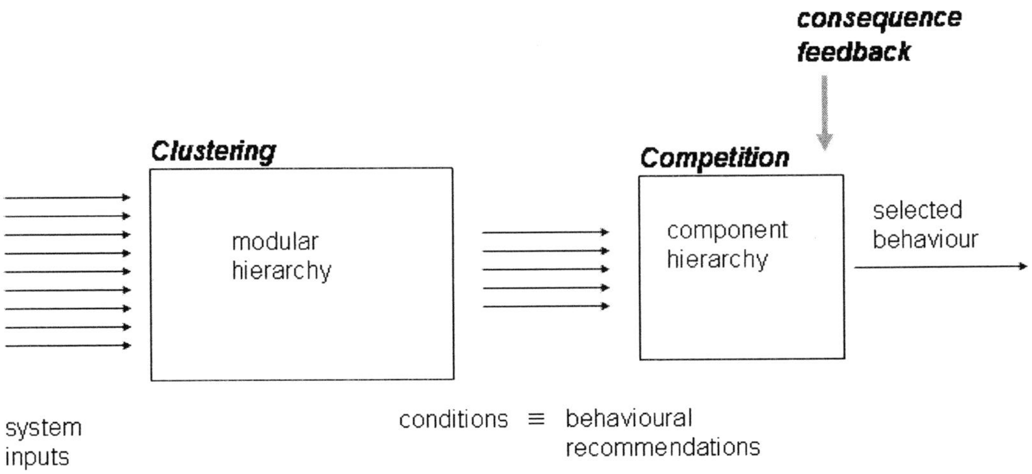

Figure 3.4 Recommendation architecture primary separation between clustering and competition. The clustering subsystem defines and detects behaviourally ambiguous conditions in system inputs. Some conditions are used to determine when and where other conditions will be added or modified within clustering. Other conditions are communicated to the competition subsystem. The competition subsystem uses consequence feedback (sometimes associated with imitation of an external teacher) to interpret conditions detected by clustering as behavioural recommendations and to select the most appropriate behaviour under current circumstances. Conditions detected by clustering cannot be directly changed by competition or by the consequence information available to competition.

If the conditions detected by clustering modules are defined heuristically, the operational ambiguity of information makes it possible to limit the undesirable side effects of heuristic changes. In general terms this can be understood as an effect of multiple recommendation generation. A wide range of conditions will be detected in the typical overall input state. Overall input states corresponding with an already learned behaviour will in general contain a high proportion of conditions which have been associated with the behaviour by the competition subsystem. Slight changes to some conditions will therefore generally not change the behaviour selected. However, a change to a condition will also change any higher complexity condition which has incorporated that condition. The critical issue is therefore the definition of "slight" and "some", as discussed in the next section.

If consequence feedback were applied directly to clustering it would need to be applied to the currently active modules. Unfavourable consequence feedback could be interpreted as indicating that the wrong conditions were being detected by those modules, but would give no indication of how the conditions should be changed. Thus any change could not be guaranteed to improve the selection of the current behaviour in the future, and would have uncontrolled side effects on any other behaviours influenced by the changed modules. In an operationally complex system such changes would be very unlikely to converge on a high integrity set of behaviours. Consequence feedback therefore cannot be used directly to guide the definition of conditions. The primary information available to guide heuristic definition of conditions is therefore limited to frequent occurrence of a condition. This temporal information must be utilized to define clusters of input conditions on many levels of complexity. The clustering process is conceptually similar to the statistical clustering algorithms used in ANN unsupervised learning, as discussed below. External instruction could of course influence the definition of conditions by influencing the environment from which the system derived its inputs.

The use of consequence feedback constrains the form of the competition subsystem. Competition cannot depend on internal exchange of information with complex operational meanings. Components in competition must each be associated with just one behaviour, specific sequence of behaviours invoked as a whole, or general type of behaviour. Competition therefore corresponds with the component hierarchy discussed earlier. The output from such a component to other competition components can be interpreted as a recommendation in favour of the behaviour and against all other behaviours, and an output which goes outside competition can be interpreted as a command to perform the behaviour. Such components can interpret inputs from other clustering modules or competition components as either encouraging or discouraging their one behaviour. The relative weights of such inputs can be adjusted by consequence feedback in response to that one behaviour. As discussed below, this approach is similar to reinforcement learning in ANNs.

Heuristic Changes to Conditions

The output of a module within clustering indicates the detection of an information condition within the portfolio of such conditions programmed on the module. In an operationally complex system such an output may be an input to many other clustering modules. The operational meaning of such an input is the recommendation of a set of behaviours which will in general be different for each recipient. The major issue is therefore how to preserve the multiple meanings which the output of a module has acquired when the conditions programmed on the module which generated the output are changing. This is the issue of context management.

To illustrate the issue, consider a simple human example. Suppose that I have an understanding with my wife that when I leave a telephone message asking her to call me at the office, an emergency has occurred. Suppose also that this understanding is completely unambiguous, and I will never leave such a message in the absence of an emergency. In this situation, if I leave a message she will stop whatever she is doing and act as if there is an emergency. If one day I just wanted to chat, leaving the message will result in an inappropriate response. In other words there is no scope for learning. However, if leaving a message was partially ambiguous, then the tone of my voice, the exact wording, my mood when I left that morning, and her knowledge of the probability of an emergency at that particular time could all contribute to selection of a different response. In other words, learning is possible. However, there are limits to the degree of change. I cannot leave a message in a foreign language and expect a call, and I cannot leave the usual message and expect to be called at the new number I have just been assigned. So the critical issues for context maintenance are how much the form of the message generated under the same conditions can be changed, and how much the conditions under which the same message is generated can be changed without destroying the usefulness of the message to recipients.

As operational complexity increases, the average degree of change which can be tolerated without unacceptable loss of meaning will decrease. For example, at the most detailed level of the modular hierarchy there must be devices which can select, record, and detect conditions which are combinations of their inputs. Suppose that one such device recorded its first combination (or condition) and indicated its presence by producing an output. Devices in many other modules might then select that output to be part of their own conditions. In general the source device has no "knowledge" of the operational meanings derived from its outputs by its targets. If that source device subsequently changed its programmed condition, the meaning assigned to the output by its recipients would be less valid. Degree of change to a device must therefore be "slight" and located on as "few" devices as possible. Determination of when and where change should occur is itself an operationally complex problem which must be managed by the system. In other words, some of the conditions detected in some modules must determine when change can occur to the conditions programmed on other modules.

Degree of Change

The minimum change situation would be if a device could only be programmed with one condition, and only produce an output in response to an exact repetition of that condition. Such a change algorithm would be extremely expensive in terms of device resources needed. A slightly greater degree of change would be if such a device could subsequently be adjusted to produce an output in response to any large subset of its original condition. A greater degree of change would be if a small proportion of inputs could be added to an existing condition, for example if the additional inputs occurred at the same time as a high proportion of the condition. A yet greater degree of change would be if the device could be programmed with multiple semi-independent conditions with inputs drawn from a population of possible inputs which tended to occur frequently at the same time. Such a device would not produce an output in response to the presence of a subset of one of its independent conditions, but might respond to the simultaneous presence of many such subsets. In all these change algorithms, the generation of an output in response to a specific condition results in a permanently increased tendency to produce an output in response to an exact repetion of the condition. Repeated exposure could be required to guarantee that an output will always occur, or a single exposure could be adequate. The key characteristic of the algorithms is that change only occurs in one direction, always resulting in expansion of the portfolio. This characteristic will be referred to as “permanent condition recording” although multiple exposures may in some cases be needed as also discussed when physiological mechanisms are reviewed in chapter 8. To avoid portfolios becoming too large and therefore indiscriminate, as a portfolio expands it will become more difficult to add new conditions. A portfolio which became too broad would be supplemented by new portfolios in a similar input space as described in the section on Resource Management Processes in the next chapter.

There is nevertheless an information context issue even with permanent condition recording algorithms: two identical device outputs at different times may indicate slightly different conditions. Change algorithms allowing higher degrees of change may require additional mechanisms to maintain context. For example, there could be structure in a device output which provided some information about where in the device portfolio the current condition was located. In addition there could be management processes to determine the consistency by some additional criterion between conditions detected by different devices. A source of information for such a criterion could be the timing of past changes to devices. Suppose that whenever a relatively low complexity condition was recorded on a device, feedback connections were established to that device from devices which were currently detecting higher complexity conditions. Then suppose that subsequently if a low complexity device produced an output indicating detection of a condition, it would cease producing the output unless some of these feedback connections were active. If some low complexity devices ceased activity, some higher complexity devices would in general cease activity because of loss of their inputs. The population of active devices would therefore be reduced towards a set with higher consistency in past recording activity. Release of information outside the overall device population could be delayed until the period required for feedback had elapsed. Conditions recorded or changed on devices at all levels in that overall population would only become permanent if device activity continued beyond the feedback period [Coward 2000].

Artificial neural networks use change algorithms based on adjustment to relative input weights. Such algorithms mean that the device does not necessarily produce an output in response to conditions which previously produced an output. As operational complexity increases, the problem of maintaining adequate context within the clustering will become more and more severe. These types of algorithm are therefore unlikely to be adequate under such conditions. However, under the operationally simpler information contexts within the competition subsystem, such relative weight change algorithms will be ubiquitous.

The selection of the change algorithm is a compromise between resource economy and context management adequacy. If there were an input domain within which relatively simple conditions could be defined in early system experience, and the definition completed before these conditions were incorporated extensively in higher complexity conditions, then the change algorithms for the simple conditions could be more flexible, even to the point of using relative weight adjustments. However, no such changes could occur after the early learning period. For ongoing learning of a very complex set of operations subject to resource constraints the most plausible device algorithm is the multiple semi-independent condition recording model with mechanisms to enhance context maintenance. This device is described in detail in the next chapter. The same compromise between resources and context maintenance could also mean that if a condition was recorded but never repeated for a long period of time, elimination of that condition and reuse of the resources would be relatively low risk, perhaps provided that the condition was not recorded at the time of a behaviour which could be viewed as highly important.

Location of Change

A further consideration for context management within clustering is the determination of when and where changes should occur. To give competition something to work with, even to make the decision to do nothing, clustering must generate some output in response to every overall input state. If output is inadequate, clustering must add conditions or modify conditions until output is adequate. The selection of where such changes should occur is itself a operationally complex problem which must therefore be managed by clustering. In other words, as described in the next chapter, some detected conditions must determine whether or not changes are appropriate in specific modules under current conditions. At the device level this implies that there will be change management inputs in addition to the inputs defining conditions. These change management inputs will excite or inhibit changes to conditions, but will not form part of any condition.

Organization of Change

As described in more detail in chapters 4 and 6, clustering organizes system input states into groups of similar conditions detected within those states. These groups are called portfolios, and conditions can be added to a portfolio but not deleted or changed. Portfolios are added and expanded to meet a number of objectives. Firstly, every input state must contain conditions within at least a minimum number of different portfolios. Secondly, the overlap between portfolios must be as small as possible.

In general terms, the process for definition of a set of portfolios is as follows. A series of input states to the portfolio set are monitored to identify groups of individual inputs which tend to be active at the same time but at different times for each group. Different groups of this type are defined as the initial inputs to different portfolios. Conditions which are subsets of a group are then defined in the corresponding portfolio whenever the input state contains a high level of activity in the input group. Additional inputs can be added to the group if required. However, as the number of different input states containing conditions within one portfolio increases, the ability of that portfolio to add conditions decreases. An a priori bias placed on the initial input group definitions can improve portfolio definition speed and effectiveness.

The set of portfolios defined in this way will divide up the input space into similarity groups which will tend to be different from each other but to occur with similar frequeny in input states. Individual portfolios will not correspond exactly with, for example, cognitive categories, but will permit discrimination between such categories. Thus different categories will tend to activate different subsets of the set of portfolios, and although the subsets may partially overlap, the subset for one category will be unique to that category. There are mechanisms for detecting if such subsets are not unique and adding portfolios in such circumstances.

As described in more detail in chapter 4, an example would be if a set of portfolios was exposed to a sequence of input states derived from different pieces of fruit. Portfolios might be defined that happened to correspond roughly with similarity groups like "red and stalk", "green and round", or "smooth surface and round" etc. Although no one such portfolio would always be detected when one type of fruit was present and never any other fruit, the specific combination of portfolios detected could provide a high integrity identification of the type. Each portfolio can be given a set of weights corresponding with how strongly it indicates different fruit types, and the most strongly indicated fruit across the currently detected portfolio population determined.

Portfolios defined on different levels of condition complexity could provide discrimination between different cognitive feature, objects, groups of objects etc.

Similarities between Modular Hierarchies in Operational Systems and Description Hierarchies in Scientific Theories

There are some strong similarities between modular hierarchies in operationally complex systems and the hierarchies of description which form theories in the physical sciences as discussed in chapter 2. In both cases the units of description on one level are defined in such a way that their internal construction can largely be neglected in the description of phenomena or processes at that level, and human understanding is made possible by the capability to shift freely between levels.

The modules into which the brain will tend to be constrained by practical resource and other considerations can therefore be the basis for a scientific understanding of cognition in terms of processes between identifiable physiological structures.

References

Bass, L., Clements, P. and Kazman, R. (1998). . Addison-Wesley.

Chastek, G. and Brownsword, L. (1996). . Carnegie Mellon University Technical Report CMU/SEI-96-TR-035. www.sei.cmu.edu/pub/documents/96.reports/pdf/tr035.96.pdf

Coward, L. A. (1990). , New York: Praeger.

Fodor, J. A. (1983). . Cambridge, MA: MIT Press.

French, R.M. (1999). Catastrophic Forgetting in Connectionist Networks. 3(4), 128-135.

Garlan, D., Allen, R. and Ockerbloom, J. (1995). Architectural Mismatch or Why it's hard to build systems out of existing parts. . New York: ACM.

Hirschfeld, L. and Gelman, S. (1994). . Cambridge University Press; Karmiloff-Smith, A. (1992). Beyond Modularity. MIT Press.

Kamel, R. (1987). Effect of Modularity on System Evolution. , January 1987, 48 – 54.

Leslie, A. (1994). ToMM, ToBY and Agency: Core architecture and domain specificity. In , L. Hirschfeld and S. Gelman, eds. Cambridge University Press.

Nortel Networks (2000). . http://a2032.g.akamai.net/7/2032/5107/20030925230957/www.nortelnetworks.com/products/library/collateral/50004.11-02-00.pdf

Chapter IV

The Recommendation Architecture

In the previous chapter the reasons that practical considerations constrain a system which learns a complex combination of features within specific architectural bounds were described. These bounds are called the recommendation architecture, because all information generated within the modular hierarchy which performs the operationally complex processing is partially ambiguous and can only be interpreted as behavioural recommendations. This is in contrast with the von Neumann architecture ubiquitous in commercial electronic systems in which such information is unambiguous and can be interpreted as behavioural commands. In this chapter the detailed forms of a system within the recommendation architecture bounds are described.

As discussed in much more detail in chapters 7 and 9, the recommendation architecture models the human and other mammal brains as control systems which receive sensory inputs from their environments and use these inputs plus a range of information recorded in response to past sensory inputs, past internal brain states, and past behaviours to determine the most appropriate current behaviour. The focus of this chapter will be on the detailed forms of any system with comparable operational complexity to the human brain.

As described in chapter 3, at the highest architectural level there is a primary separation between clustering and competition. Clustering is a modular hierarchy which defines and detects conditions within the input information space available to the system. Competition is a component hierarchy which defines frequently used behaviours and behaviour types, and associates conditions detected by clustering with recommendation strengths in favour of different such behaviours and types. Competition determines and implements strongly recommended behaviours, and adjusts recommendation strengths on the basis of consequence feedback following a selected behaviour. Consequence feedback does not directly affect the conditions detected by clustering.

The Structure of Clustering

A range of considerations influence the detailed structure of clustering. These considerations include the needs to maintain synchronization of related information as discussed in the previous chapter; to define conditions which are as behaviourally useful as possible; to manage the circumstances under which conditions are recorded to minimize unnecessary recording; to define groups of similar conditions which can be detected by resource sharing modules; to limit the use of connectivity resources; and to minimize the information exchange between modules as far as is consistent with available resources. The way in which these considerations influence the detailed structure of clustering is the topic of this section.

Conditions and Recording Substrates

Because conditions are defined heuristically, the most appropriate conditions for guiding behaviour are not known to the system a priori. The population of possible conditions within a large available input space is immense, and in the absence of a priori guidance there will be a random element to the definition process. Conditions cannot be changed (with some limited exceptions discussed below), and therefore cannot be evolved to correspond exactly with behaviourally "ideal" conditions. Various strategies can reduce the random element to some degree but cannot eliminate it.

Two general strategies limit the type of conditions and the circumstances in which recording can occur. The general limit on type is that only conditions which actually occur in the course of system experience can be recorded. This limit on type recognizes that a condition which actually occurs has a higher chance of being useful than a condition selected completely at random. The general limit on circumstances is that conditions will only be recorded when the population of currently detected conditions is below a minimum. This limit on circumstances is equivalent to a requirement that some minimum level of behavioural recommendations must be present in response to every system experience, and conditions will be defined and recorded only until this minimum is achieved. These two general strategies limit the range of condition recording to some degree, but additional strategies to further reduce the random element are essential and will be discussed later.

An implication of the heuristic definition of conditions and the unavoidable random element to condition selection is that condition recording will occur throughout the life of the system, with the rate of recording depending on the degree of novelty in the current system inputs.

As discussed in more detail in chapter 8, the ultimate substrate on which conditions are recorded in the recommendation architecture model for the human brain is the pyramidal neuron in the cortex. A condition is defined by any group of inputs which can cause the neuron to fire, so many conditions are recorded on one neuron. To give a rough order of magnitude for the condition recording process, if it is assumed that there are $\sim 10^{10}$ neurons in the brain, each neuron has $\sim 10^4$ synapses, and ~ 50 active synapses cause a neuron to fire (and therefore simplistically correspond with one condition), there could be total of 2×10^{12} conditions recorded, or for a lifetime of ~ 70 years, an average recording rate of ~ 1000 conditions per second. The actual condition recording rate will vary considerably depending on the degree of novelty in current experience, and the lifetime average rate may be different from the estimate by an order of magnitude or more. The purpose of this very rough estimate is to emphasize the contrast between the condition recording algorithm and artificial neural network models in which neuron response is by continuous adjustment to input weights. It should be noted, however, that algorithms similar to classical artificial neural network algorithms are used in competition to define and adjust recommendation weights. As discussed in chapter 7, in the recommendation architecture model for the brain, subcortical structures including the thalamus, basal ganglia and cerebellum correspond with competition.

The presence of a random element in condition definition means that conditions will not correspond exactly with, for example, cognitive features. In other words, one condition may be present when a number of different features are present, and its absence does not conclusively demonstrate the absence of any one feature. However, the presence of a specific condition increases the probability that features in a particular group could be present, and the most probable features are indicated by the overall population of detected conditions. This lack of exact correspondence with cognitive features is equivalent to operational ambiguity as discussed in the previous chapter.

Preprocessing of Raw Sensory Information

A vast range of raw inputs containing information related to the external environment and the internal states of both the brain and the body are available to the brain from various senses. Individual raw inputs correlate very weakly with the presence of behaviourally relevant objects and circumstances. These raw inputs must be processed to detect conditions which correlate more strongly with such objects and circumstances.

In the early stages, retinal outputs may be processed into signals correlating with the presence of boundary elements in the visual field. The detection of boundaries may then be used to generate information on objects which is independent of the position and size of the image of the object on the retina. Experimental evidence indicates that early visual processing had this capability [see e.g. Biederman et al 1992; Furmanski et al 2000; Tovee et al 1994; Ito et al 1995]. Some of this early processing, or preprocessing, could use algorithms which differ from permanent condition recording for a number of reasons.

Firstly, the type of conditions could be genetically specified (i.e. a priori) to some degree. For example, predefined connectivity could be biased in favour of the detection of boundary elements within raw inputs, and could also favour the definition of conditions in the information derived from within a continuous boundary which are independent of the position and size of the image of the object on the retina. Secondly, conditions could be changed after initial definition, provided that such changes only occurred in early learning before the conditions were distributed to and incorporated in a wide range of different higher level conditions. Limiting change to the period before the conditions are generally utilized means that the changing conditions have not acquired multiple behavioural meanings prior to change, and the requirement to maintain an adequate context for these prior meanings is not present.

Although the use of a priori information from genetic sources and of condition change in the period before the conditions are widely distributed will tend to be most applicable in the early processing of raw sensory inputs, it could be present to some degree in later clustering as discussed below.

Definition of a Condition within Clustering

The outputs from the preprocessing of raw inputs are the sensory inputs to clustering proper. A somewhat oversimplified way of understanding the definition of information conditions within clustering is that a particular condition is present if each of a specific set of these sensory inputs is in a specific state. Conditions may be defined by the presence of a combination of other conditions, in which case the presence of the contributing set of sensory inputs reaches the condition via its component conditions.

Conditions are recorded on substrates, and the recording of a condition is indicated by activation of the substrate on which it is recorded. Any subsequent exact repetition of the condition will again activate the same substrate. Multiple similar conditions are recorded on the same substrate, and activation of the substrate therefore indicates the current presence of one or more of the conditions within the recorded set. Activations last for a brief time after the occurrence of a condition and then cease. This activation time is the period during which the detection of a condition is available to the system. If there is operational value in making this information available for a longer period, the activation time of the substrate can be extended as discussed below.

The definition of similarity in this context is that all the conditions on one substrate are subsets of a group of component conditions which have tended to be active at similar times in the past. For two conditions on the same substrate, in general some of their defining component conditions will be the same, and others will be different for the two conditions but will often have been present in the past at the same time. The larger the correlations of this type between component conditions, the more similar the conditions and the more probable that the two conditions will be present when a similar range of behaviours are appropriate, in other words that they will have a relatively consistent behavioural meaning.

This simple view of conditions is made considerably more complex because a substrate can be activated not only by the presence of its conditions within sensory inputs, but also indirectly by two other types of mechanism. One mechanism is that a substrate can be activated if a number of other substrates are already active which have often been active in the past at the same time as the substrate. The other mechanism is that a substrate can be activated if a number of other substrates are already active which have recorded conditions at the same time in the past as the substrate.

Immediately after the simultaneous activity of two substrates, there is a strong tendency for the activity of one substrate to activate the other, but this tendency declines fairly rapidly with time. However, if an indirect activation actually results from the tendency and occurs in the course of generating a useful behaviour (i.e. one followed by positive consequence feedback), the decline is reduced. If such indirect activations occur a number of times, the tendency stabilizes long term and may even increase again.

If there is simultaneous recording on two substrates, the substrates must have been simultaneously active. The tendency to activate on the basis of simultaneous recording is therefore in addition to the tendency on the basis of simultaneous activity, but is stronger, declines less rapidly, and is supported by independent connectivity. Again, a behaviourally useful indirect activation on the basis of simultaneous recording reduces the decline, and repeated such activations stabilizes the tendency long term. For both past activity and past condition recording, there is also a tendency for a substrate to activate a second substrate if one substrate was active or recorded conditions slightly before or slightly after the other.

These indirect activation mechanisms can be viewed as supplementing the conditions present in current sensory inputs with other conditions which have a significant probability of being relevant to determining the most appropriate current behaviour. For example, conditions which have been active in the past at the same time as currently present conditions may contain information about the current environment which cannot currently be observed.

A newly recorded condition is made up of a set of currently active component conditions. Some of these component conditions may be combinations of currently present sensory inputs, and some could have been activated by one of the two indirect mechanisms. Both the definition of conditions in terms of sensory inputs and the relationship between sensory inputs and the resultant pattern of condition activation can therefore become very complex.

The Condition Recording Device

At a detailed physical level there must be devices which record and detect conditions. A condition is physically defined by a set of inputs on to its device from other devices or from sensory inputs. Any recording or detection of conditions results in activation of the device, where an activated device generates an output indicating the detection to other devices.

A condition recording device is illustrated in figure 4.1. A number of different recorded conditions are implemented by different groups of inputs. Depending on the total level of activity within a group required to activate the device, one group may correspond with many very similar conditions, each of the conditions being a large enough subset of the group to activate the device. The output of a device does not indicate which conditions within its recorded set are currently present, but the structure of the output can provide some additional information. For example, if outputs are sequences of spikes, the average rate at which the device generates output spikes can indicate the number of different conditions currently being detected. As discussed in detail later, this average rate can also be modulated in various ways to indicate different input domains within which the conditions are being detected.

The recording of a condition on a specific device is itself a behaviour which must be adequately recommended before it can occur. A condition recording device therefore has inputs exciting and inhibiting such condition recording. These inputs indicate the overall level of activity in specific source modules as discussed below, but do not form parts of conditions. An input as illustrated in figure 4.1 may in fact be a group of physical inputs from different sources with a relatively high level of activity across the group being required to support the corresponding change behaviour. Because it would be difficult to achieve meaningful consequence feedback on a decision to record conditions on a particular device, these change management connections will not generally involve competition.

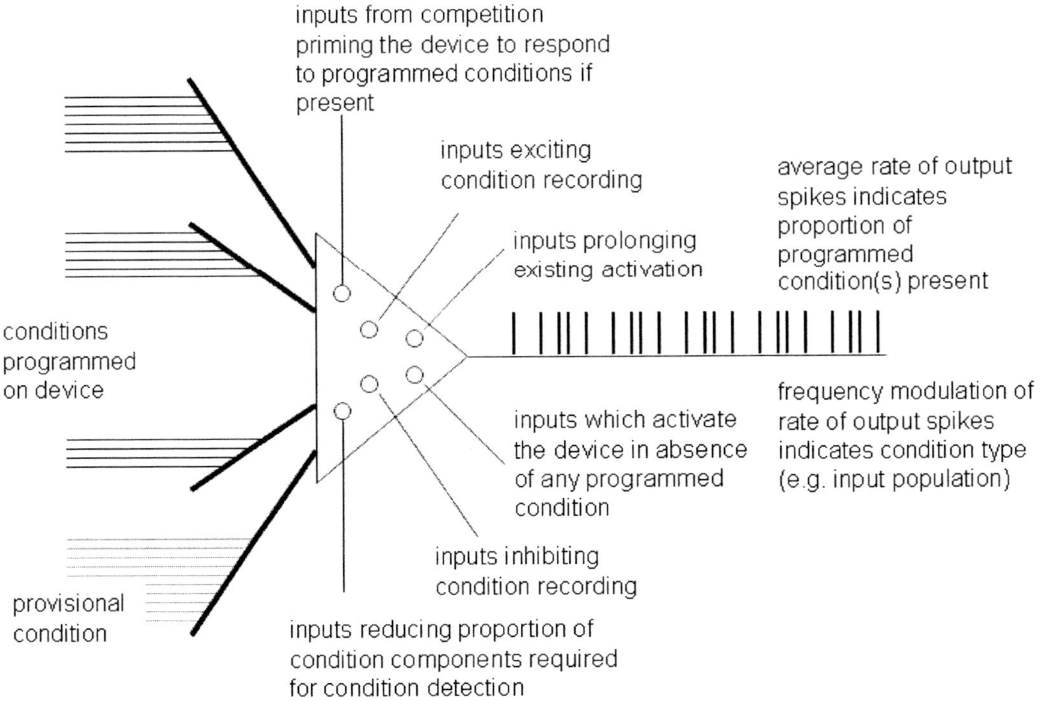

Figure 4.1. The condition recording device in the recommendation architecture. These devices have different groups of inputs defining different conditions or very similar sets of conditions. The presence of one or more recorded conditions activates the device. Other inputs determine whether or not a device will record an additional condition at any point in time, but these inputs do not form parts of conditions. Yet other inputs prolong the activity of an active device or activate it in the absence of any of its recorded conditions, or activate it if substantial proportions of conditions are present but not full conditions. An activated device generates a series of voltage spikes. The average spike rate indicates the number of recorded conditions currently present, and the phase of a frequency modulation imposed on the average rate indicates the input space within which the conditions have been detected.

Given that conditions are defined by physical connectivity, the inputs which define a new condition will in general need to be established in advance of the occurrence of the condition to be recorded. In practice therefore, provisional conditions defined by a group of provisional inputs will need to be physically created on devices. To ensure adequate similarity to existing conditions on the same device or within the same module, the selection of provisional inputs must be biased in favour of inputs which have often been active in the past when the device has been active, or when other devices in the module to which the device belongs have been active.

Recording of a condition can occur if several requirements are met simultaneously. Firstly, the level of device activation must be low. Secondly, there must be strong inputs exciting the recording of conditions. Thirdly, inputs inhibiting the recording of conditions must be weak. Fourthly, a provisional condition must have significant activity in a high proportion of its inputs. If these four requirements are met, the active proportion of the provisional condition becomes a recorded condition and the device will increase its output activity to indicate detection of the condition. Any future repetition of the condition will activate the device independent of the state of the change management inputs.

A condition recording device also has some special purpose inputs as illustrated in figure 4.1. One type of input can activate the device in the absence of any of its recorded conditions. This type of input is required to implement the capability to activate conditions on the basis of past temporally correlated activity as discussed earlier. A second type of input prolongs the activity of an already active device. This type of input makes the indication of condition detection available for a longer period if required. Both of these types of input can only be active as a result of acceptance of recommendations in favour of such activation behaviour by competition.

Two more types of special purpose input influence the ability of the device to respond to recorded conditions. One type increases the level of output generated in response to a given level of condition detection, the other type decreases the level of output. The operational role of these inputs is discussed later.

Not all of the special purpose inputs need to be present in every device. The presence of such inputs will be determined by the module in which the device is located. Furthermore, the physical implementation of such inputs may differ depending on, for example, the time scale over which they need to operate.

Role of Simultaneous Activity in Condition Definition and Detection

When consequence feedback and a priori knowledge are excluded, simultaneous activity is the only source of information available to guide efficient organization of information derived from sensory inputs. The organization of conditions in the recommendation architecture uses information on simultaneous activity in a number of ways. Firstly, provisional conditions are defined by random selection within a population of appropriate inputs, but with a bias in favour of inputs which have tended to be active at similar times in the past. Secondly, permanent conditions are created by a combination of activity in all the inputs which define them, strong overall activity in some specific modules, and weak activity in other specific modules, all occurring simultaneously. Thirdly, simultaneous activity of two condition recording substrates creates a tendency for one substrate to activate the other at a later point in time. This third way is similar to the mechanism suggested by Hebb [1949] in which the ability of one neuron to activate another increased when the two neurons were active at the same time. However, unlike Hebb, in the recommendation architecture the tendency decays unless it results in a behaviourally useful activation, and as discussed later, such activations cannot occur unless other active substrates also encourage secondary activations of the type. Fourthly, simultaneous condition recording on two condition recording substrates creates a tendency for one substrate to activate the other at a later point in time.

Device Layering

To ensure that conditions are detected within a set of sensory inputs with adequate temporal consistency, condition recording devices are arranged in layers, with most condition defining inputs to one layer coming from devices in the preceding layer. Change management and special purpose inputs may be derived from other layers, but do not form parts of conditions.

The layering of devices also means that any one layer detects conditions which are within a specific range of complexity, where the complexity of a condition can be defined as the total number of sensory inputs which contribute to the condition, either directly or via intermediate conditions. The condition complexity of a layer is the product of the average number of inputs to condition recording devices in the layer and in each of the preceding layers.

Conditions within different ranges of complexity may be more appropriate for recommending different types of behaviour. In this context, "appropriate" refers to the degree of ambiguity of conditions with respect to a behaviour type. For example, suppose that an average condition in one layer can sometimes occur when N different behaviours of a type are appropriate, while an average condition in a second layer can sometimes occur when n different behaviours of a type are appropriate. If $n \ll N$, then conditions in the second layer are less ambiguous with respect to the behaviour type than conditions in the first layer. The lower ambiguity will make the second layer conditions more effective for selecting a behaviour within the type. Such differences in behavioural ambiguity between randomly defined conditions have been demonstrated by electronic simulation (see chapter 6). As an illustration of the concept, conditions with the same order of magnitude of complexity as the number of sensory inputs making up a typical visual object may be less ambiguous with respect to categorization of such objects than conditions with other degrees of complexity, even though no conditions on any level correlate unambiguously with such categories.

The Modular Hierarchy in Clustering

The modular hierarchy is overlaid on the layers of condition detection devices. Modules are defined to economize on condition recording resources, and must therefore detect groups of conditions which are relatively similar in an information sense, although less similar than the group detected by one device. As discussed earlier, this information similarity does not guarantee an operational meaning that is unambiguously the same. However, useful modules must detect conditions with a limited range of operationally useful meanings.

A critical task of the modular hierarchy is to manage the definition and detection of conditions. This task must meet several objectives. One is that the number of conditions recorded must be minimized as far as possible, implying careful management of when and where conditions will be recorded. A second objective is that outputs must be generated to competition within all behaviourally useful ranges of complexity. A third objective is that the conditions detected within different ranges of complexity must be as consistent with each other as possible. The meaning of consistency is discussed below. Management to achieve these different objectives is primarily performed by modules on different levels in the hierarchy.

A significant level of module above the device level is the column, made up of devices in each of a sequence of layers. Most of the conditions detected within a column are communicated only within the same column. In other words, most condition defining connectivity is within columns. This connectivity limitation is one aspect of the minimization of information exchange between modules as far as possible. The column module has a primary operational role of determining when and where conditions will be recorded. A higher level of module is the array made up of a set of parallel columns across the same layers. The array module has a primary operational role of detecting some conditions on a behaviourally useful level of complexity in every input state from a particular domain, in such a way that the conditions detected provide adequate discrimination between circumstances in which different behaviours are appropriate. The next highest level of module is a sequence of arrays in which each array receives its condition defining inputs from the preceding array in the sequence. The sequence is called an area module for reasons to be discussed later. Area modules manage the consistency of conditions detected within different ranges of complexity by its constituent arrays, and releases a consistent set to competition. An area module can also be evolved to detect conditions optimized for managing one general type of behaviour. A modular hierarchy with the modules on each level is illustrated in figure 4.2.

Figure 4.2. An area module spanning twelve device layers. Each layer detects conditions on a different level of complexity within a roughly synchronous set of system inputs. The area module is made up of three array modules, each spanning four of the device layers. An array module contains a number of columns which all span the four layers. The layer modules within a column are made up of a group of condition recording devices on the same layer.

Column Modules and Condition Portfolios

A column detects the presence of all the conditions detected by its component devices, but only communicates the presence of a small subset of these conditions outside the column. This small subset will be called the portfolio of conditions detected by the column. Most of the conditions detected by a column are used only to determine when and where new conditions wiii be recorded within the column or in other columns within the same array.

The physical instantiation of a portfolio is a specific subset of column devices which have outputs targeted outside the column. The conditions detected by these devices are the portfolio. These devices will in general be located within a specific layer of the column (or perhaps more than one) and will detect conditions within a specific range of complexity which is most appropriate for their targets.

Because column portfolios are instantiated by specific condition recording devices, portfolios can expand but not contract, and portfolios will not in general correspond exactly with cognitive features or with specific behaviours. Rather, the presence of an active portfolio can be interpreted as a recommendation in favour of a limited range of category identifications and behaviours.

The primary role of one column alone is to manage the recording and detection of its portfolio of conditions. In addition, the column must support the roles of the higher level modules within which it is located. Each role requires a number of different functions which must be managed by the detection of conditions within some appropriate range of complexity.

Different layers within a column are sub-modules that perform different management functions in support of one or more roles. The first function is to ensure that if the column is already producing an output, no changes are made to its portfolio. A second function determines if the column is the appropriate place within the array for such changes to occur and initiates such changes within the column if required. The second function will depend upon determination of the relative degree of similarity between the array input state and the current portfolios of each existing column, as indicated by the number of conditions detected on some level of complexity in each column. A third function determines whether an additional column is required, generally because no existing column is sufficiently similar to the current input state. These three functions make up condition change management.

A fourth function ensures that activity within the column only continues if it is consistent in information terms with activity in higher level arrays. A fifth function generates activity within the column corresponding with information which is not present but may be relevant (i.e. indirect activations as discussed above). A sixth function prolongs the activity of a column when the information is required for a longer period of time. For example, if conditions made up of inputs derived from several objects viewed consecutively are required to determine behaviour, some conditions from the object viewed first must be kept active until their activity overlaps with the object viewed last. Such a prolongation would only occur on some levels of condition complexity. In addition, prolongation of activity is required for consistency management as discussed below. These three functions make up condition detection management.

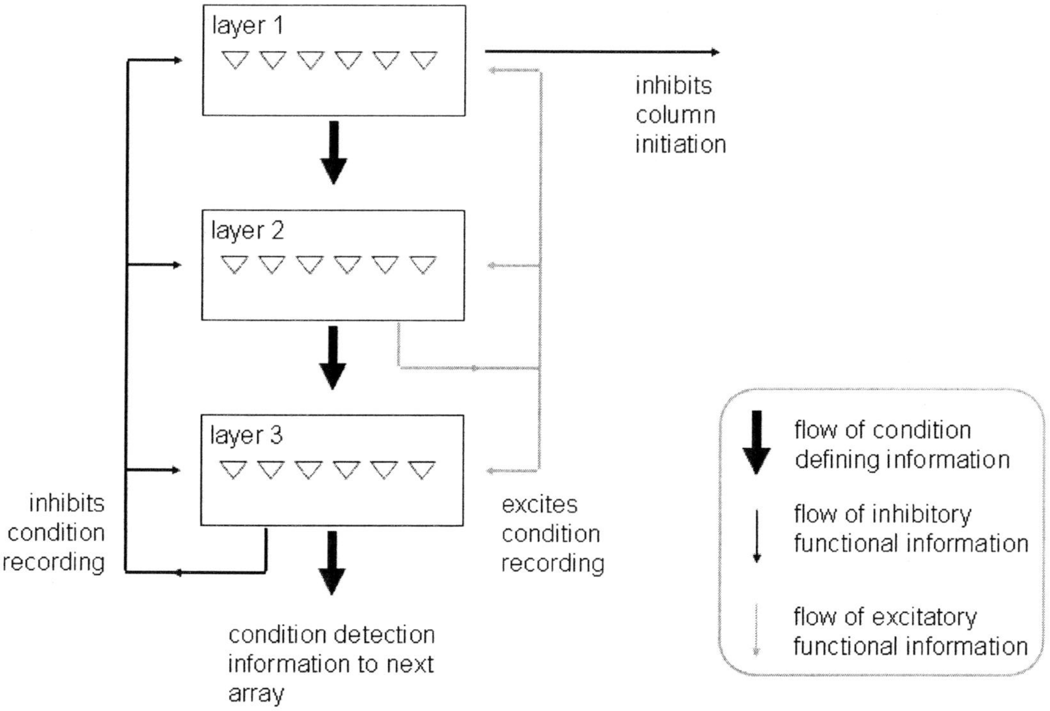

Figure 4.3. A simple three layer column made up of groups of condition recording devices on three consecutive layers. Most condition defining connectivity into a group on layer 2 or 3 comes from the group in the same column in the preceding layer. Condition defining outputs go from layer 3 to columns in arrays detecting conditions in higher ranges of complexity. Non condition defining connectivity enters and leaves specific layer-modules as shown. Much of this connectivity indicates the general level of activity in the source layer-modules, and may derive from special purpose devices which receive inputs from many condition recording devices in their layer-module and generate an output that averages those inputs.

A seventh function manages the input space to the column. The objective here is to add to that input space (by adding provisional inputs which are not part of any existing condition) only if current provisional conditions are inadequate to generate required additions to the column portfolio. An eighth function manages the column output space. The activation of different devices in the output layer corresponds with the presence of different subsets of the portfolio of the column. The specific combination of active output devices therefore provides some information on the currently detected subset. The objective of output space management is to ensure that outputs are added only if additional discrimination is required between different sets of conditions already in the portfolio. Functions seven and eight contribute to minimizing information exchange between modules. These two functions contribute to condition selection management.

The eight functions must all be managed by detection of information conditions at some level of complexity. If the optimum level of complexity for performing two functions is similar, the functions could be performed by one layer. As the difference between the optimum levels of complexity for two functions increases, the more the functional advantages of two layers will tend to outweigh the resource costs of the additional layer. Hence the optimum number of layers is a compromise between the resource cost of and operational discrimination provided by additional layers. Even some of the eight functions may require subdivision of corresponding layers in an operationally complex system. For example, generation of outputs to different targets could require different layers. The arrangement of the layers will therefore depend on the operational problems which the system needs to solve.

Change Management in a Three Layer Column

The simple three layer column illustrated in figure 4.3 can be used to explain the operation of change management. A column of this type has been used in electronic implementations of the recommendation architecture (see chapter 6). The change management functions of the layers are that activity in layer 1 inhibits creation of new columns, activity in layer 2 excites the recording of conditions throughout the column, and activity in layer 3 inhibits such recording.

Inputs derived from an input domain are available to condition recording devices in layer 1, and some of these inputs are incorporated into layer 1 conditions. Detections of conditions in layer 1 are the available inputs to layer 2 and so on. Conditions detected in layer 3 are the portfolio of the column.

Conditions are detected within an input state which is a set of inputs derived at the same time from the input domain. The complexity of the detected conditions increases from layer 1 through layer 2 to layer 3. The higher the complexity, the more specific the condition is to an input state, and the smaller the proportion of input states which will contain the condition. Detection of some layer 1 conditions will therefore occur in response to a relatively wide range of input states, detection of some layer 2 conditions in response to a narrower range, and detection of layer 3 conditions in response to the very narrow range of input states which contain the column portfolio.

A small degree of activation in layer 1 therefore indicates a small degree of resemblance between the input state and the column portfolio. A small degree of activation in layer 2 indicates a significantly larger degree of resemblance. Any activation in layer 3 indicates the presence of the column portfolio.

If there is activation in layer 3, the presence of the portfolio has been detected. The greater the activation, the lower the operational value which would be derived from recording additional conditions, and the more such recording is inhibited. If there is a strong activation in layer 2 but no activation in layer 3, there is a strong resemblance between the current input state and the column portfolio, and condition recording to expand the portfolio slightly to include the input state is appropriate. This condition recording occurs until there is enough activation in layer 3 to inhibit further recording. If there is strong activation in layer 1, but not enough activation in layer 2 to trigger recording, it could be that other input states occurring in the future might expand the portfolio in such a way that an input state identical with the current state would contain the expanded portfolio. Such a strong activation in layer 1 therefore tends to delay a decision to create a new column with a portfolio corresponding with the current input state by inhibiting such creation.

The effect of these layer functions is that an input state which is completely different from a column portfolio may result in creation of a new portfolio. An input state which is slightly similar to the column portfolio will result in inhibition of new portfolio creation, and an input which is sufficiently similar will result in expansion of the portfolio to include the input state by condition recording throughout the column.

The definitions of "slightly" and "sufficiently" similar are managed at the array module level as discussed below.

Activity Detection Devices

The change management inputs to a condition recording device must indicate the overall level of activity in an appropriate layer of an appropriate column. This overall level could be communicated by connectivity from all condition recording devices in the source layer to all such devices in the recipient layers, but a more effective use of connectivity resources would be a device or set of devices within the source layer which receive inputs from the condition recording devices in their layer and generate outputs proportional to the degree of activity of those devices.

Minimized Information Exchange Column

As an example of how the arrangement of layers can be affected by the relative priority placed on different functions, figure 4.4 illustrates an arrangement optimized for reducing required information exchange. The figure and the discussion are limited to the different change management process.

Condition recording devices exist in all of the layers, and condition recording connectivity goes from column input space to layer 1, through 2, 3 and 4. Layer 4 generates column outputs.

Within layer 3 there are activity detection devices which detect moderate activity within the layer and generate excitatory change management outputs to the condition recording devices in layers 2 and 3. Within layer 3 there are also activity detection devices which detect strong activity within the layer and generate inhibitory change management outputs also to the condition recording devices in layers 2 and 3. The effect of this change management connectivity is that if there is moderate activity in layer 3 there will be recording of additional conditions in layers 2 and 3 until activity reaches a strong level, at which point further recording will be prevented. There is an analogous device and connectivity arrangement between layers 1 and 4.

If such a column receives an input state which only generates a weak level of activity in layer 3, the column will produce no output. If the input is strong enough to generate a moderate level of activity in layer 3, recording will occur first in layers 2 and 3 and only if this does not generate enough activity in layer 4 to produce an output without recording in layers 1 and 4 will such recording occur in those layers.

The effect of this change algorithm is that the column will only add to its input space (by recording in layer 1) and its outputs (by recording in layer 4) if it has not succeeded in producing an output by internal recording in layers 2 and 3. This column therefore reduces information exchange at the resource cost of extra layers and connectivity.

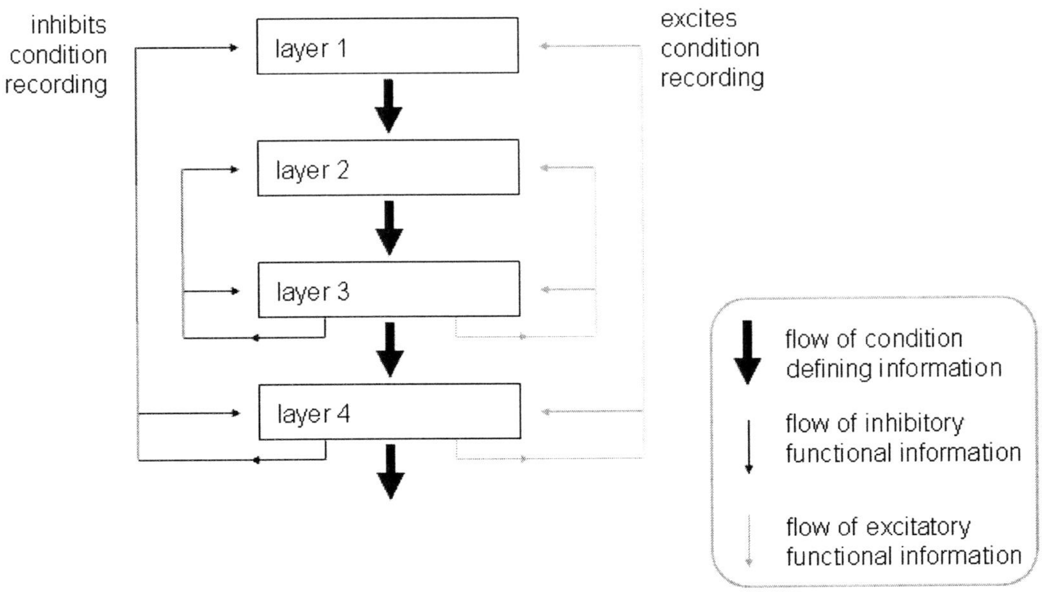

Figure 4.4. A column module which separates the eight functions performed into four layers in such a way that information exchange between the column and other modules is reduced. Comparison with cortex layers and connectivity is made in chapter 7.

The Array Module

An array module is made up of a set of parallel columns across the same device layers. Such an array module defines and detects a set of condition portfolios in an input space. The input space is the sensory inputs from a particular sensory domain (visual, auditory etc.) or the outputs from an earlier array.

An array module manages the creation and expansion of its constituent column portfolios in such a way that each portfolio defines a set of conditions which tend to occur at similar times, the overlap between portfolios is minimized as far as possible, one or more portfolios are detected in every input state, and portfolios are added if the existing set provides inadequate discrimination between informationally similar but behaviourally different input states.

The change management connectivity between two of the columns in an array module is illustrated in figure 4.5. The columns are the three layer type discussed in section 4.1.3.1.1, but the mechanisms would be similar for more complex columns.

Layer 1 receives inputs from outside the column which indicate the presence or absence of relatively simple conditions detected in an earlier array. Conditions which are combinations of these inputs are defined and detected in layer 1 and form the inputs to layer 2. Layer 2 defines and detects conditions which are combinations of its inputs and these conditions are the inputs to layer 3. Combination conditions defined and detected in layer 3 are the outputs from the column indicating the presence of the column portfolio. The identity of the specific devices providing the output can provide some discrimination between different types of conditions within the column portfolio.

An array begins with randomly defined provisional conditions programmed on devices in every layer of every column. The provisional conditions on layer 1 devices are random selections from the input space to the column, but biases are placed on the random selection process for these conditions to ensure that the provisional conditions in one column are relatively similar to each other and fairly different from those in other columns. As discussed earlier, similarity is defined by selection of provisional inputs with a bias in favour of inputs with a tendency to be active at the same time. The very early experience of the system therefore does not record conditions in any column, but is used to identify different groups of array inputs which tend to be active at the same time. The random definition of provisional conditions in the first layer of different columns is biased in favour of inputs from different such groups.

To understand the development of an array, suppose that the array of which two columns are illustrated in figure 4.5 is exposed to sensory inputs derived from a series of different instances of a number of different categories of visual object. The system never experiences exactly the same object more than once. For clarity of explanation, suppose that the categories are different types of fruit: apples, pears, plums, peaches etc. Suppose also that the minimum level of condition detection which the system must achieve in response to every input state, as discussed earlier, is outputs from two columns.

Initially the array is exposed, with no condition recording, to a series of sensory inputs derived from different instances of different types of fruit. Any one input will tend to be active more often when some inputs are also active, and less often when other inputs are active. Groups of inputs which tend to be active at the same time are defined, and provisional conditions for layer 1 in different columns are biased in favour of different groups. Condition recording in response to instances then begins.

Suppose the first instance is a red apple. At first no column produces an output, because no column has any regular conditions. As illustrated in figure 4.5, input activity in layer 1 excites initiation of columns. The two columns with the highest level of input activity will be triggered by that activity to record conditions on every level until a column output is present. Initiation of other columns will then be inhibited by that activity. The initial portfolios of the two columns will correspond with two sets of similar conditions which were all present in the apple instance. One set might correspond with red, but such an exact correspondence with a cognitive condition is very unlikely given the random definition of the provisional conditions. More plausibly, one set might correspond with some mix of redness and presence of stalk, the other might correspond with smoothness and "spherical-ness". Even more plausibly, the portfolios might be sets of conditions which correlated even less precisely with well defined cognitive features. However, each portfolio corresponds with some set of similar conditions in the input space derived from the specific apple instance. Future apple instances would be somewhat more likely to contain conditions within these portfolios, future instances of other types of fruit would be somewhat less likely to contain such conditions.

When the next fruit instance is presented, there are a number of different possibilities. If this instance is almost identical with the first, there will be activity in layer 3 of both columns, and the condition recording inhibition generated by those layers will prevent further recording. If the next instance is different but sufficiently similar to the first apple, there will be significant activity in layer 2 of both of the columns but not layer 3, and condition recording will occur in all layers until there is activity in layer 3. This additional recording slightly expands the definition of condition similarity for the column portfolios. If the next instance is somewhat less similar, one or other of the columns will not have activity in layer 2, and a third column will be initiated with conditions different from those in the first two columns. A very dissimilar instance might result in initiation of two new columns.

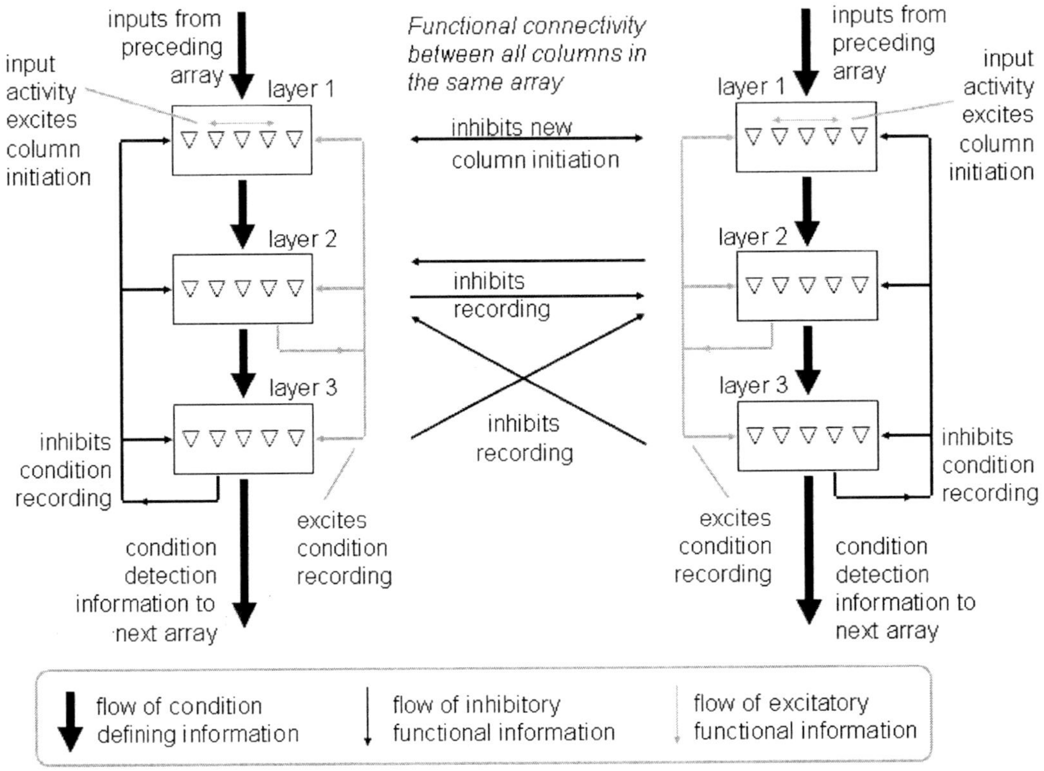

Figure 4.5. Condition defining and condition change management connectivity between and within two columns in the same array. The simple columns illustrated have three consecutive layers of condition recording devices. Condition defining connectivity enters layer 1 from a previous array, condition defining connectivity into layer 2 comes from layer 1 in the same column and so on. Conditions in one layer are therefore combinations of conditions in the preceding layer. Condition defining outputs go from layer 3 to columns in arrays detecting conditions in higher ranges of complexity. Non condition defining connectivity (functional connectivity) enters and leaves specific layers as shown. Much of this connectivity indicates the general level of activity in the source layer of the column, and may derive from special purpose devices which receive inputs from many condition recording devices in the same layer and column, and generate an output that averages those inputs. Inputs supporting extended activation and indirect activation of the column enter or leave specific layers but are not illustrated.

Experience of more and more instances of different types of fruit gradually expands the number of portfolios and the similarity definition for each portfolio until there are enough portfolios to generate output (with or without condition recording) in response to every fruit instance. However, at any time a novel type of fruit could initiate new portfolios.

Any one portfolio will be activated (i.e. its column generate an output) in response to a number of different kinds of fruit, but not all types of fruit. Members of a small subset of the portfolios will each tend to be often (but not always) active in response to fruits of one type, and a different but sometimes partially overlapping subset in response to fruits of a different type.

The array definition process thus processes a large number of input portfolios which discriminate weakly between different categories into a (generally smaller) number of portfolios which discriminate more strongly between the categories. In other words, the output portfolios have less operational ambiguity with respect to fruit categories than the input portfolios as discussed in the Device Layering section earlier in this chapter.

The way in which competition can learn to associate such portfolios with appropriate behaviours using only consequence feedback and a genetically defined tendency towards imitation is described in detail later.

A problem which could occur in an array is that a small group of portfolios might be defined which were activated in response to two different categories, in such a way that the activations in response to instances of the two categories were indistinguishable. A symptom of this problem would be if after learning the association between portfolios and a behaviour in competition, the portfolios participating in the selection of the behaviour were often present when both positive and negative consequences follow the behaviour. The implication of this contradictory consequence feedback situation is that the portfolios do not adequately discriminate between similar input conditions with behaviourally different implications.

To solve this problem, some indication that contradictory consequence feedback has been detected over a period of time must be provided to the array. This indication is used to force the array to add new outputs to the outputs already being produced when the operationally inadequate portfolios were again active. Such outputs could be generated by additional devices in the output layer of the problem columns, or by additional columns provided with inputs often active when the problem columns were active. These new outputs would provide additional discrimination. Such indirect use of consequence information generates additional outputs which do not affect the existing outputs managing other behaviours.

Alternative Array Definition Processes

There are a number of detailed processes by which portfolio definition could occur, which differ in the tradeoffs made between learning time and resources required to achieve a given level of discrimination effectiveness. At one extreme, one column could be configured by random selection of provisional conditions within a population of inputs which tended to occur at the same time over a period of experience. The corresponding portfolio would be initiated by the first input state which contained a significant proportion of the favoured inputs. The initiation would be by recording of conditions which all occurred in the one input state, at various levels of complexity in the new column instantiating the portfolio. The portfolio would be expanded by addition of conditions present within sufficiently similar input states until the portfolio was detected in a significant proportion of input states. Only at that point would an additional column be configured by random selection of provisional conditions within a population of inputs which tended to occur at the same time over a period of experience within input states generating no response in the existing column. The new column would then be initiated by random selection of conditions within an input state which contained a significant proportion of the defining inputs and generated minimal activity in the existing column, and the new portfolio expanded. New portfolios would continue to be added, but only once the previously added portfolio responded to a significant proportion of input states, until every input state resulted in detection of one or more portfolios.

At the other extreme, after a period of early experience a large number of columns could be configured by random selection of provisional conditions within different groups of inputs. Each group would be made up of inputs which tended to occur at the same time in the early experience period, but with groups defined to have as little overlap as possible. A column would be initiated by the first input state which contained a significant proportion of its favoured inputs at a time when no other column was detecting its portfolio.

Once an array responding to all input states was achieved, both approaches could evolve portfolios by indirect consequence feedback as required. The first approach would tend to result in fewer, better separated portfolios but a longer learning period. The second approach defines a full array of portfolios more rapidly, but with greater risk of discrimination problems requiring additional column resources. In practice, some process intermediate between the two extremes would be most effective.

Genetic bias could also guide the selection of the starting points for some portfolios, which could therefore all be initiated at the start of experience. Such use of genetic information is discussed later.

Sequences of Arrays

Once an array has been configured using sensory inputs, the outputs of the array could then be inputs to another array defining portfolios at a higher level of condition complexity and so on. For example, the first array could detect condition portfolios in a range of complexity comparable with that for features of visual objects, and the next array could detect condition portfolios in a complexity range comparable with visual objects. Such conditions would have similar numbers of sensory inputs contributing to them as features or objects, but would not correlate unambiguously with such features or objects. Portfolios in intermediate ranges of complexity could also be useful for behavioural purposes.

Later arrays could receive inputs derived from a sequence of outputs from earlier arrays, and therefore detect conditions comparable in information content and complexity with groups of visual objects. Yet later arrays could receive a sequence of outputs from such arrays and detect conditions comparable in information content and complexity with groups of groups of visual objects. Such portfolios would be useful for guiding behaviours in response to the presence of such groups.

Parallel Arrays

In general, unless information is segregated at the input level, sensory input will generate a single array of portfolios across the entire input space. Visual and auditory inputs are separate, and could generate separate arrays, although higher level arrays might detect conditions incorporating information from both sources. Colour and shape information could be derive from separable sources and therefore generate separate arrays with their separate outputs incorporated into one later array. Thus in the example discussed earlier, a practical array of portfolios will generate responses to all possible categories of visual object, not just types of fruit.

However, genetic information could result in partial physical separation between portfolios responding to different broad types of category. For example, if a physically separate array of portfolios was genetically biased in favour of the types of condition which tended to occur in human faces, the result could be faster learning of faces and partial separation between the columns responding to faces and the columns responding to other visual objects. The separation would only be partial because the biased random element in the process for condition definition means that some portfolios in the "face" array could respond to other objects and vice versa.

Area Module

An area module is made up of a series of arrays and generates outputs indicating the presence of conditions in behaviourally useful ranges of complexity. There are two functions most efficiently managed at the area level. One is management of the consistency across populations of conditions detected at different levels of complexity, the other is the optimization of conditions for different general types of behaviour.

Management of consistency is required to reduce inconsistencies between the input state and the conditions detected. These inconsistencies arise because multiple conditions can be recorded on one device. To understand these inconsistencies and how they can be reduced, consider the connectivity illustrated in figure 4.6.

In the figure, the target device C1 in layer C has recorded a condition which includes condition b1-1 recorded by device B1 in layer B. This condition is in turn made up of simultaneous presence of conditions recorded by devices A1, A2, and A3 in layer A. However, device B1 has also recorded condition b1-2 which is made up of a different set of devices A4, A5, A6 and A7 in layer A. Because the inputs to device B1 will have been limited to inputs which tend to be active at the same time, there is likely to be some similarity in information meaning between b1-1 and b1-2. However, as more and more complex conditions are defined across many layers, with multiple conditions defined on each device, the degree of similarity may become inadequate for guiding behaviour. In other words, some of the higher complexity conditions may be inconsistent with the simpler conditions.

A mechanism which can improve the consistency uses additional functional connectivity between devices in layer C and devices in layer A. Such connections would be established provisionally and made operational if the device in layer A was active when the device in layer C recorded a condition. The strength of the connection would increase each time the device in C recorded a condition when the device in layer A was active. Analogous connectivity could be established in the opposite direction. Such connectivity would not form part of conditions, but would have the functional effect of prolonging the activity of the target device.

The effect of this additional connectivity would be that when a population of devices was activated on all levels of complexity, the initial population would decay towards a population made up of devices on different levels of complexity which tended to have recorded conditions in the past at the same time. Such a population would have a somewhat higher internal consistency than the original population. Depending on the number of conditions recorded on each device, such mechanisms might or might not be required to achieve an acceptable level of consistency.

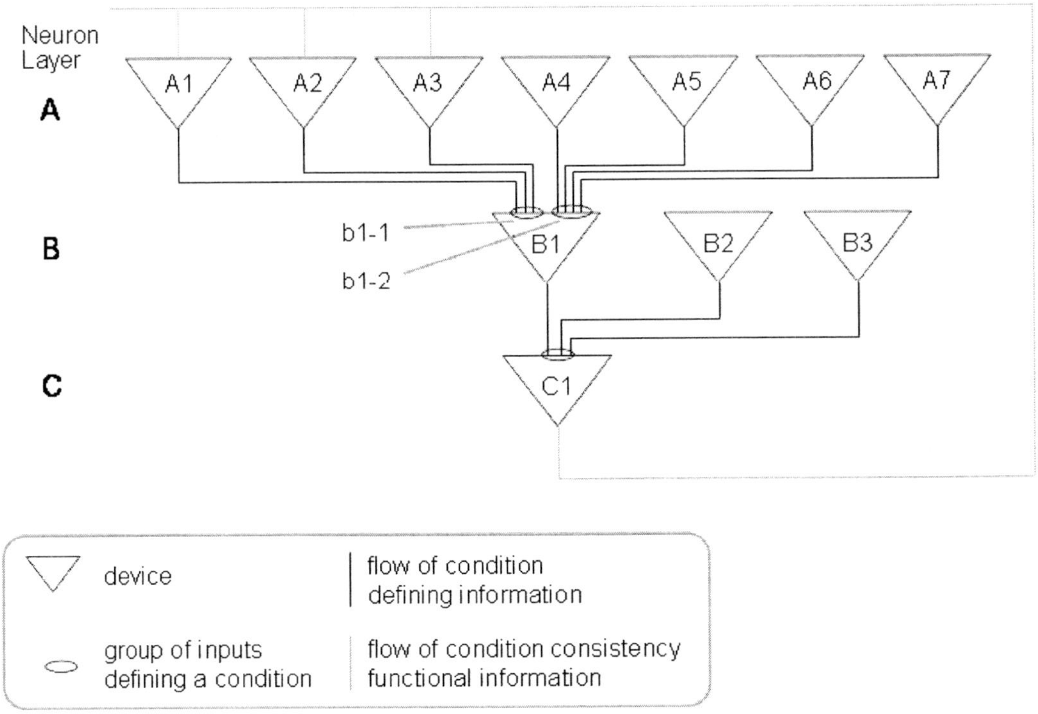

Figure 4.6. Condition consistency problem and solutions. If a device records multiple repetitions, devices receiving its outputs need to have some capability to distinguish between outputs due to relevant conditions and accidental repetitions. For example, condition p1 on B1 is relevant for C1, but not p2 on B1. One mechanism is to forbid sharing of inputs between different repetitions on the same device and allow some inputs to reach the target from earlier levels. A less restrictive mechanism is to combine partial input orthogonality with feedback to lower complexity devices, which results in continuing output from the lower level only if the higher level is present. The system will then converge on a consistent set of device activations

Minimization of information exchange and connectivity resources would suggest that the required connectivity between devices detecting conditions at higher and lower complexity levels originate in one specific device layer and target devices in another specific device layer. The optimum distance between the layers in complexity terms will depend on the operational problem and cannot be determined a priori, but will tend to be a compromise between the improvement in consistency and longer times required to reach a consistent activation as the distance increases.

The optimization of an area module for a general type of behaviour uses indirect feedback of consequence information to arrays as discussed in the previous section. The types of condition portfolios which provide the best discrimination for one type of behaviour may be different from those for a different type of behaviour. For example, to ability to distinguish between two visual conditions may be critical for one type but irrelevant for another type. There could therefore be operational advantages to separate areas for different types of behaviour. However, there is a high resource cost to such duplication, and in practice a compromise with separate areas just for different major types of behaviour is more probable.

Such a separation into areas generating recommendations of different types of behaviour has a further operational advantage. The relative probability of recommendations of different behavioural types being generated can be modulated by modulating the level of device output generated in response to a given level of condition detection across the area. The device inputs required are illustrated in figure 4.1.

For example, if there were separate areas for food seeking, aggressive, and fearfui recommendations, then specific triggers (e.g. low blood sugar) could increase the probability of one type of behaviour (e.g. food seeking). It is also plausible that areas optimized for internally directed behaviours such as indirect activations of portfolios would exist.

Resource Management Processes

Within clustering, learning occurs by permanent addition of conditions to modules on many levels including device, column, array and area. There are a number of considerations which demand careful management of the resources used for this condition recording. The first consideration is that once recorded, conditions are (with minimal exceptions) not subsequently changed. It is therefore important to ensure that a recorded condition has a reasonable probability of being operationally useful to the system and thus avoid excessive resource consumption. The second consideration is that conditions and portfolios are defined heuristically, and there is no detailed a priori knowledge of which higher level modules such as arrays will require many column portfolios, or which column portfolios will need many device level portfolios etc. Hence assignment of column and device resources to higher modules must be performed heuristically on the basis of need. The third consideration is that the greater the connectivity between different modules, the greater the risk of undesirable side effects on some behaviours when conditions are added. Such connectivity should therefore only be established if there is some indication that it is both needed and likely to be useful.

The fourth consideration is that the need for a behaviour type in which one portfolio activates another portfolio on the basis of temporally correlated condition recording in the past implies the need for creation of provisional connectivity to support such activations. This provisional connectivity will become permanent if temporally correlated activity actually occurs. Although all-to-all provisional connectivity would work in principle, such a level of connectivity would use excessive resources and raise the risk of undesirable side effects. A means to establish only provisional connectivity with a reasonable probability of being useful is needed.

Resource management must therefore perform a number of functions. Firstly, it must assign provisional conditions to devices, devices to columns and columns to arrays on the basis of current need. Secondly, it must ensure that provisional conditions are appropriate to the device portfolio and to the higher level portfolios within which the device is situated. Because conditions added to a portfolio must be generally similar to existing conditions, provisional conditions must be made up mainly of inputs which are part of existing portfolio conditions. Thirdly, because newly assigned resources have a higher probability of simultaneous condition recording in the immediate future, it should bias the assignment of provisional connectivity to support activation on the basis of simultaneous recording in favour of connectivity between resources assigned at the same time to different modules. Fourthly, to minimize information exchange, inputs not contributing to the current portfolio must be included in provisional conditions only if there is some indication that such additional inputs are needed, and if the additional inputs have a reasonable probability of being relevant to the portfolio.

One example of an indication that additional inputs are needed would be if the presence of the portfolio on a number of occasions resulted in a particular behaviour, but the consequences of the behaviour were sometimes positive and sometimes negative. As discussed earlier, this contradictory consequences feedback indicates that the presence of the portfolio is not discriminating between similar but behaviourally different circumstances. The solution to the problem is to add devices to the output layer of the corresponding column or to add columns with the same input space as the problematical column. Either of these changes would provide additional discrimination without affecting existing portfolio outputs to other behaviours. The changes may well require a proportion of inputs not contributing to the original portfolio.

An important issue is how to select appropriate provisional connectivity when the only available information is past experience. As discussed earlier, an indicator which increases the probability of relevance is if a potential source of inputs has often been active in the past at the same time as a potential target. Identification of such inputs could in principle be performed by collecting and recording data on simultaneous activity of all possible pairs of modules on all levels, but such a solution would be very resource intensive. A more efficient alternative would be to perform an input driven partial rerun of recent experience. This would only require storage of a record of input activity on some level of condition complexity. Portfolios identified as requiring additional inputs could then accept inputs from other modules frequently active at the same time during this partial rerun.

Resource management therefore requires two components. One is a map of condition recording resources identifying which are unassigned, the other is a partial rerun process for identifying appropriate connectivity. Resource management would then consist of a periodic process during which requirements for new resources would be identified, resources would be assigned, and appropriate provisional connectivity provided. Connectivity to support indirect portfolio activation on the basis of simultaneous condition recording could be efficiently provided via the resource map, at least initially. Connectivity to support indirect portfolio activation on the basis of prior or subsequent condition recording would for efficiency reasons tend to continue to be dependent on the map.

As discussed in detail in chapter 8, if resource management is identified as a function of the Hippocampal complex and the partial rerun to optimize connectivity identified with REM sleep, a plausible account for the deficits observed as a result of damage to the Hippocampus system in amnesias such as Korsakoff's syndrome develops.

Management of the Random Element to Condition Definition

The more that random element can be reduced, the lower the resources required to define enough conditions to adequately guide behaviour. However, the absence of external intellectual guidance means that there is no way to avoid some random element in the definition of conditions. Reduction of the random element requires some external information. Such information could be genetic, or could be the result of an experienced outside system influencing the environment of the inexperienced system to promote condition recording of particular types. These possibilities are discussed in more detail in chapter 9. Another approach is to use past experience as an indicator of the types of condition required in the future as discussed in the previous section. Other approaches to minimizing the random element require resource tradeoffs against the rate at which effective learning is achieved. For example, as discussed in the earlier section on Alternative Array Definition Processes, the use of indirect consequence feedback to indicate when additional conditions are required implies that a relatively low level of condition recording will take place initially, resulting in inadequate initial learning.

The Structure of Competition

A range of considerations influence the detailed structure of competition. These considerations include the needs to limit information exchange meanings to simple levels compatible with the uses of consequence feedback; to limit the overall flow of information within competition; and to limit the number of input weights which must be managed. The way in which these considerations influence the detailed structure of competition is the topic of this section.

Definition of Behaviours within Competition

The competition system is made up of components corresponding on a one-for-one basis with all possible system behaviours. Each component is a device or group of devices. Some components correspond with behaviours which are "atomic" in the sense that the system can implement the behaviour or not implement it, can vary the speed and perhaps the degree with which the behaviour is performed, but cannot change the nature of the behaviour. An atomic behaviour could be the contraction of an individual muscle, or genetically programmed groups and sequences of such contractions. Other components correspond with higher level behaviours such as groups and sequences of atomic behaviours, and yet higher behaviours which are groups and sequences of such groups. At the highest cognitive levels, behaviour is achieved by outputs from clustering driving a sequence of competition components which in turn activate more specific competition components.

In principle even a high level behaviour could be invoked by outputs from clustering driving a sequence of atomic behaviours. However, the connectivity required and the complexity of managing the huge number of connection weights would be excessive. Any very frequently occurring group or sequence of atomic behaviours will therefore tend to result in one intermediate level component in competition which receives most of the inputs from clustering and drives behaviour more directly into atomic behaviours. Yet higher level behaviour components will be defined as frequently occurring groups or sequences of intermediate behaviours.

The outputs of a component are the outputs of specific devices within the component. As discussed earlier, such component outputs can have two types of operational meaning. One is a recommendation to perform the behaviour corresponding with the component. If such an output exits competition, it becomes a command to perform the corresponding behaviour. Otherwise it is directed at a range of components corresponding with more detailed or specific behaviours within the recommended type, and increases the probability of such behaviours being accepted. These more detailed components will also receive inputs directly from clustering.

The other type of operational meaning is a recommendation against performing any behaviour other than the component behaviour. Such outputs are directed at competition components corresponding with different behaviours. A high proportion of these outputs are directed at peer components, in other words components corresponding with different behaviours on roughly the same level of detail.

The Behavioural Component Hierarchy

There are two functions which must be performed by the behavioural component hierarchy which constitutes the competition subsystem. One is to determine the strength with which each behaviour is recommended by current clustering outputs. The second is to select one behaviour or a group of consistent behaviours based on comparison of recommendation strengths. Competition must ensure that attempts to implement inconsistent behaviours (such as trying to move an arm in two different ways at the same time) are prevented.

Devices in Competition

At the most detailed level, competition is made up of devices with the structure illustrated in figure 4.7. These devices have excitatory and inhibitory inputs with different weights, and a threshold. A device generates an output proportional to the degree by which the total of all active input strengths multiplied by their corresponding weights averaged over the recent past exceeds the threshold. These inputs are derived from devices in specific clustering modules and competition components. Averaging can be viewed as a gradual decay of the strength of an input after it first arrives. However, different inputs may decay at different rates. For example, some inhibitory strengths decay more slowly than excitatory strengths. The functional importance of this difference is discussed below.

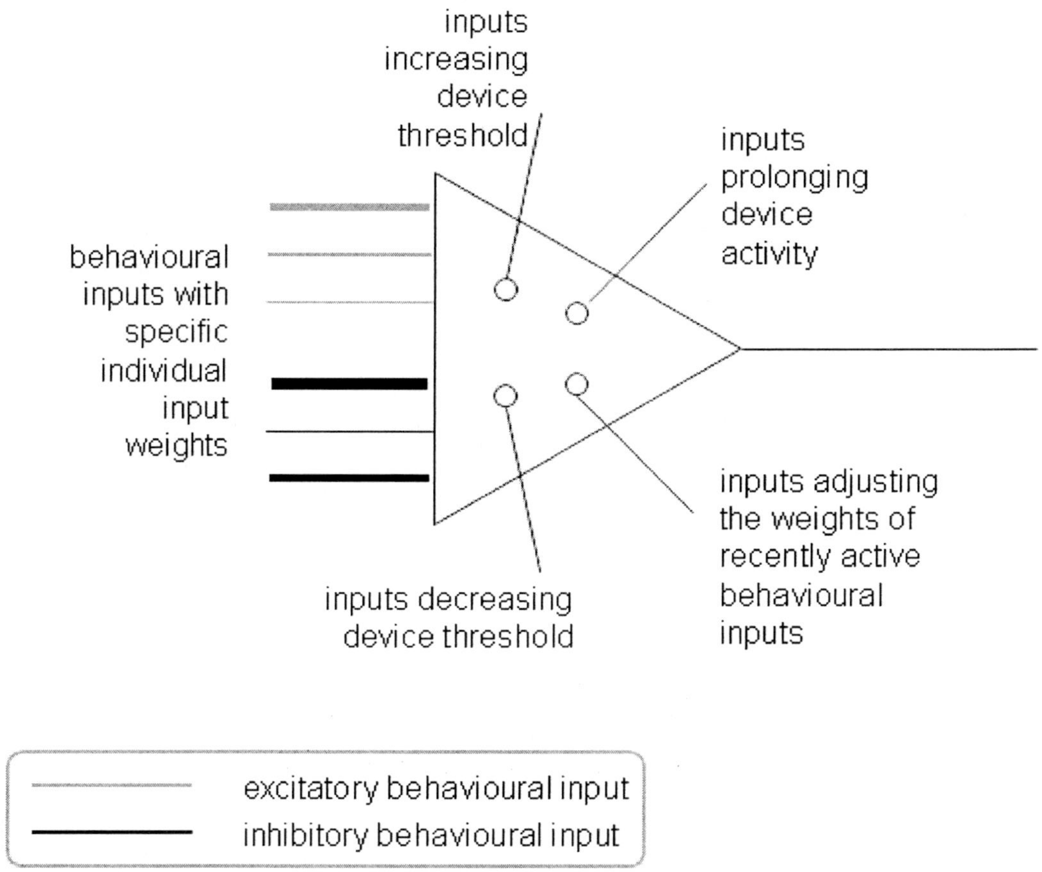

Figure 4.7. A device within competition. The device has primary inputs from different sources in clustering and competition, each with a different excitatory or inhibitory weight. The device produces an output proportional to the degree to which the total of the input level multiplied by the weight for each input exceeds its current threshold. The threshold can be modulated by inputs derived from specific other sources. There are also inputs which can prolong the activity of a device.

Devices also have inputs with different specific functions. Some inputs can increase the current threshold, others can decrease that threshold. Yet other inputs can prolong the period of time during which the device produces an output. The reason for these different input types is differences in the type of information implicit in different input sources.

Unlike the devices used in clustering, these competition devices adjust their weights in response to consequence feedback. There is therefore some general similarity to algorithms of the perceptron type used in artificial neural networks [see e.g. Jain, Mao & Mohiuddin, 1996; Ballard, 1997].

Higher Level Components in Competition

Components in competition are made up of groups of these devices. Each group receives a range of inputs: from clustering recommending for or against their corresponding behaviour; and from other competition components recommending in favour of the behaviour or behaviour type corresponding with the source component and against any other behaviour or type. The target component must integrate these inputs to generate appropriate outputs. If a source component corresponds with a general type of behaviour and the target with a specific behaviour of that type, the input can be interpreted as a recommendation in favour of the specific type. However, because the use of consequence feedback within competition limits the operational complexity of information meanings, the input cannot be interpreted as a stronger recommendation for one specific behaviour of the general type than for another. These inputs therefore lower the thresholds of devices in all appropriate specific behaviour components using the different input type discussed in the previous section. Otherwise, any input from a competition component corresponding with a different type of behaviour can be interpreted at a recommendation against the behaviour corresponding with the target component.

If a component corresponding with a sequence of behaviour types is selected, the effect is that the component lowers the thresholds for the component corresponding with the first behaviour type. When that behaviour type has been performed, the sequence component lowers the thresholds for the component corresponding with the next type and so on. Selection of behaviours within the types will depend on recommendation strengths provided directly to the individual type components by clustering portfolios activated as the sequence proceeds.

Thus outputs from a component which are directed at any other component either recommend in favour of the corresponding behaviour or against any behaviour different from the corresponding behaviour. Outputs targeted outside of competition can be interpreted as commands to implement the corresponding behaviour.

There are therefore a number of different functions performed by a competition component, which may for operational efficiency reasons be performed by separate subgroups of components within the group making up the component.

As a simple example, consider the three device layer parallel pipe implementation illustrated in figure 4.8. Each pipe corresponds with a different specific behaviour. Two pipes are illustrated.

The first layer of each pipe receives inputs from clustering. These inputs may be a subset of the total clustering output set, where the identity of the subset has been selected by a higher level behavioural component as discussed below. Each input excites or inhibits its target devices in the first layer. Devices in the second layer receive excitatory inputs from the first layer of the same column and inhibitory inputs from the second layer of other columns. In the second layer, the effect of a given level of inhibitory input persists longer than excitation. Hence activity in the second layers will tend to decline until only activity in one pipe remains. There is excitatory connectivity from the second layer to the third layer within each pipe, but any inhibitory connectivity from the second layer of a pipe to the third layer in all other pipes is strong enough to completely prevents any third layer activity. Such third layer (output) activity will therefore be generated in only one pipe.

The inputs from behavioural components corresponding with different general types of behaviour act upon all first layer devices to modulate their thresholds.

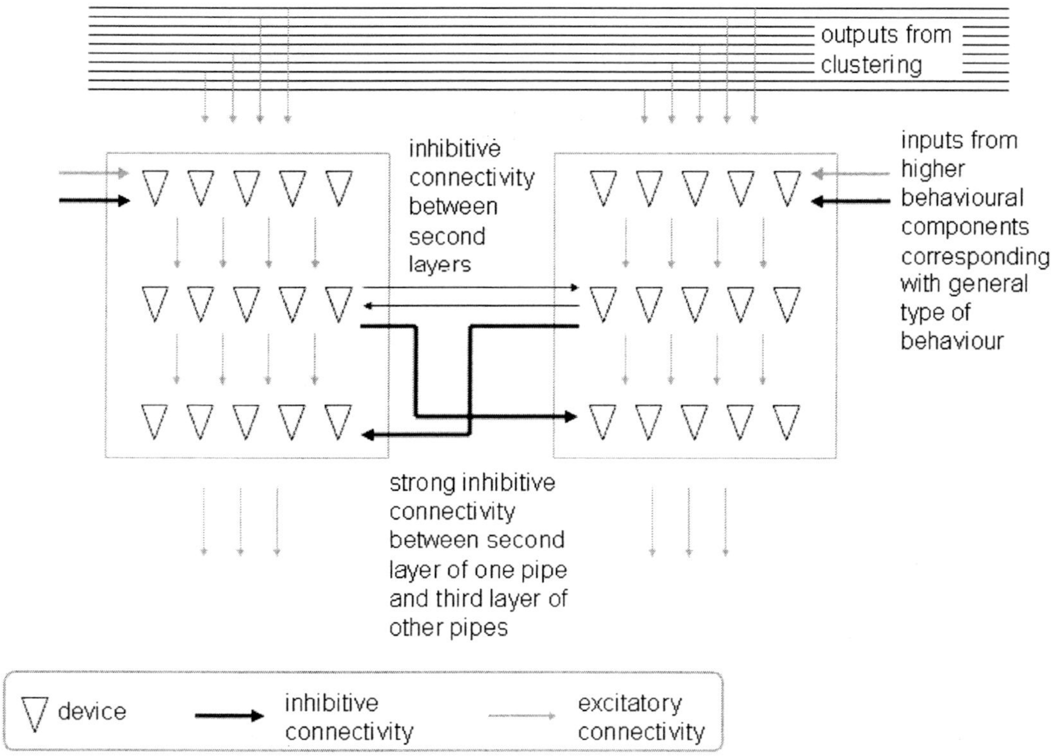

Figure 4.8. Two pipe components in competition within a wide range of peer pipes managing different behaviors. The three device layers manage integration of recommendation weights from clustering, comparison of total weights between alternative behaviours, and ensuring that only one behaviour is implemented.

The Highest Level Competition Architecture

Clustering provides detections of conditions on many different levels of complexity to competition. These condition detections could in principle target all possible behavioural components, but connectivity resource requirements would be excessive. There are architectural arrangements which can reduce both the degree of condition detection needed in clustering and the flow of condition detection information to competition. These arrangements separate clustering into a number of levels, each level providing most of its outputs to the next level of clustering and to one corresponding competition subsystem. In this arrangement, the sequence of competition subsystems select behaviour with an increasing degree of specificity.

For example, suppose that there were three clustering levels and three corresponding competition subsystems. The first competition subsystem could select the input domain (for example, one visual object) which is allowed to influence behaviour. In other words, the subsytem performs an attention function. The second subsystem could select a general type of behaviour in response to the selected input domain, and the third could select a specific behaviour of the selected general type. Competition cannot change clustering outputs, it can only select a subset to be allowed to go to processing by the next clustering level. However, this architectural arrangement reduces both the volume of information which must be searched for conditions by clustering, and the volume of information which must be evaluated by competition. The actual number of levels and subsystems will be a compromise between resouce efficiency gains and increased architectural complexity.

Important Behavioural Types

Behaviours directing muscle movements such as limb motions, eye movements, or speech generation are important. However, there are a number of behaviours directed back into clustering which are also important.

Such behaviours include activation of portfolios often active in the past at similar times to currently active portfolios, or activation of portfolios which recorded conditions at the same time in the past as currently active portfolios. Other such behaviours are prolonging the activation of currently active portfolios or reactivating recently active portfolios. Yet others are increasing or decreasing the thresholds of devices throughout an area module corresponding with a particular behavioural type, to decrease or increase the probability of recommendations of the type being generated in response to current input states. There must be competition between all these types and with externally directed behaviours resulting in selection of a consistent set.

Because outputs from competition cannot acquire operationally complex meanings, these outputs cannot directly drive activations of specific modules in clustering. Competition can only select subsets of currently active clustering portfolios to drive such activations. For example, the effect of acceptance of a recommendation to activate portfolios often active in the past at the same time as a set of currently active portfolios will be release of outputs from those portfolios to targets along connectivity established by the presence of simultaneous past activity.

Clustering and competition structures to support such behaviour management processes are illustrated in simplified form in figure 4.9. A set of condition portfolios are detected within current sensory inputs by first level clustering, and a subset of these portfolios is selected by a competition subsystem corresponding with attention. This attention competition is not illustrated, but operates in the same way as the illustrated competition subsystems and is discussed in more detail later.

The selected portfolios are communicated to a set of competition units corresponding with different general types of behaviour. Such general types could include aggressive, fearful, food-seeking or speech generation behaviours directed externally, or internally directed behaviours such as prolonging activation of current conditions or activating new conditions often active in the past at the same time as current conditions. Internally directed behaviours could also be managed by the attention competition. A competition module corresponding with one general type of behaviour wins a competition of cross inhibition between all the modules at this level, and the effect is that the subset of outputs from clustering which recommended that type are released to the third level of clustering. Portfolios detected within the third level of clustering are released to the third level of competition, but only to those modules corresponding with specific behaviours within the selected general type. All these modules compete by cross inhibition and one module wins the competition. The effect is that the third level clustering outputs which recommended the selected specific behaviour are released to control behaviour.

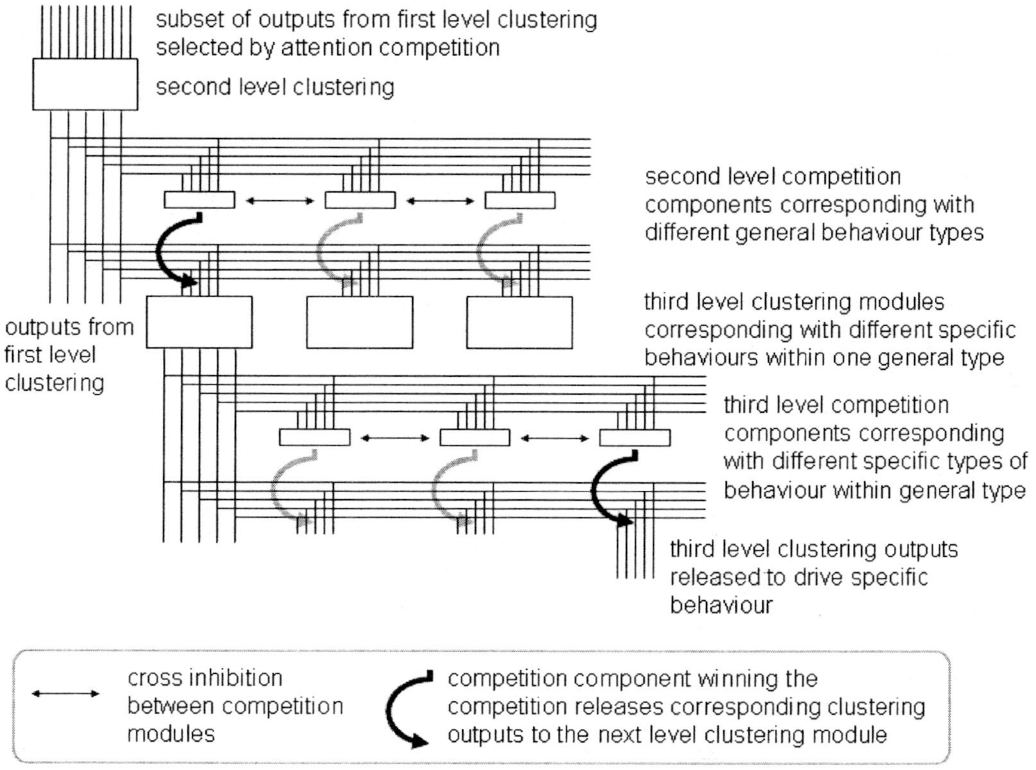

Figure 4.9. Management of release of second level clustering outputs to third level clustering and of third level clustering outputs to control of behaviour. Second level clustering outputs first go to a set of competition modules corresponding with different types of behaviour. The outputs are interpreted as recommendations for or against the competition module behaviour type. The competition module with the strongest recommendation releases the outputs from second level clustering which are in favour of that behaviour to the third level clustering module generating specific recommendations of the general type. A similar process in third level competition releases the appropriate subset of third level clustering outputs to control behavior.

It is important to note that device outputs within competition cannot acquire complex behavioural meanings. Hence although a component output can command a general type of behaviour, any processing to determine the specific behaviour within a range of options requires information from clustering. Competition can therefore select the subset of information from clustering which will be used to determine behaviour in more detail, but cannot modify the content of that information. This process requires close interaction between competition modules and clustering modules to ensure that outputs from the preceding clustering level are only released to appropriate targets, and only if they recommended the selected behaviour or behaviour type. One way to manage this interaction is illustrated schematically in figure 4.10.

In figure 4.10, there is connectivity from specific modules in clustering to specific components in competition, but also connectivity from those components in competition back to the source modules in competition. The competition component with the strongest recommendation generates activity back into clustering towards modules which were the sources of the excitatory inputs making up the accepted recommendation. This activity results in the release of those module outputs to the next level of clustering.

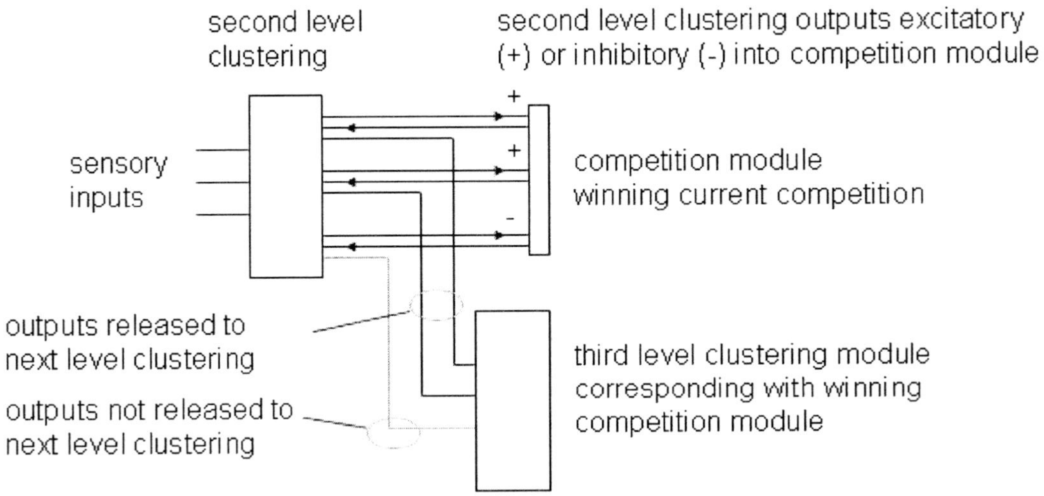

Figure 4.10. Connectivity supporting management of release of second level clustering outputs to third level clustering. Outputs go first to each competition module, only one is illustrated. If that competition module wins the competition with its peer modules, connectivity back to first level clustering triggers release of outputs to second level clustering if the output was excitatory into the competition module.

Major Information Flows and Highest Level Architectural Separations

A number of factors drive the form of the recommendation architecture at the highest levels. The resultant architectural form is conceptually illustrated in figure 4.11, and the primary flows of condition detection information are illustrated in figure 4.12 for the same form. To make the explanation of the architectural form more comprehensible, the integration of information from different sensory domains is not illustrated. This integration will be discussed in detail in chapter 9.

The typical complexity ranges of conditions detected at different points have been labeled in figure 4.11 by the type of unambiguous cognitive condition which would be in a similar range of complexity. Such unambiguous conditions would be cognitive features, objects, groups of objects etc. However, this labeling is conceptual and only intended as an aid to explanation. In practice, conditions will be defined within many different ranges of complexity, and different ranges will correlate to different degrees with different types of unambiguous cognitive conditions. Conditions within a number of different ranges of complexity may be useful for guiding any one behaviour type.

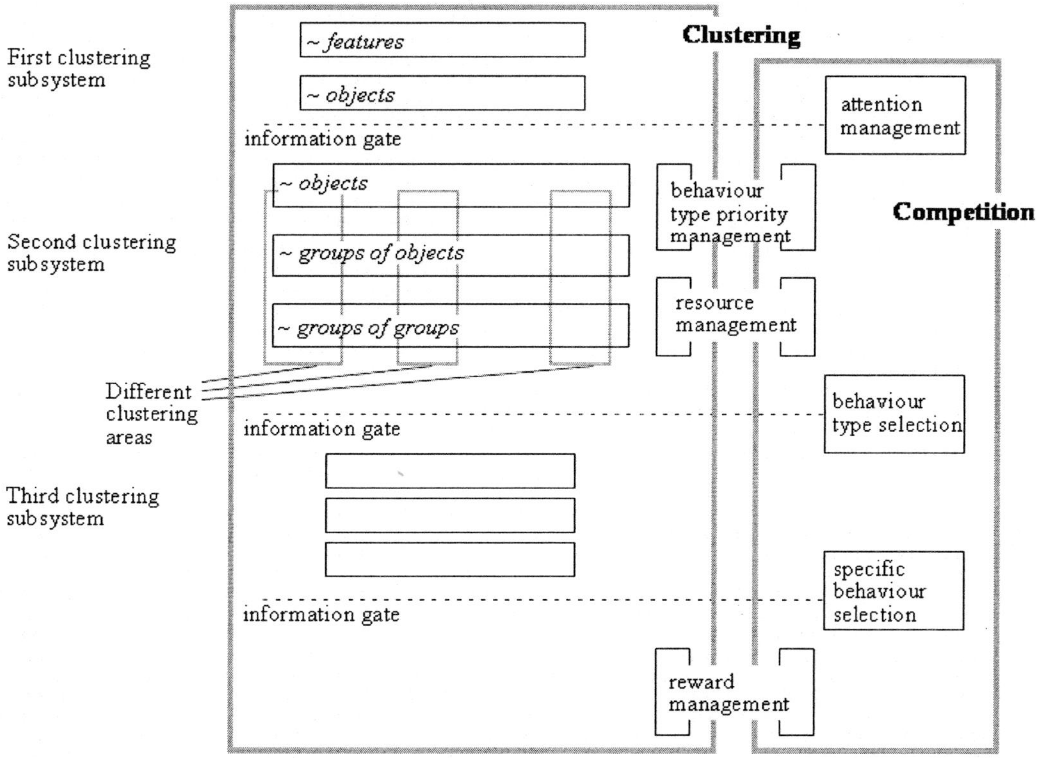

Figure 4.11. Highest level architectural separations of the recommendation architecture. Clustering is separated into three major subsystems which detect conditions used to manage attention, the selection of a general type of behaviour, and the selection of a detailed behaviour within an already selected general type. The clustering subsystem managing selection of behaviour type is further separated into areas detecting conditions on levels of complexity comparable with object features, objects, groups of objects and groups of groups. This subsystem may have a further separation into areas addressing different general behaviour types which is partially overlaid on the condition complexity separations. The corresponding competition subsystems manage the release of the appropriate subsets of condition detections to the next stage of clustering or to muscle control in the case of the final clustering subsystem. Management of the relative priority of different behaviour types and management of clustering resources are both performed by separate subsystems which each have clustering and competition separations.

One of the factors driving the detailed form of a system with the recommendation architecture is the need to detect conditions which are optimized as far as possible for behavioural types. This factor results in separate areas for different major types of behaviour as discussed in the earlier section on Area Modules.

The second factor is the need to generate appropriate behaviours not only in response to perceived objects but also in response to groups of objects and even groups of groups. To illustrate this point, consider speech comprehension. The understanding of speech requires detection of conditions within single objects like words, within groups of words making up phrases, and within groups of phrases making up sentences. Conditions must be detected which incorporate information from all the members of the appropriate group at each point in time. Although speech comprehension is a clear illustration of the need, most behavioural selections must take account of multiple objects in the perceived environment, even if the behaviour selected is primarily with respect to just one of the objects. The selection of the behaviour of sitting on a chair will be influenced by the locations and identities of people and other objects in the environment, and the way the people (and perhaps the objects) are organized into groups. The second factor results in separation of clustering into areas which detect conditions correlating with objects, groups of objects and groups of groups, and also conditions which can be interpreted by competition as recommendations for which information sources from lower level areas (e.g. those detecting conditions in objects) will be combined as inputs to higher level areas (e.g. those detecting conditions in groups of objects).

The third factor is the need to limit the volumes of information flowing within clustering and between clustering and competition. This factor tends to result in the separation of clustering into three major subsystems, each with a corresponding competition subsystem. The first clustering subsystem detects conditions within the entire sensory domain, and the first competition subsystem interprets detections of some of these conditions as recommendations to release specific subsets of currently detected conditions to the second clustering subsystem. The primary role of these first clustering and competition subsystems can thus be viewed as managing attention. The second clustering subsystem detects conditions within the subset, and the second competition subsystem interprets these conditions as recommendations in favour of different general types of behaviour. Acceptance of one such recommendation results in release of the subset of condition detections corresponding with the accepted recommendation type to the third clustering subsystem. The third clustering subsystem interprets these conditions as recommendations in favour of different specific behaviours of the general type. Acceptance of one such recommendation results in release of the subset of conditions corresponding with the accepted behaviour to drive that behaviour.

The primary flows of condition detection information are therefore as illustrated in figure 4.12. The first major flow is from the first clustering subsystem to the attention management competition subsystem. This flow results in release of a subset of the conditions detected in the first clustering subsystem to the second clustering subsystem. Conditions detected in the second clustering subsystem flow to both the attention management and the behaviour type selection competition subsystems, where they are interpreted as recommendations to release particular subsets of currently detected conditions to further clustering, and so on.

All these practical factors mean the major separations of the architecture illustrated in figure 4.11 and 4.12 will occur in any system which learns a sufficiently complex combination of features. However, the heuristic definition of both conditions and information flows means that although this architecture will be very useful for understanding the operations of such a system, particular operational needs may lead to some limited variations at a detailed level in information flow from the primary flows in the general architectural form.

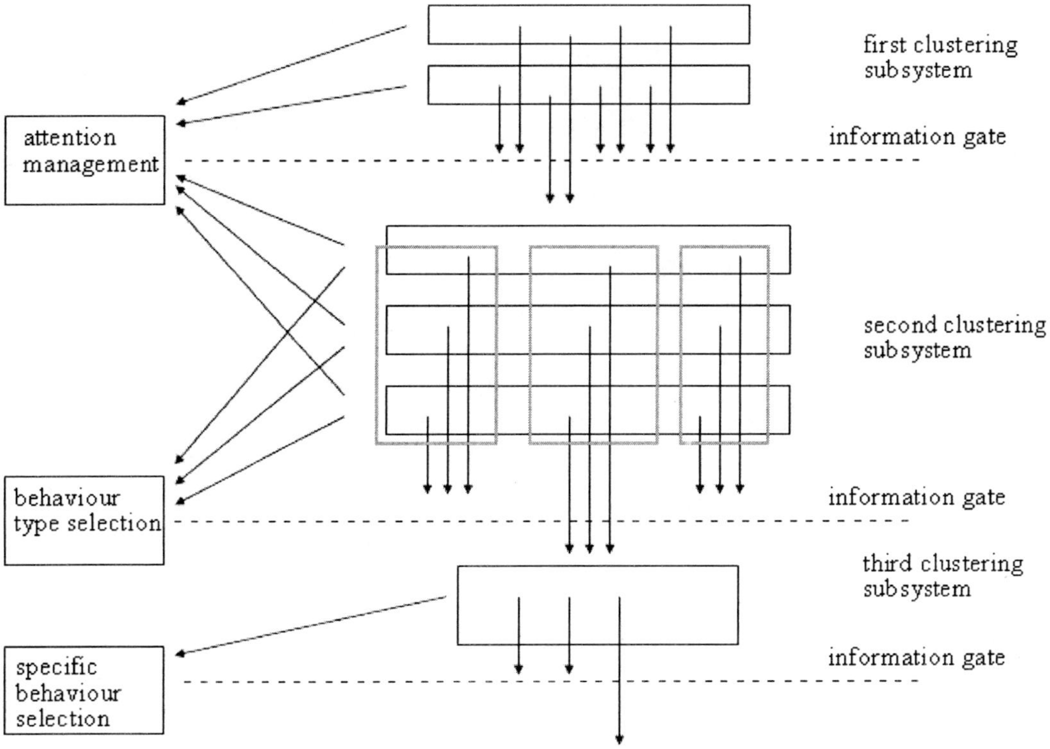

Figure 4.12. Major condition defining information flows in the high level architecture illustrated in figure 4.11.

For example, even in figure 4.11, both clustering subsystems 1 and 2 detects conditions with complexity comparable with perceptual objects. However, the specific conditions may differ in the exact range of complexity and in the way in which definition is influenced by indirect inconsistent consequence feedback or genetic programming. The outputs from subsystem 1 to competition will be optimized for managing selection of attention objects, while the outputs from level 1 to level 2 will be optimized to be most likely to be useful to level 2, and may not be limited to, or even include, the highest level of complexity in level 1.

Furthermore, although the three sets of clustering and competition subsystems correspond primarily with attention management, selection of behaviour type, and selection of specific behaviours within an already selected type, in practice some conditions detected within one clustering subsystem may acquire recommendation strength at a different behavioural level if the operational benefit is high enough. For example, conditions detected within the attention subsystem could be interpreted as recommendations for an avoidance behaviour (such as ducking a flying object) where response time is critical. Another example would be if conditions detected within a group of objects need to influence the selection of the next attention object. It may be more efficient to use detections of such conditions in the second clustering subsystem rather than duplicating the condition recording in the first subsystem. The resultant flow from the second clustering subsystem to both attention management and behaviour type selection is illustrated in figure 4.12. Another option would be for some complex attention behaviours to be located in the behaviour type selection subsystem.

It is of course possible that structuring into four or even more sets of subsystems implementing an even more gradual selection of behaviour could have advantages. The actual architecture would be a compromise between reduced information flow, increased complexity, and generally longer selection times as the number of sets increased.

A fourth factor driving detailed recommendation architecture forms is the need to manage resources. As discussed in the earlier section on Resource Management Processes, this factor results in a subsystem which includes a map of condition recording resources throughout clustering. The subsystem detects conditions correlating with a need for resources in specific modules, assigns unused resources, and manages the creation of appropriate connectivity.

A fifth factor is the need to prioritize behaviour type. As discussed in the earlier section on Area Modules, different areas may be configured to detect conditions optimized for different types of behaviour. Given these configurations, it is possible to manage the relative probability of the different behaviour types. Suppose that there were three areas optimized for avoidance, aggressive and food seeking behaviours. A separate subsystem which detected (perhaps genetically specified) conditions which generally indicated that avoidance, aggressive or food seeking behaviours were appropriate could provide a signal which increased the output activity of devices in the appropriate area which those devices generated in response to a given level of input. The device inputs supporting this capability were discussed in the earlier section. The conditions supporting food seeking behaviour could correlate with, for example, low blood sugar levels. The conditions supporting avoidance behaviour could correlate with life threatening circumstances, for example the shapes of dangerous animals. The conditions supporting aggressive behaviour could correlate, for example, with circumstances in which loss of property was threatened. The signals generated by this subsystem would therefore resemble biological signals labeled hunger, fear and anger.

A sixth factor is management of reward in the competition system. There are two aspects to this management process. One is identifying the targets of reward within competition, the other is determination of the conditions which attract a reward. The selection of behaviour involves a series of selections on different levels of detail as described earlier. Hence the reward can be applied independently to the behaviour decisions on these different levels. For, example, a negative consequence following an inept muscle movement will not necessarily apply to the general type of behaviour under way at the time. A reward targets recent competitive selections. However, the definition of recent may cover a longer period for general types of behaviour than for detailed muscle movements. The conditions under which reward will be assigned will vary with the level of the behaviour. There will be specific types of conditions defined a priori as attracting rewards, but conditions which are frequently present when a reward occurs could be added to the a priori conditions. In the case of more general behaviour type decisions, a priori conditions could be similar to those which rewarded more detailed behavioural decisions, but conditions which are of the level of complexity of the clustering outputs which determined the more general behaviour could come to influence such reward decisions. A separate subsystem is therefore required to manage the generation of reward signals.

Furthermore, there could be behavioural value in some conditions which precede a reward themselves acquiring recommendation strength in favour of generating a reward. For example, suppose that a sensory input results in a sequence of behaviours, and the final behaviour in the sequence results in a reward provided by some external supervisor. Then if conditions correlating with the presence of the sensory input followed by the behaviour acquire recommendation strength in favour of generating a reward, the effect will be that a reward will be generated if the same behaviour follows the same sensory input in the future, regardless of the presence or absence of the supervisor and independent of the intermediate sequence of behaviours by which the final behaviour was reached. Alternative sequences which might be more efficient will then be rewarded in the absence of the supervisor.

The use of consequence feedback will be considered in more detail when the bootstrapping of behaviour is discussed in chapter 9.

Summary of Characteristics of a System with the Recommendation Architecture

Any sufficiently operationally complex system which heuristically defines or modifies its own operations is forced to adopt a form within a set of constraints called the recommendation architecture.

In the recommendation architecture there is a primary separation between a clustering subsystem and a competition subsystem. Clustering is a modular hierarchy which defines and detects different characteristics of system inputs. Competition uses consequence feedback to associate different sets of characteristics detected by clustering with different behaviours.

Practical considerations also constrain the form of clustering and competition at a more detailed level. For clustering these considerations include the needs to maintain synchronization of related information; to define conditions which are as behaviourally useful as possible; to manage the circumstances under which conditions are recorded to minimize unnecessary recording; to define groups of similar conditions which can be detected by resource sharing modules; to limit the use of connectivity resources and to minimize the information exchange between modules as far as is consistent with available resources. For competition these considerations include the needs to limit information exchange meanings to simple levels compatible with the uses of consequence feedback; to limit the overall flow of information within competition; and to limit the number of input weights which must be managed.

These practical considerations result in a range of constraints on the detailed form of any system with the recommendation architecture. It is important to note that just as many different types of system could be implemented within the constraints of the von Neumann architecture, so many different types of system could be implemented within the constraints of the recommendation architecture. However, just as all von Neumann systems tend to exhibit some specific forms, so any system with the recommendation architecture will tend to exhibit the forms of that architecture.

References

Ballard, D. H. (1997). . MIT Press.

Biederman, I. and Cooper, E. E. (1992). Size invariance in visual object priming. Journal of Experimental Psychology: 18(1), 121 - 126.

Furmanski, C. S. and S. A. Engel (2000). Perceptual learning in object recognition: Object specificity and size invariance. 40(5), 473 - 484.

Hebb, D. O. (1949). . New York: Wiley.

Ito, M., H. Tamura, I. Fujita, and K. Tanaka (1995). Size and position invariance of neural responses in monkey inferotemporal cortex. 73 (1), 218 - 226.

Jain, A. K., Mao, J. and Mohiuddin, K. (1996). Artificial Neural Networks: A Tutorial. 29(3), 56 - 63.

Tovee, M. J., E. T. Rolls, and P. Azzopardi (1994). Translation invariance in the responses to faces of single neurons in the temporal visual cortical areas of the alert macaque. 72 (3), 1049-1060.

Chapter V

Attention Management

As discussed in the previous chapters, determination of behaviour depends upon detection of conditions within the information available to the system. Such information, derived from the entire available sensory domains, is provided to the system at all points in time. However, a sensory domain may contain many different objects which are potentially relevant to behaviour. For example, the visual domain may contain many physical objects, and the auditory domain may contain words derived from multiple conversations. It must therefore be possible to detect conditions within information limited to specific subsets of these domains corresponding with sensory objects of different kinds. There are therefore a number of information management issues.

Firstly, it must be possible for the detection of conditions within one sensory object to get priority over detection of conditions within other objects in the determination of behaviour. Furthermore, it must be possible for conditions detected within different sensory domains (such as colour and shape) but corresponding with the same object to be identified and combined into higher level conditions. This is essentially the classical "binding problem".

Secondly, activity indicating conditions detected within different objects must be simultaneously present in order to be combined into conditions detected within groups of objects and groups of such groups. However, such combinations must be performed in a controlled fashion. It must be possible to distinguish between conditions derived from different objects at levels of condition complexity where they have not been combined. There must be appropriate selections of the specific objects making up a group and of the level of complexity of the conditions derived from different objects at which they will be combined into conditions within the group as a whole. Understanding speech provides a good example of this issue. Such understanding requires detection of conditions within different words and within groups of words making up phrases. Populations of conditions detected within each word must be separable, and conditions within a phrase must be detected within the appropriate combination of activated condition populations.

Thirdly, although conditions may be activated on the basis of frequent past activity at the same time as currently active conditions, it must be possible to distinguish between conditions which are actually present and conditions which have been indirectly activated.

All the information available to the system is ultimately coded in the activity of devices like neurons. This activity must therefore make it possible to solve these information management issues. The solution to these issues requires a dynamic model for the condition recording device. Discussion of this dynamic model makes it necessary to address some detailed neuron physiological considerations in advance of the more complete physiological discussion in chapter 7, and some recommendation architecture simulation results in advance of the more complete description in chapter 6.

Neuron Network Dynamics

There have been a number of studies of the dynamic behaviour of cortical neurons in vivo [e.g. Shadlen and Newsome 1998], in electronic simulation [e.g. Diesmann, Gewaltig, and Aertsen, 1999], and even using one biological neuron to model many different neurons in a network [Reyes 2003].

A typical cortex neuron receives 3000 – 10,000 synaptic contacts, of which 85% are believed to be excitatory. The post synaptic potential generated by one input action potential depolarizes the membrane by about 3 – 10%, so neglecting inhibition a neuron would be expected to produce a spike whenever 10 – 40 input spikes arrive within 10 – 20 milliseconds [Shadlen and Newsome 1998]. Average neuron firing rates can be up to 100 spikes/second, but within an average rate the actual spike generation timing is very irregular [Tomko and Crapper 1974].

Simulations have typically employed the leaky integrator model neuron, in which a spike input generates a post synaptic potential which then decays with some time constant. Examples of such simulations include Brody [1998], Shadlen and Newsome [1998], Diesmann, Gewaltig, and Aertsen [1999] and van Rossum, Turrigiano, and Nelson [2002]. The simulated neural networks have generally been successive feedforward layers of neurons with connectivity between layers all to all (e.g. Diesmann et al [1999]) or randomly selected more sparse (e.g. 10% in Reyes [2003]). In general, only excitatory connectivity has been included.

One general result is that synchrony tends to develop across layers. This synchrony has been labeled pulse packets by Diesmann et al [1999], and it appears that a pattern of input activity either propagates synchronously across a series of layers or dies away [Reyes 2003]. Another interesting result is that a relatively slow interaction between neurons can result in synchronization of their outputs on a much shorter timescale [Brody 1998].

Information Content of Neuron Action Potentials

An early proposal for the way in which the information content of neuron output action potentials or spikes in the cortex could address the binding problem was that temporal synchronization of neuron outputs could be used to tag the activity of neurons with a signal indicating the perceived object to which its activity relates. This temporal binding hypothesis [von der Malsburg 1981; 1999] was seen as requiring neurons capable of detecting coincidences of spikes to within a few milliseconds, and therefore led to controversy over whether coincidence detection is a viable model for cortex neurons [Konig et al 1996]. The hypothesis was severely criticized by Shadlen and Movshen [1999] on a number of grounds, including the incompleteness of the theory in that it describes the indicator of binding without describing how binding is computed; the presence of the proposed binding indicator in areas of the visual cortex where binding is not cognitively present; and the lack of evidence for the required coincidence detection neurons.

Another aspect of the debate has been discussion of which aspect of action potentials carried the information content. There have been two major positions on this issue: rate coding and temporal coding [Gautrais and Thorpe 1998]. In rate coding, information is represented by average firing rates, consistent with the observations that firing rates in sensory neurons correlate with the intensity of an encoded feature [van Rossum, Turrigiano, and Nelson 2002]. Synchrony is not a factor with the model, and may even compromise rate coding [Reyes 2003]. An issue with the rate coding view is that the actual generation of action potentials by biological neurons is very erratic, and it has therefore been argued that 10 – 50 millisecond integration times would be required in each neuron layer to read out rate coded signals [Gautrais and Thorpe 1998; Shadlen and Newsome 1998]. Such integration times would be in conflict with the human ability to categorize a complex scene in ~150 milliseconds [van Rossum, Turrigiano, and Nelson, 2002].

In temporal coding, some aspect of the relative spike times plays a crucial role in the information content of the spikes. One example is the proposal that information is contained in the order in which neurons spike [Gautrais and Thorpe, 1998]. The temporal binding hypothesis is another example.

Simulations of the responses of cortex neurons using leaky integrator models with a range of biological realism indicate that in layered networks of such neurons, propagation of activity across multiple layers tends either to become synchronized or to be extinguished [Diesmann, Gewaltig, and Aertsen, 1999; van Rossum, Turrigiano, and Nelson, 2002; Reyes 2003].

The way in which information is coded in the specific population of neurons generating action potentials has received somewhat less attention, partly because the response fields of neurons do not appear to correlate with cognitively significant conditions or objects [Tanaka 1993].

Information Coding in the Recommendation Architecture Cognitive Model

In the recommendation architecture the condition recording device generates a sequence of output action potentials or spikes. The outputs of a column are the outputs of devices in a specific layer of the column. These outputs indicate the presence of the conditions detected by these devices. The learning algorithms result in a column detecting a set of fairly similar conditions, and an array of columns detecting different sets of conditions within the same input space. Because the sets can expand but not contract, a column set will not correspond exactly with a cognitive object. However, the learning algorithms result in column outputs which have better discrimination between such objects than column inputs. In other words, if an input can typically occur in instances of N cognitive objects and an output in n such objects, then n << N. An object will therefore be indicated by the presence of any large subset of a limited group of columns. There will be some overlap between the groups for different objects. The output from one column can be regarded as a recommendation in favour of the presence of a small number of objects, and the largest total recommendation weight found by summation of the recommendation weights for all active columns across all recommended features indicates the object currently present. This summation and comparison takes place in subcortical structures.

A condition recording device can be dynamically modeled as a leaky integrator neuron. Outputs from the device are sequences of spikes. Any output indicates the current presence of conditions within the neuron set, and the average spike generation rate indicates the number of such conditions. A frequency modulation of the rate indicates the input population within which the conditions were detected. Frequency modulation means that although the average rate over some period of time is constant, spikes are more likely to occur bunched near peaks in some modulation signal as illustrated in figure 5.1. Strong bunching is equivalent to packetization as defined by Deisman et al [1999]. If such a modulation were imposed on the inputs to a layer of neurons, then for higher thresholds the activity in the layer will be much higher than in the absence of modulation as illustrated in figure 5.1. Furthermore, the difference in activity level between modulated and unmodulated inputs would be maintained or even amplified in subsequent layers, because of the tendency for neurons to produce outputs in phase with the modulation.

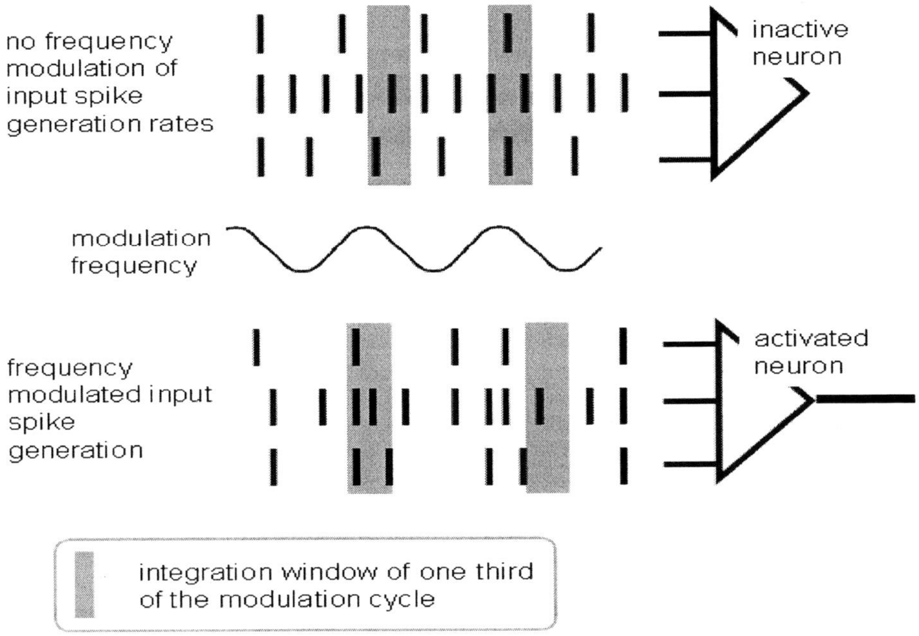

Figure 5.1. Frequency modulation slightly shifts each individual action potential spike towards the nearest peak in a modulation signal. If three inputs to a neuron are unmodulated, the number of spikes within one integration window varies from two to four, but from one to five with modulation. If the neuron threshold was equivalent to five spikes, only the modulated inputs would generate any output.

The functional effect of modulation can be understood as follows. Suppose that inputs were derived from a visual field, and within that field was an object defined by a continuous closed boundary. Suppose further that detection of the presence of such a boundary could trigger imposition of a frequency modulation only on inputs from within the boundary. The result would be preferential detection of conditions within the visual object relative to the rest of the visual field. Independent detection of simple conditions of different types (colour, shape etc.) could be associated by a common frequency modulation phase and therefore used to detect combination conditions limited to information derived from the one object.

The recommendation architecture approach thus employs a combination of population, rate and temporal coding to perform information processing. The frequency modulation coding has some resemblances with von der Malsburg's original proposal that temporal coherence in neuron spike generation could contribute to identifying populations of activity related to individual objects, but there are some important differences. Von der Malsburg and later workers [see Shadlen and Movshon 1999 for extensive references] have argued that synchronicity between spikes within a few milliseconds is the determining factor indicating neuron activity relating to a particular visual object, and the 40 Hz signal found in the EEG gamma band is a manifestation of such synchronization processes [Llinas et al 1994]. In the recommendation architecture model, spike synchronization can be much less precise.

The Von der Malsburg temporal binding model has been strongly criticized by Shadlen and Movshon [1999] on a number of grounds. Firstly, they point out that in the theory as proposed, temporal correlation is not a theory about how binding is computed, only a theory of how binding is signaled. In other words, it does not indicate how the visual system decides which elements belong to different objects. They suggest that the binding problem could more readily be solved by successive elaboration of progressively more complex representations of visual scenes as explored by Tsotsos [1995]. Secondly, they point out that the study of visual deficits indicates that binding only exists in the higher levels of the visual cortex hierarchy, in which case the observations of synchrony in the primary visual cortex is unexpected. Thirdly they argue that the implied coincidence detector model for cortex neurons is implausible, and that given a more plausible integration window of 5 - 10 milliseconds, all spikes are in synchrony with other spikes. There has been extensive debate over whether cortical neurons are integrators or coincidence detectors. Konig et al [1996] point out that the essential difference between these views is the period of time over which synaptic inputs are integrated. This period of time is of the order of the synaptic potential decay time, or comparable with the membrane constant which is estimated to be in the range 8 - 16 milliseconds [Konig et al, 1996], much longer than the few milliseconds suggested for temporal coincidence detectors.

Effect of Synaptic Integration Time on Frequency Modulation Mechanism

Simulation results demonstrate that frequency modulation coding in the recommendation architecture does not place unrealistic requirements on neuron parameters such as synaptic integration times. The neuron organization used in the various simulations described in this chapter is conceptually illustrated in figure 5.2. Six hundred neurons in three layers of two hundred were simulated by a software implementation. Two hundred externally generated signals were inputs to the first neuron layer, outputs from the first layer were inputs to the second layer and so on. The externally generated signals were structurally identical with neuron outputs, being made up of sequences of spikes modulated and delayed in various ways. For each run the performance of the model was simulated over a period of 500 milliseconds broken down into 0.33 millisecond time slots. The status of each neuron was determined from its past condition, its current inputs, and some noise related uncertainties once every time slot.

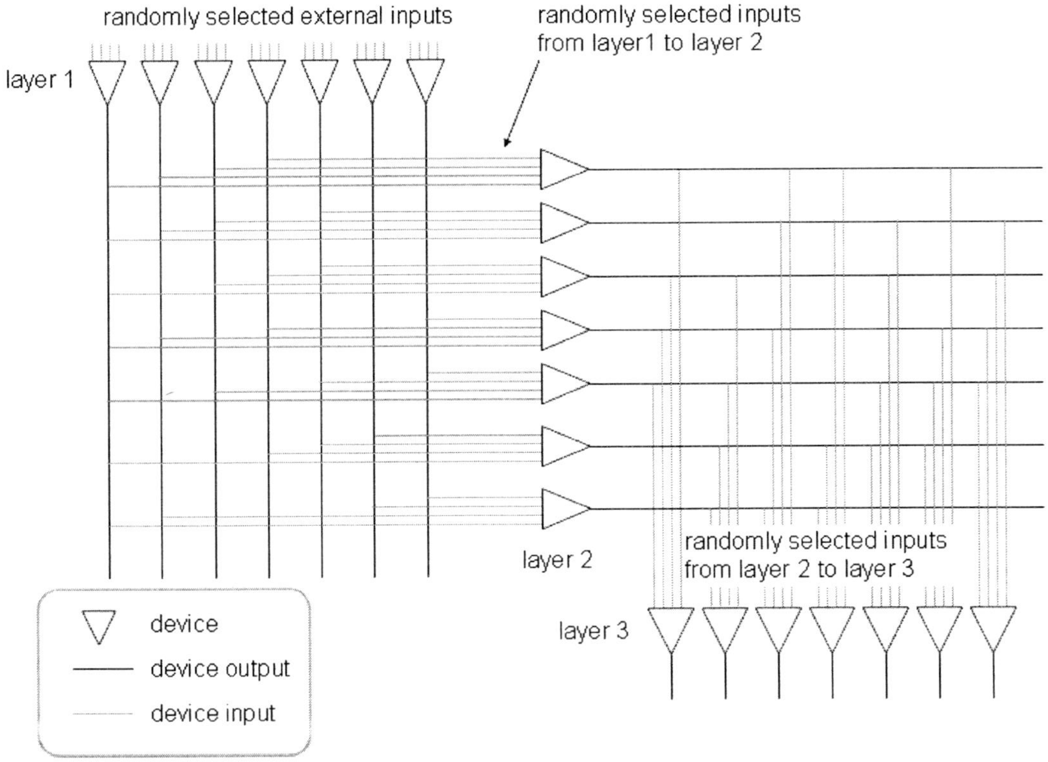

Figure 5.2 Neuron organization in simulations. Three layers of neurons were used. The inputs to the first layer were excitatory inputs randomly selected from an external input set. The inputs to the second layer were excitatory inputs randomly selected from first layer outputs. The inputs to the third layer were excitatory inputs randomly selected from second layer outputs. Not illustrated are the inhibitory inputs from layer 1 to layer 2 and from layer 2 to layer 3. In some simulations there was a bias placed upon the various random selection processes.

Each neuron had a set of randomly selected inputs from the appropriate input population, and the size of that input set was randomly selected in the range 40 - 60 for each neuron. Duplicate inputs to one neuron were not permitted. Each input from the appropriate population had the same chance of being selected for any neuron. The externally generated inputs to this model had spike rates for which the average interval between spikes was randomly selected for each input within the range 20 - 100 milliseconds, corresponding with rates in the range 10 - 50 spikes/second. Each of the 241 possible intervals at 0.33 millisecond resolution has the same selection probability. The overall average of spike rates across a set of 200 inputs was 20 spikes/second.

A spike occupied one time slot of 0.33 milliseconds. Once the interval had been selected for an input, the initial spike was delayed from the start of the 500 millisecond simulation period by a randomly selected phase of up to one interval. Generation rates were noisy in the sense that the time slot for each spike varied by a randomly selected interval of zero to two time slots from the slot defined by the average rate. The noisy average spike generation rates were then modulated at 40 Hz. This modulation was implemented by defining a phase for the 40 Hz signal relative to simulation start. Each spike was then shifted towards the nearest modulation peak. This shift was randomly selected for each spike, with a 50% probability of selecting a primary modulation shift, a 40% probability of selecting one time slot less that that shift, and a 10% chance of two slots less. The primary value could be varied in the range 3 to 5 time slots or 1 - 1.67 milliseconds, and the average shift (or the strength of modulation) therefore varied in the range 0.8 - 1.5 milliseconds. In addition, it was possible to introduce delays which could be randomly selected for each input in the range 0 - 2 milliseconds. The purpose of the delays was to test the tolerance of the process for inconsistent transmission delays. The various random noise factors test the robustness of the observed phenomena.

A neuron model had a set of inputs, and detected any significant total activity in those inputs. In information terms it detects the set of conditions defined by activity of any large subset of its inputs. A single spike generated a post synaptic potential which varied as Ate^{-at} where t is the time after arrival of the spike and A and a are constants which were selected so that the peak in the post synaptic potential curve was 100 arbitrary units and the time from arrival of the spike until the potential declined to 33% of its peak value was a controllable parameter called the integration time.

The values of this integration time were selected in the range 2 - 25 milliseconds, to cover the periods postulated for synchronization detection (a few milliseconds) through the observed membrane constants in cortical neurons (8 - 16 milliseconds, see Konig et al [1996]), up to the period defined by the modulation frequency (i.e. 25 milliseconds for 40 Hz). Examples of the synaptic potential decay curves used are shown in figure 5.3. The neuron model employed was thus a conventional leaky integrate and fire mechanism [Dayan and Abbott 2001].

A neuron generated an action potential or spike if at any point in time the total potential generated by past input spikes exceeded a threshold T, provided that no output spike had been generated within an immediately preceding absolute refractory period set at 3 milliseconds to correspond with such periods observed in cortex neurons [Dayan and Abbott 2001]. The output spike was delayed by a randomly selected interval of 0.33 - 1 millisecond after threshold was exceeded to model the observed reliability of spike timing in neocortical neurons [Mainen and Sejnowski 1995]. When a spike was generated, the post synaptic potential was reduced to zero. The value of the threshold T was the same for all neurons in one layer, but set at different values for each synaptic integration time. The values were set so that a modulated set of inputs with an average spike rate of 20 spikes/second generated a similar level of output activity in each layer.

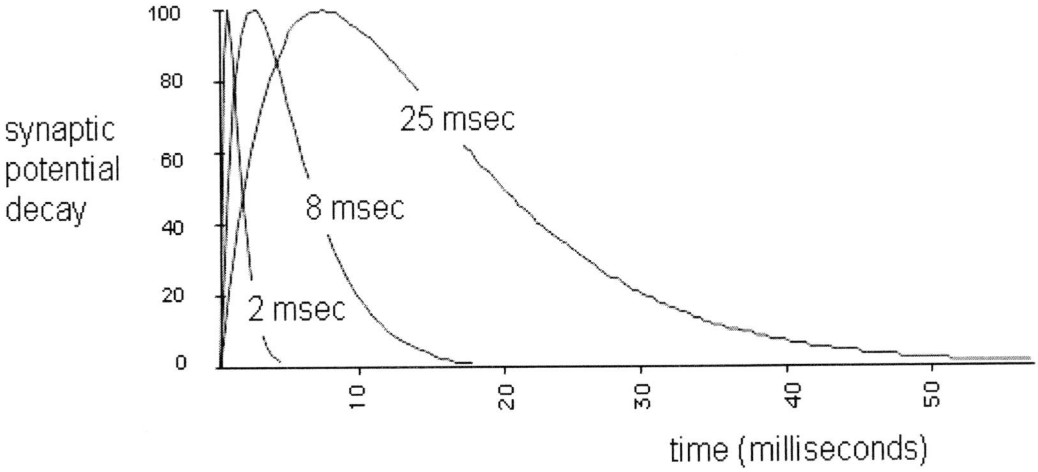

Figure 5.3 Synaptic potential decay curves for neuron model. The synaptic potential resulting from one input spike at time zero is shown for a range of integration time parameters.

A number of different system models were generated and a number of different randomly constructed input states were applied to these models. Each model construction resulted in a different pattern of connectivity, each input state construction resulted in different spike rates, phases, noise, delays and modulation shifts, and each presentation of the same input state to the same model would generate slightly different results because of random delays in spike generation. The results described in this section are from over 100 different runs with different parametric settings.

The spikes in each timeslot in the input and in each layer in the first 40 milliseconds of a typical simulation with modulated inputs is shown in figure 5.4, with an unmodulated input for comparison. As can be seen from the diagram, response is packetized in phase with the modulation signal, and packets are shorter and slightly delayed (by an interval of the order of the time between arrival of spike and peak in synaptic potential) as they progress through the layers. Response to an unmodulated signal also packetized, but was smaller and unrelated to a modulation signal.

The ratio of the sums of the average spike rates of all the neurons in each layer between when the same external inputs are unmodulated and modulated is illustrated in figure 5.5, with a modulation strength of 0.8 milliseconds and as the integration time varied over the range 2 - 25 milliseconds. The variability between runs for layer 2 is also illustrated. Random delays of up to 2 milliseconds in external inputs, had very little effect on the activity ratios. The variation in activity ratio with modulation strength in the range 0.8 - 1.5 milliseconds is shown for an integration time of 8 milliseconds in figure 5.6.

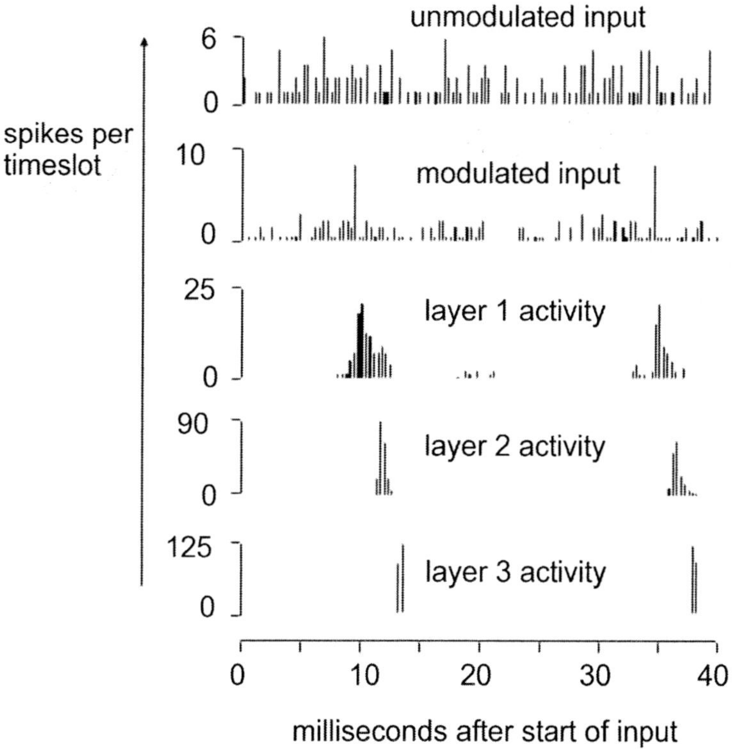

Figure 5.4 Total spikes per timeslot in a typical unmodulated and corresponding modulated input, and activity in the three layers in response to the modulated input. Response to an unmodulated input was also packetized but with lower total packet sizes.

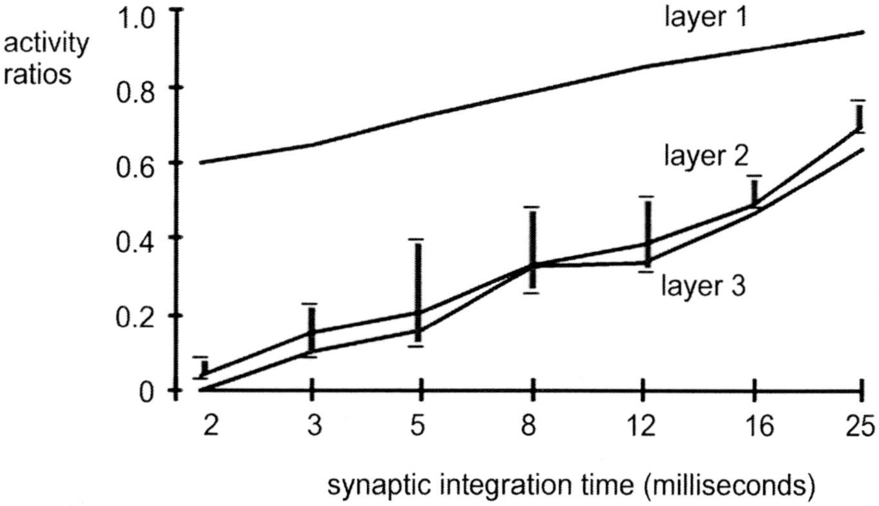

Figure 5.5. Effects of modulation on neuron activity levels as a function of synaptic integration time. Unmodulated sets of inputs and the same sets modulated were presented to the first layer. The diagram shows the average ratios of activity between when the input layer 1 is presented an unmodulated set of inputs and when the same input rates are modulated at 40 Hz with an average modulation shift of 0.8 milliseconds. Ratios are shown for the input layer of neurons and two successive layers. The bars indicate the variation in the ratio found for layer 2 between different runs.

A number of implications follow from these results. Firstly, consistent modulation at 40 Hz across a set of inputs considerably enhances the activity of the neurons targeted by those inputs. Although the enhancement is greater for integration times in the range 2 - 3 milliseconds (i.e. the "coincidence detector" realm), it remains strong at much longer integration times, even times comparable with the modulation period. Secondly, the enhancement is transmitted between layers, even though the modulation is only applied directly to the inputs and the neurons have no special tuning for 40 Hz. Thirdly, small, inconsistent delays of less than 2 milliseconds in the arrival times of the external inputs reduce but do not eliminate the enhancement. Fourthly, the enhancement is present despite random delays of the order of 1 millisecond in the generation of a spike following achievement of threshold. Fifthly, because for all spike generation rates there is a tendency for spikes to occur shifted towards the nearest modulation peak, the generation of spike trains will appear irregular, as observed for in vitro cortex neurons [Stevens and Zador, 1998].

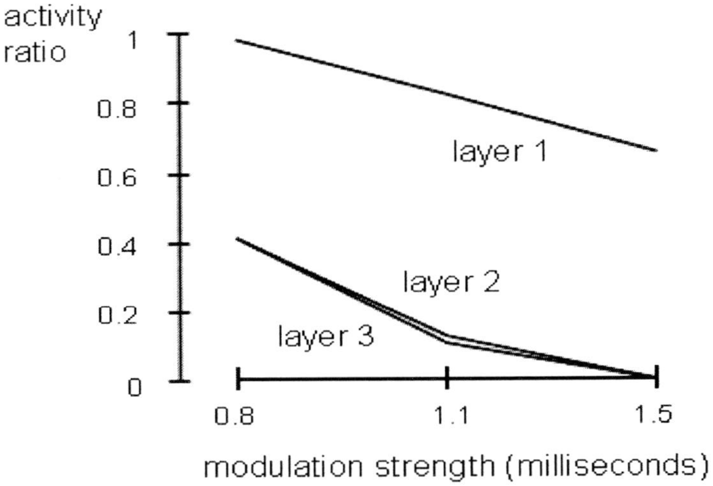

Figure 5.6. Effect of degree of modulation on activity in three successive layers of neurons with an 8 millisecond integration time. As modulation strength increases, the ratio by which response to a modulated set of inputs exceeds the response to the same inputs unmodulated increases.

The simulations thus demonstrate that if the inputs to a population of leaky integrator neurons are given a consistent frequency modulation of 40 Hz with a modulation amplitude which shifts spikes towards the peaks in a modulation frequency by a time ~ 1 millisecond, the activity of the population increases substantially, and the modulation is propagated across the population. The effect exists for a wide range of synaptic integration times and is robust against a range of noise parameters.

Recommendation Architecture Binding Model

The model for binding of visual objects in the recommendation architecture is therefore as follows. A visual domain corresponding with an object defined as the area within a continuous enclosing boundary is selected within the total visual field. The definition of this attention domain would be the result of recommendations for domain changes generated by conditions including those corresponding with the existence of boundary elements. The outputs from the domain will in general be at many different average spike generation rates and phases. The selection of the attention domain results in all the rates from within that domain being modulated at a common frequency and phase.

Outputs from the entire visual field are then processed, but the frequency modulation at the same phase places a bias in favor of detecting conditions derived only from information within the attention domain, and these conditions would be tagged by the same spike rate frequency modulation because the synaptic potentials of the devices tend to exceed threshold in phase with the modulation. The set of domain outputs could be processed separately, for example to detect conditions correlating with color and shape, and when the conditions came together they would be tagged by a common frequency modulation phase which would favor detection of higher level conditions derived only from the original attention domain. Because the higher level activity populations indicate conditions predominantly derived for the object within the attention domain, the corresponding behavioral recommendations will be predominantly appropriate to that object.

This binding model addresses the issues raised by Shadlen and Movshon [1999]. It is a model for how binding is computed, it explains why a binding related synchronization should be present at levels where binding is not detected cognitively, and it does not require unrealistically narrow integration windows.

Simulation of the Binding Model

To demonstrate the operation of the combination of rate, frequency and population coding in the recommendation architecture model, simulation results with a system model will be described. Leaky integrator neurons were organized into three layers as illustrated in figure 5.2. Four parallel columns spanned the three layers. Each column contained 100 neurons in each layer and detected different types of conditions within external input states. There were external excitatory inputs to the first layer from 200 separate sequences of action potential spikes. The excitatory inputs to the second layer were outputs from the first layer and so on.

Four types of input state were defined by the most probable range of values for the average spike generation rate of each of the 200 external inputs. For one input state type, each input was randomly assigned an integer bias parameter in the range 1 to 10, with equal selection probability for each value. A bias parameter value had an associated relative probability for each possible average spike generation rate. A bias parameter of 1 corresponded with a high probability of a high average spike rate, 5 with a high probability of an intermediate spike rate, and 10 with a high probability of a low or zero spike rate. Different instances of an input state type were constructed by random selection of average spike rates for each input, with the selection biased by the corresponding bias parameter. Instances were then normalized to an overall average of 20 spikes/second. The presence of a type is therefore only indicated by a trend across all inputs.

Neurons were the leaky integrators described earlier, except that each neuron also had inhibitory inputs derived from the other neurons in the same layer. The synaptic potential decay curve in response to both excitatory and inhibitory spikes was for an 8 millisecond synaptic decay time as illustrated in figure 5.3. However, for inhibitive inputs, the curve could be scaled so that the maximum synaptic potential was adjusted by a factor from 0 (i.e. no inhibition) to 8. The time at which a spike was generated was a randomly selected interval in the range 1 - 5 milliseconds after post synaptic potential exceeded threshold. The threshold for all devices in all layers was the same, but could be varied in the range 20 - 40 times the peak magnitude of the excitatory synaptic potential in response to one spike.

Recommendation architecture learning results in inputs to first layer neurons in a column tending to be inputs which have had high activity at similar times in the past, with a corresponding bias placed on second and third layer devices in the column. Detailed algorithms are described in chapter 6. Learning was not modeled directly for the current simulations, but the results of such learning were modeled by biasing the selection of external inputs to the first layer of each column in favour of inputs with a low bias parameter (i.e. a high probability of high activity) in a different input state type. Excitatory inputs to the second layer were biased in favour of first layer devices often active at the same time and so on. One effect of the bias was that 90% of excitatory connectivity was within columns. The number of excitatory inputs to a device was randomly selected in the range 40 - 60.

The inhibitory connectivity to a device from other devices in the same layer was randomly selected with a 3:1 bias in favour of inputs from outside the column of the device, and set at 15% of the number of excitatory inputs. The intended effect of the inhibitive connectivity was that if spike packets of comparable size were initiated in more than one column, the earliest packet would suppress the packets in other columns.

Instances of one input state type would typically generate the strongest activity in the column corresponding with the type. In a practical learning situation there would be many columns corresponding with different condition types, each defined by different groups of system inputs which tended to have high activity at the same times. A cognitive feature would be indicated by significant activity in many columns in a particular subset, different features being indicated by different (but overlapping) subsets. However, to simplify the description of the simulations and results it will be assumed that the input state types correspond with different cognitive objects.

The scenario was that two cognitive objects were within the visual field, and attention was focused on one of the objects. The attention focus was assumed to result in outputs from one object being frequency modulated at 40 Hz. The modulation of inputs was imposed by shifting each spike towards the nearest modulation peak. The average shift was a controllable parameter varying between 1 and 3.7 milliseconds. The actual shift for a spike was varied randomly from the selected average by □0.7 millisecond. Eight sets of instances were generated, each made up of one instance of each input state type. Mixed inputs for each of the twelve possible combinations of modulated instance of one type and unmodulated of the other type were constructed as a 200 input mixed set by adding the modulated spikes for one (attention) input state to the unmodulated spikes for a different type of input state. A supplementary modulation, in phase with the input modulation, could be applied to the first level neurons. This modulation was intended to test if such a modulation could be used to adjust to different overall levels of input activity.

Simulations were performed for 200 millisecond input period with a timeslot of 0.3 milliseconds. The total activity in each layer of each column was determined over the period, and also the timing of the spikes in each column.

Simulation Results

Spike counts were measured in each layer of each column. Spikes generated by layers were again packetized. The first packet in layer 1 of the column corresponding with the modulated input was usually generated ~ 8 milliseconds after the start of inputs, in response to the first modulation peak in the inputs. Packets were initiated in layer 1 of other columns, particularly in the column corresponding with the unmodulated input, with a tendency to be initiated later than in the column corresponding with the modulated input.

In the absence of inhibition, a packet with less than a minimum number of spikes in layer 1 decayed through the other layers. The minimum depended on threshold. At the highest thresholds, packets in all columns tended to be extinguished by the third layer. For low thresholds there tended to be activity in the third layer of the columns corresponding with both the modulated and the unmodulated input or even other columns at the lowest threshold levels.

For intermediate threshold levels, there was a strong tendency for the packet in the column corresponding with the modulated input to be earlier and larger than packets in any other column in layer 1. Often the smaller packets decayed in later levels. When inhibition was introduced, the decay of packets in other than the column corresponding with the modulated inputs was more frequent and faster. However, at certain threshold levels the introduction of inhibition would result in decay of all packets by layer 3.

If the packet in layer 3 of the column corresponding with the modulated condition was earlier than and had at least twice the number of spikes as any packets in any other columns, system response was regarded as correct. A typical example of correct spike activity in layer 3 in the two columns corresponding with the modulated and unmodulated condition is illustrated in figure 5.7. Under these conditions activity in layer 3 of other columns was minimal.

Table 5.1 gives the number out of the 12 possible modulated/unmodulated instance combinations of a set of four instances, one of each input state type in which spike activity was in the correct column. The variables in table 5.1 are the degree of input modulation, device threshold, inhibition level and first level modulation. The variation for different sets of instances and different system connectivity selections was of the order of ± 1. For low thresholds, errors were activity in the wrong column, for high thresholds, errors were lack of activity in any column.

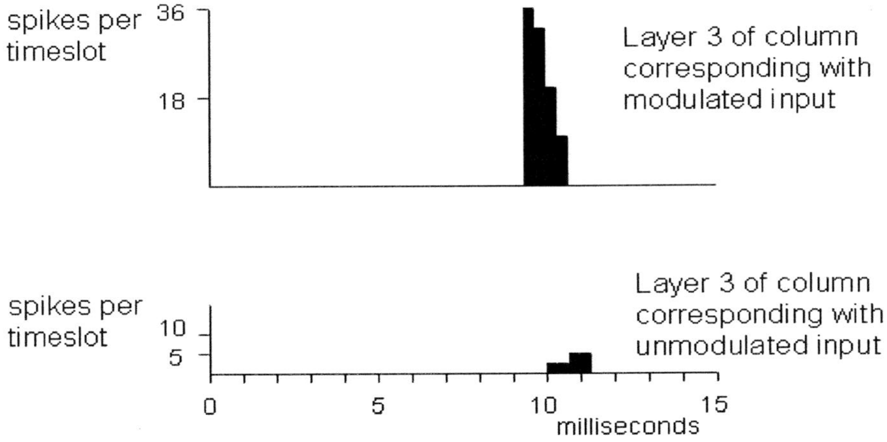

Figure 5.7 Typical spike generation by layer 3 neurons in two columns in response to a mixed input with modulated spikes for the condition corresponding with one column and unmodulated spikes for the condition corresponding with the other column. Activity in the rest of the columns was negligible.

		First level modulation = 0			First level modulation = 4			First level modulation = 8		
Inhibition	0	2	4	6	2	4	6	2	4	6
Average modulation = 3.7 milliseconds										
Threshold										
3000	0	2	8	8	0	4	3	0	0	2
3250	1	10	11	11	2	8	9	0	4	7
3500	6	12	11	10	9	12	11	3	8	9
3750	9	7	6	6	12	11	11	10	11	12
4000	0	0	1	1	8	6	6	11	10	10
Average modulation = 2.3 milliseconds										
Threshold										
3000	1	8	9	8	2	6	6	0	3	3
3250	10	10	9	8	8	9	9	2	6	6
3500	7	5	4	4	10	9	7	5	10	10
3750	0	0	0	0	5	5	3	10	9	9
4000	0	0	0	0	0	0	0	4	4	3
Average modulation = 1.0 millisecond										
Threshold										
3000	7	8	5	3	4	7	7	2	4	6
3250	0	0	0	0	6	6	4	4	7	7
3500	0	0	0	0	0	0	0	9	5	5
3750	0	0	0	0	0	0	0	0	0	0
4000	0	0	0	0	0	0	0	0	0	0

Variables are the device threshold, degree of modulation of external inputs, and the degree of additional modulation to first level devices. Grey indicates bands in which maximum accuracy occurs.

A number of conclusions can be drawn from figure 5.7 and table 5.1. Firstly, column activity discriminates between modulated and unmodulated inputs with a high degree of accuracy, reaching over 90% for the highest modulation level and optimized threshold range. Secondly, first level modulation changes the threshold range within which the highest accuracy is achieved, but does not affect the level achieved. This means that for a fixed threshold level, first level modulation could be increased until there was column activity, and at this point the accuracy would be high. Thirdly, accuracy is improved by inhibition in the threshold ranges in which peak accuracy is achieved, but the improvement in accuracy is not sensitive to the degree of inhibition in the range studied. At higher threshold levels inhibition sometimes made a lower accuracy worse, because the effect of inhibition was to eliminate activity in all columns by layer 3. Fourthly, figure 5.8 indicates that discrimination is achieved within ~ 10 milliseconds for 3 layers and requires an extra 2 milliseconds per layer, consistent with observed human categorization times ~ 150 milliseconds.

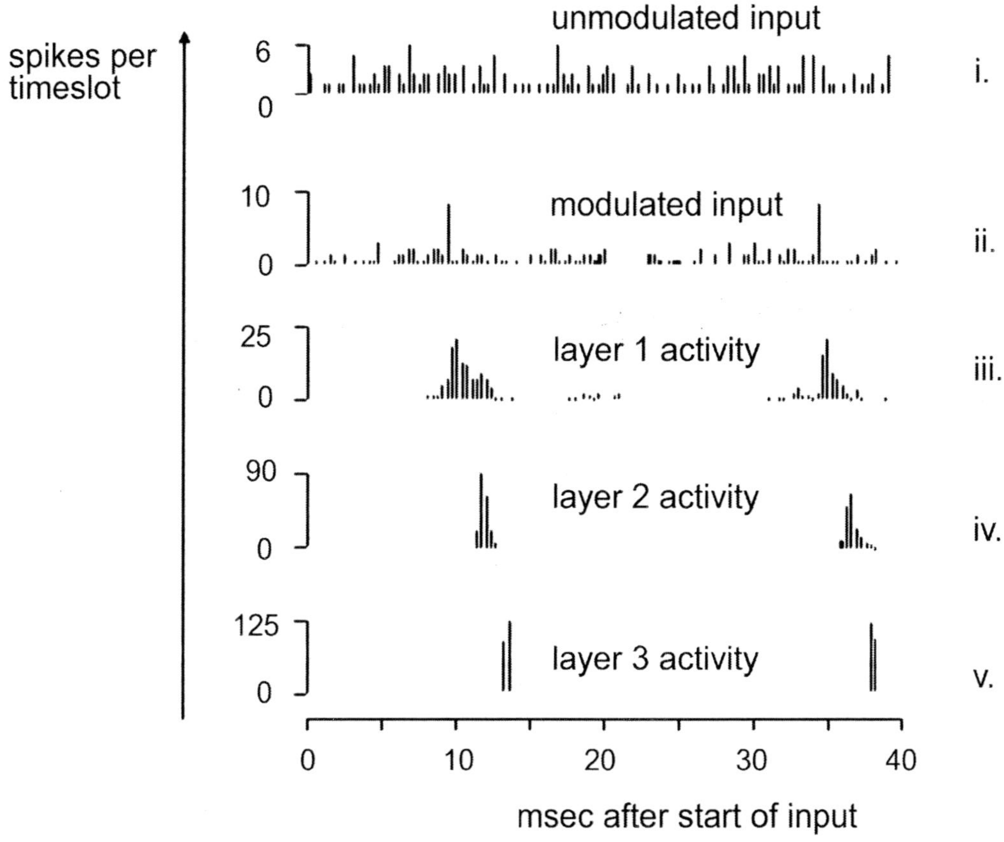

Figure 5.8. Spikes per third of a millisecond timeslot across all external inputs when unmodulated (i) and modulated (ii), and total output spikes per millisecond in response to the modulated inputs in layer 1 (iii), layer 2 (iv) and layer 3 (v).

Management of Information Derived from Multiple Objects

Two other information management issues were identified in the first section of this chapter. One is the need to be able to distinguish between condition detections within different perceived objects, and to combine conditions within specific different objects as required. The other is to distinguish between conditions detected within current direct sensory inputs and conditions indirectly activated on the basis of frequent past simultaneous activity.

Control of the phase of frequency modulation provides a solution to these issues. As illustrated in figure 5.9, suppose that there are inputs from a visual domain which correspond with areas within two different boundaries. Suppose further that inputs from both areas have been frequency modulated at 40 Hz, but with a different phase, and the phases are separated by 8 ms, or about a third of the modulation interval. In figure 5.9 the inputs with the two different frequency modulation phases are coded as black and gray spikes.

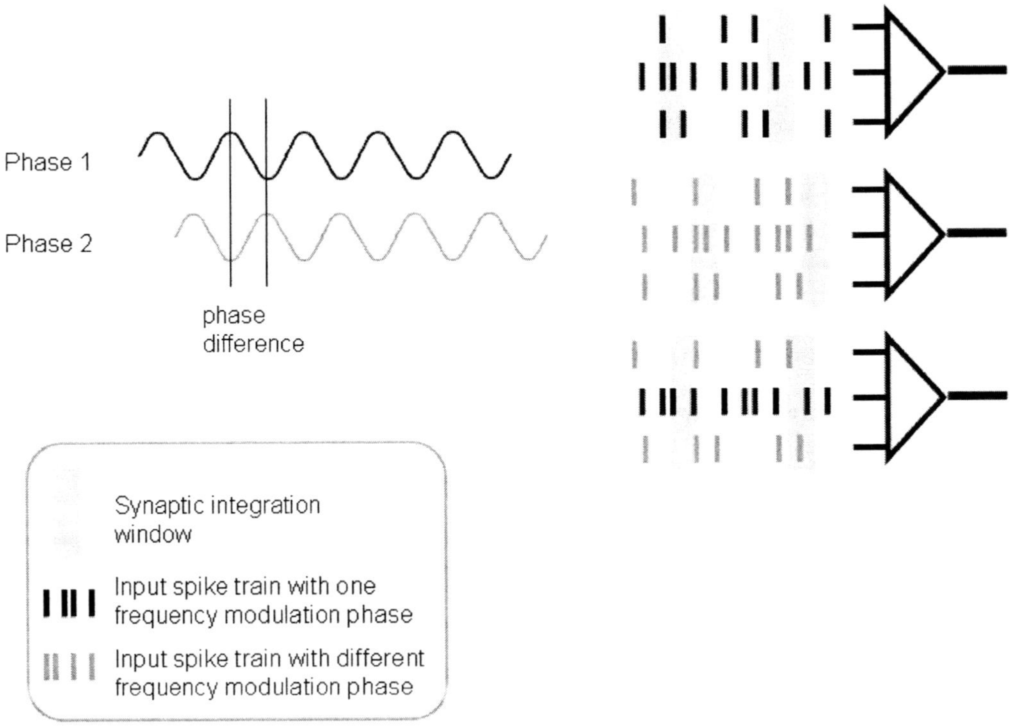

Figure 5.9. Management of separate information populations using the phase of frequency modulation. There are two populations of inputs to a layer of devices, one population frequency modulated with one phase, the other modulated at the same frequency but a different phase. Devices with most of their active inputs from one population will tend to have enough inputs within the synaptic integration window centred at the peak of the corresponding modulation to produce outputs. Devices with inputs from both populations will not have integration windows with as many input spikes and will tend not to produce outputs. The output of all the devices responding to one input population will tend to be modulated in phase with each other, with a slight delay over the input phase. In a layer of devices there will therefore tend to be two populations of active devices distinguished by the frequency modulation phase of their outputs. This population difference will be propagated through subsequent layers.

A neuron which receives three active inputs with the phase coded black experiences a range of spikes within its integration window which varies from two to five, and similarly with the neuron receiving three active inputs with the phase coded gray. If the threshold level was equivalent to five spikes, these neurons would activate in response to the activity of their three inputs. The third neuron in figure 5.9 has one black input and two gray inputs. With this combination the maximum number of spikes within the integration window is lower at four spikes and less likely to activate. As discussed earlier, the outputs of activated neurons will be in phase with their inputs, with a slight delay which will be the same for all neurons in a layer. Hence the effect of the two groups of inputs with two different phases of frequency modulation is the activation of two separate condition populations containing conditions within the two separate groups. Conditions made up of mixtures of information from the two different groups will tend not to be activated. This situation is illustrated in figure 5.10.

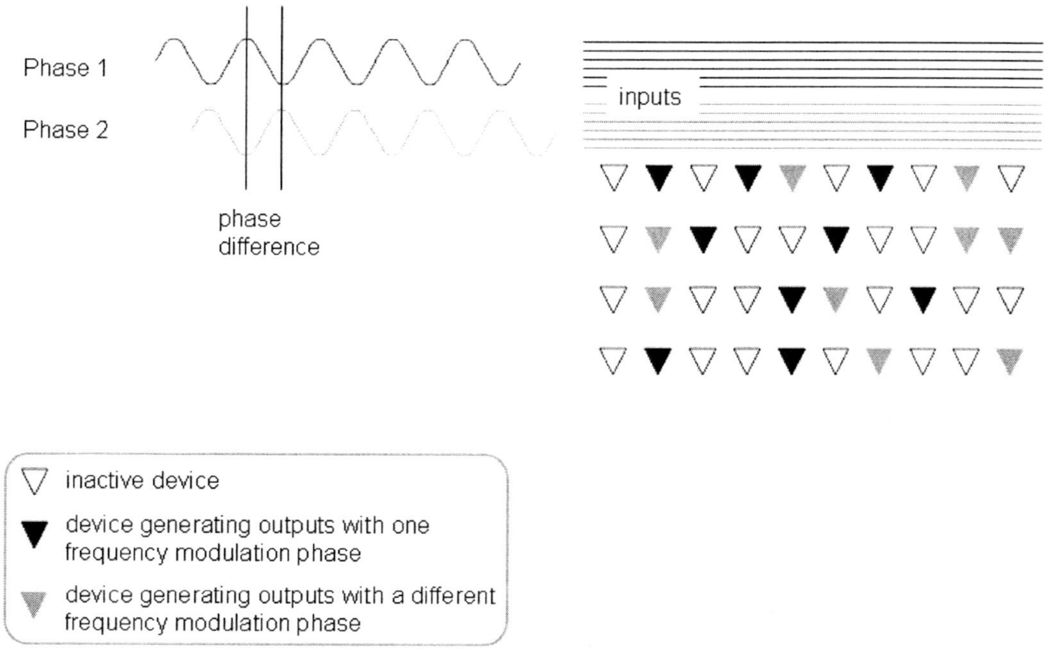

Figure 5.10. Separate populations of active devices in the same multilayered set of devices, distinguished by the phase of frequency modulation of their outputs.

If the synaptic decay time is of the order of 8 milliseconds, a phase difference of less than 8 milliseconds would tend to result in breakdown of the separation between populations. Hence the number of separate populations which can in practice be maintained by a separate modulation frequency will be the modulation interval divided by the synaptic integration time. For example, for a 40 Hz modulation frequency the modulation interval is 25 milliseconds. Hence for an 8 milliseconds integration time, the number of separate populations would be limited to three or four. The significance of this limitation for working memory is discussed in chapter 11.

This mechanism can therefore be used to maintain populations of activated conditions derived from three or four different visual objects or indirectly activated populations in the same physical group of condition recording devices, with a frequency modulation of 40 Hz.

If conditions are required within a group of several objects, the activations corresponding with conditions at some level of complexity from the separate objects must by brought into the same phase. A simple process to achieve this would be as follows. Suppose that three objects have sequentially been the focus of attention. A specific example could be three written symbols "6 + 3". A population of conditions detected initially within "6+3" viewed as a single object has resulted in an initial recommendation to detect conditions within the three constituent objects at some level of complexity, then detect conditions at a higher level of complexity which contain information derived from all three objects. This initial recommendation is maintained active by the activity of an appropriate device population. Conditions are then first detected within the visual image of "6". The population of devices recording these conditions are activated with a common frequency modulation phase. Populations of devices corresponding with the visual images of "+" and "3" are also activated, with the same modulation frequency but different phases. The initial recommendation then results in the activity of the three populations being extinguished, and then results in all recently active devices at the appropriate level of complexity being reactivated at a rate proportional to their recent activity but with a common phase. Conditions are then detected at a higher level of complexity within the combination of the three objects. Indirectly activated populations can also be managed by a similar process as discussed in more detail in chapters 11 and 12.

Note that different frequency modulation frequencies would be less effective than different phases of the same modulation frequencies because different frequencies would periodically come into phase, or at least into close enough phase to generate erroneous activity.

Conclusions

In the recommendation architecture, the activity of neurons contains information by a combination of rate, temporal and population coding. The average spike generation rate indicates the proportion of the conditions recorded on the neuron which is currently being detected. The frequency modulation phase of the generation rate indicates the input domain within which the conditions are being detected. The identities of the activated population with a common frequency modulation phase contain information on appropriate behavioural responses to the object in the corresponding input domain.

This model indicates how binding is computed, it explains why a binding related synchronization should be present at levels where binding is not detected cognitively, and the required neuron model has physiologically plausible parameters.

Simulations have demonstrated that if the inputs to a population of leaky integrator neurons are given a consistent frequency modulation at 40 Hz which shifts individual action potential spikes by ~ 1 millisecond from the position defined by an average generation rate, the activity of neurons in the population is considerably increased, and the modulation is transmitted across the population. The enhancement has been demonstrated to be significant for synaptic potential integration times over a range from 2 milliseconds to 25 milliseconds. The enhancement is therefore not dependent on the existence of coincidence detection type mechanisms with integration times of the order of a few milliseconds.

The binding mechanism has been tested by simulations of populations of model leaky integrator neurons organized into columns in which each column learns a different, cognitively ambiguous condition in accordance with the algorithms of a proposed recommendation architecture cognitive model. Results of these simulations demonstrate that discrimination of cognitive objects can occur with high integrity on timescales consistent with human cognitive processing times.

References

Brody, C. D. (1998). Slow Covariations in Neuronal Resting Potentials Can Lead to Artefactually Fast Cross-Correlations in Their Spike Trains. 80, 3345 – 3351.

Dayan, P. and Abbott, L. F. . MIT Press: Cambridge, Mass.

Diesmann, M., Gewaltig, M-O., and Aertsen, A. (1999). Stable propagation of synchronous spiking in cortical neural networks. 402, 529 – 533.

Gautrais, J. and Thorpe, S. (1998). Rate coding versus temporal order coding: a theoretical approach. 48, 57 – 65.

Konig, P., Engel, A. K., and Singer, W. (1996). Integrator or Coincidence Detector ? The role of the cortical neuron revisited. 19(4), 130 - 137.

Llinas, R., Ribary, U., Joliot, M., and Wang, X.-J. (1994). Content and Context in Temporal Thalamocortical Binding, in Buzsaki G. et alii (eds), , Berlin: Springer.

Mainen, Z., and Sejnowski, T. (1995). Reliability of Spike Timing in Neocortical Neurons. 268 (5216), 1503 - 1506.

Reyes, A. D. (2003). Synchrony-dependent propagation of firing rate in iteratively constructed networks in vitro. 6(6), 593 – 599.

Shadlen, M. N. and Movshon, J. (1999). Synchrony Unbound: a critical evaluation of the temporal binding hypothesis. 24, 67 - 77.

Shadlen, M. N. and Newsome, W. T. (1998). The Variable Discharge of Cortical Neurons: Implications for Connectivity, Computation, and Information Coding. 18(10), 3870 – 3896.

Stevens, C.F. and Zador, A.M. (1998). Input synchrony and the irregular firing of cortical neurons. 1 (3), 210 - 217.

Tanaka, K. (1993). Neuronal Mechanisms of Object Recognition. 262, 685 – 688.

Tomko, G. J. and Crapper, D. R. (1974). Neuronal Variability: non-stationary responses to identical visual stimuli. 79, 405 – 418.

Tsotsos, J. K. (1995). Towards a Computational Model of Visual Attention. , T. V. Papathomas ed. Cambridge MA: MIT Press.

Van Rossum, M. C. W., Turrigiano, G. G. and Nelson, S. B. (2002). Fast Propagation of Firing Rates through Layered Networks of Noisy Neurons. 22(5), 1956 – 1966.

Von der Malsburg, C. (1981). 81-2, Max Planck Institute for Biophysical Chemistry, Gottingen, Germany.

Von der Malsburg, C. (1999). The What and Why of Binding: the modeler's perspective. 24, 95 - 104.

Chapter VI

Electronic Versions of the Recommendation Architecture

In order to confirm the capabilities of the recommendation architecture, various electronic versions have been implemented. Because much of the learning in the recommendation architecture involves additions to connections between devices, and physical additions of large numbers of connections are difficult to implement with current electronic technology, devices and connectivity have been implemented by software emulation.

Initial system implementations have been in the Smalltalk language. This is a very powerful language for system design, having the flexibility to make experimental changes to system architecture relatively easily. The architecture developed within Smalltalk has been reimplemented in C++. This language is much less effective for system design and experimentation, but has advantages for production versions. Implementations run on Apple Macintosh hardware platforms.

One implementation has been targeted at modeling the detailed mechanisms for maintaining a separation between conditions detected simultaneously within different input sources. This implementation was described in chapter 5. Another implementation targeted at modeling the learning processes will be described in this chapter. The various device algorithms will be described first, then the higher level module algorithms in clustering. The algorithms used in competition will then be described. Finally, results obtained from modeling of cognitive processes will be described and discussed.

Device Algorithms in Clustering

Within clustering, condition recording devices are organized into layers, columns and arrays as described in chapter 4. The condition recording device implemented in clustering is illustrated in figure 6.1. This device has regular and provisional condition-defining inputs. These inputs are from external sensory systems or from other devices in clustering. All condition-defining inputs have the same weight, and inputs are either active or inactive. Each device has a threshold, and if the total count of active weights exceeds the threshold, the device activates its output. This output has no structure, it is simply present or absent. A device therefore detects a group of conditions which is the set of all subsets of its inputs larger than its threshold.

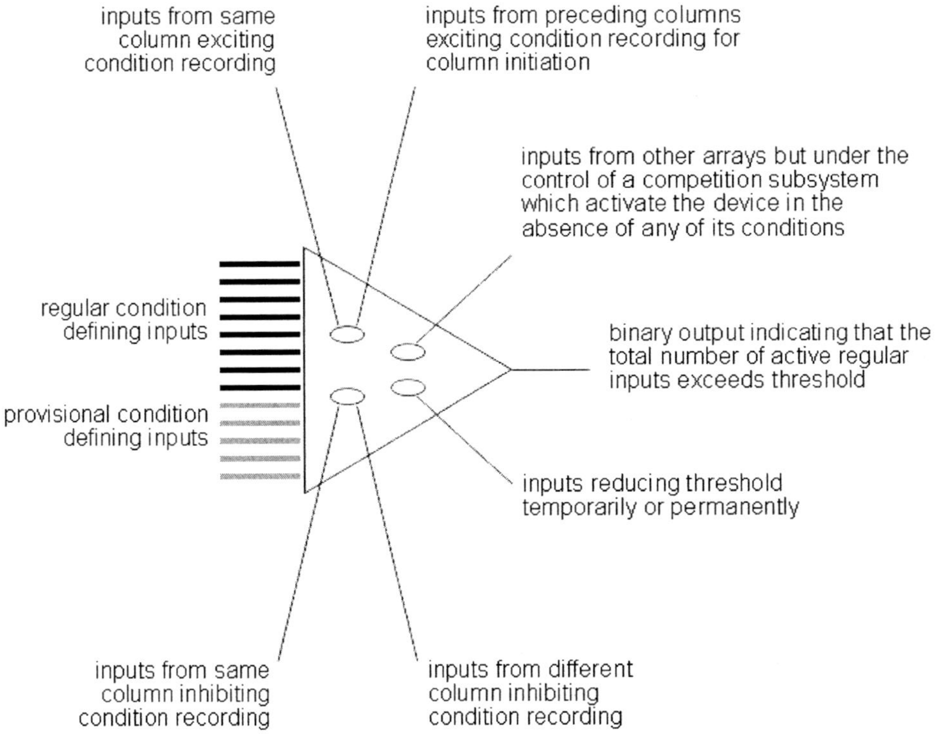

Figure 6.1. A condition recording and detecting device in clustering. The device has provisional and regular inputs which can form parts of conditions detected by the device. These inputs are derived from condition defining and detecting devices in the preceding clustering device layer. The device also has a range of change management and indirect activation management inputs which are derived from specific other locations in clustering. In the case of indirect activation inputs, the access of the inputs to the device is managed by competition.

The device also has inputs which excite condition recording. These inputs come from two types of source. One source is within the column in which the device is located. The other is the inputs to the column in which the device is located, and only operates in a column which has never previously recorded conditions, and indicates when such a column should be initiated. Inputs inhibiting the recording of conditions also come from two types of source. One is within the column in which the device is located. The other is from peer columns (i.e. columns within the same array) detecting conditions at the same level of complexity in the same input space. The sources of all these change management inputs will be described in detail below.

Each device also has inputs which can reduce its threshold, either permanently or temporarily. Again, the sources of these inputs is described below. Finally, devices have inputs from other arrays which can cause them to activate in the absence of any of their recorded conditions. These inputs can only be activated via a competition subsystem, and implement a capability to activate conditions which are not currently present but have often been active in the past at the same time as currently active conditions.

The device activation and change management processes are illustrated in figures 6.2 to 6.4. Figure 6.2 illustrates the first recording of a condition on a device. The device initially has only provisional inputs. In the absence of appropriate change management signals, no combination of these inputs will cause the device to produce an output. If there is an active input exciting condition recording and no active input inhibiting condition recording, at a time when the number of active provisional inputs is above a minimum level, the device is activated. All active provisional inputs are converted to regular, and the device threshold is set at or below the total regular input count. As illustrated in figure 6.3, if on some future occasion the number of active regular inputs exceeds the threshold, the device will produce an output independent of the state of its change management inputs.

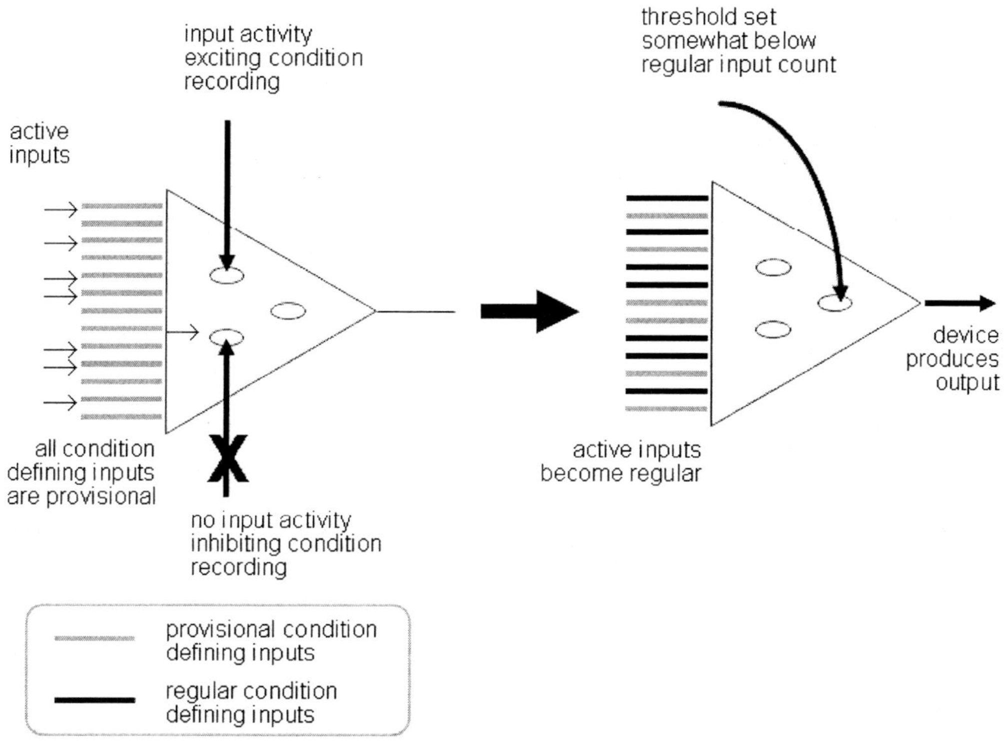

Figure 6.2. Mechanism for initial condition recording on a device. If there are active inputs exciting condition recording and no active inputs inhibiting such recording, and if the number of active provisional inputs exceeds a minimum, the device produces an output, active provisional inputs are converted to regular inputs, and the threshold is set at or slightly below the number of active inputs.

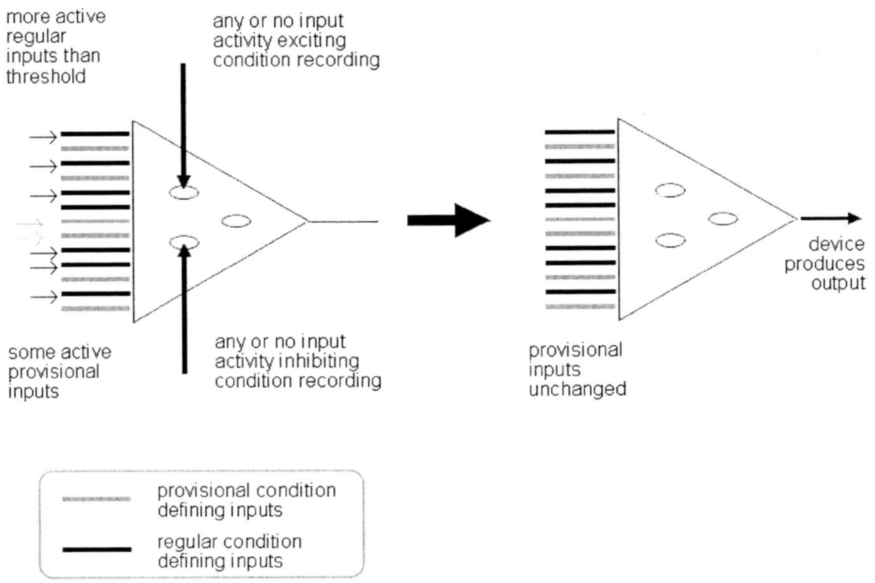

Figure 6.3. Mechanism for detected a previously recorded condition. If the number of active regular inputs exceeds the device threshold, the device produces an output regardless of the status of the change management inputs. No provisional inputs are converted to regular.

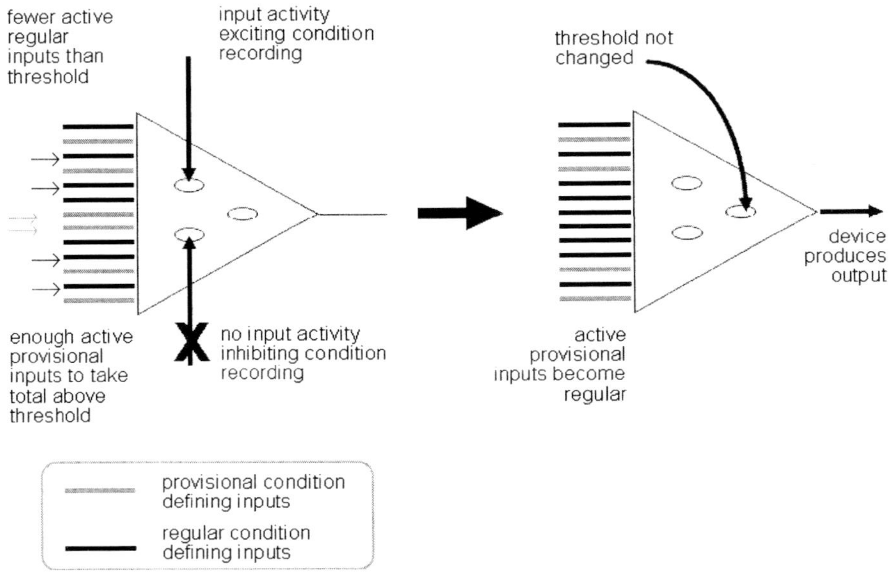

Figure 6.4. Mechanism for expansion of the device condition portfolio. If the number of active regular inputs is less than but a significant proportion of threshold and there are enough active provisional inputs that the total active input count exceeds threshold, and if there are active inputs exciting condition recording and no active inputs inhibiting such recording, the device produces an output and the active provisional inputs are converted to regular.

If a device is activated by a set of active regular inputs larger than its threshold, no provisional inputs can be converted to regular, even if they are also active. However, as illustrated in figure 6.4, if the number of active regular inputs is less than but a significant proportion of the device threshold; and the number of active provisional inputs is large enough that the total active input count exceeds the threshold; and if the appropriate combination of change management inputs is also present, then the device will activate and the active provisional inputs will be converted to regular. The device threshold is not changed, and the device will therefore detect any condition which is a subset of the larger group of regular inputs, provided the subset exceeds the same threshold.

The process by which provisional conditions are assigned to a device is illustrated in figure 6.5. This assignment process occurs when the system is taken off-line from processing its sensory inputs. During this off-line period, which is called sleep, all provisional inputs are deleted and new provisional inputs connected. These new inputs are selected randomly, but with various biases described in detail below. During sleep, the device threshold can be permanently reduced. Such reductions take place if the device or the column within which the device is located have rarely been active over some long period of experience. The number of provisional inputs assigned initially and after condition recording has commenced are parameters which depend on the number of available inputs as described below.

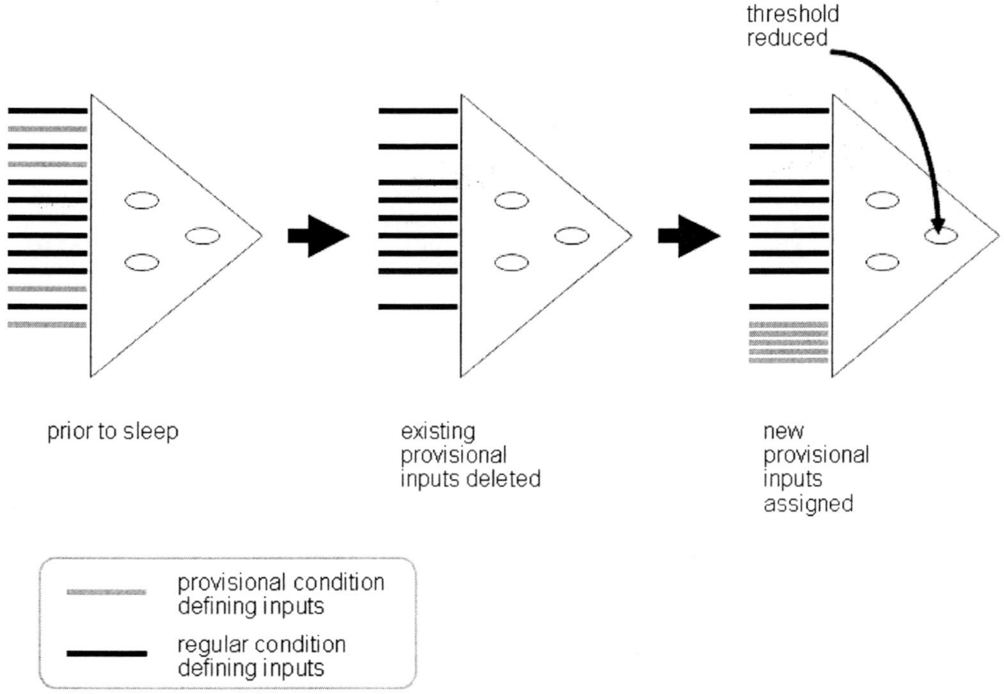

Figure 6.5. Mechanism for addition of provisional inputs. During sleep, all prior provisional inputs are deleted. New provisional inputs are selected and added. The selection process is random, with a statistical bias in favour of inputs which have often caused devices in the same layer of the same column as the target device to produce outputs in the past.

Column Organization and Algorithms

Condition detection devices are arranged in layers, and sequences of three layers are organized into parallel columns as illustrated in figure 6.6. A population of inputs are available to all the devices in the first device layer of all columns. The outputs of layer 1 of a column are available as inputs to layer 2 of the same column, and the outputs of layer 2 are available to layer 3, again only within the same column. The outputs of layer 3 devices are the outputs of the column. This connectivity defines the conditions detected in each layer. Conditions detected in one layer are therefore combinations of conditions detected by the preceding layer.

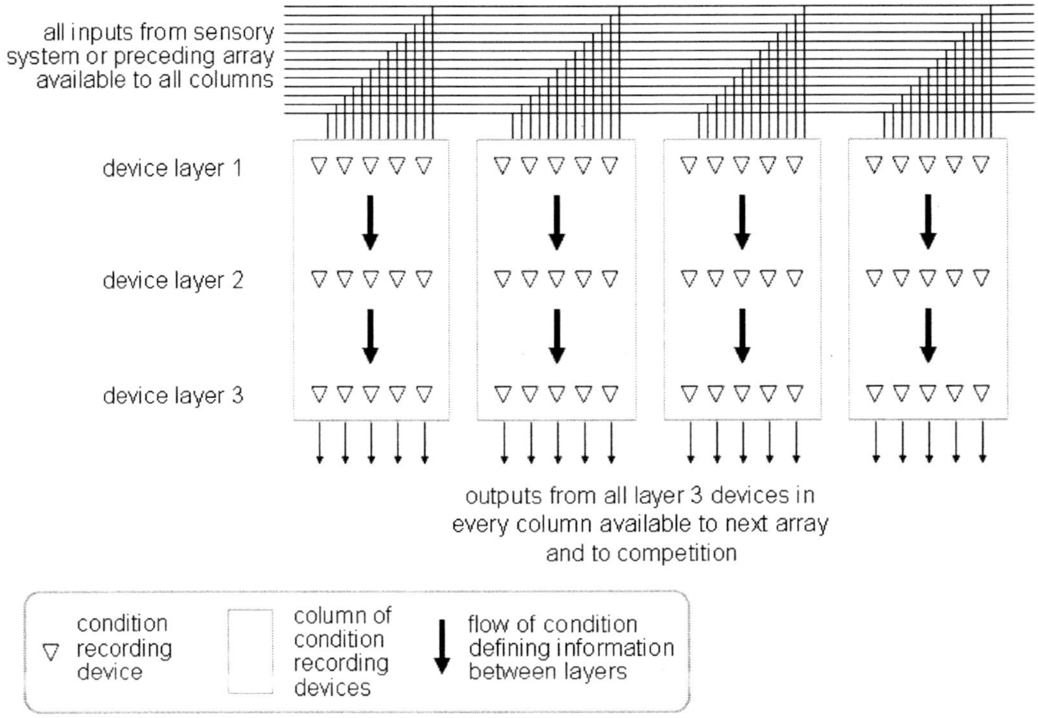

Figure 6.6. An array module illustrating the organization of condition recording devices into layers and columns. The number of columns in the array and the number of devices in each layer of a column are defined heuristically. Condition defining inputs to first layer devices are derived from inputs to the array. Condition defining inputs to devices in the second layer of a column are derived from devices in the first layer of the same column. Condition defining inputs to devices in the third layer of a column are derived from devices in the second layer of the same column.

Change management connectivity between two parallel columns is illustrated in figure 6.7. A column starts with a number of devices in each layer. The initial number of devices is a controllable parameter, but additional devices are added as required, with the requirement determined heuristically. The initial devices have only provisional condition defining inputs, but have connectivity to support all of the change management functions illustrated in figure 6.7.

Provisional inputs to layer 1 devices of a new column are selected randomly from the total population of inputs available. However, a bias is placed on the selection process. This bias is in favour of a group of inputs which have tended to be active in the same input states in the past, but only within input states which have generated minimal activity in layer 1 of any other, previously initiated column. Provisional inputs to layer 2 and layer 3 are selected randomly without bias from within the set of outputs from the preceding layer of the column. The selection of provisional inputs occurs during an off-line process called sleep.

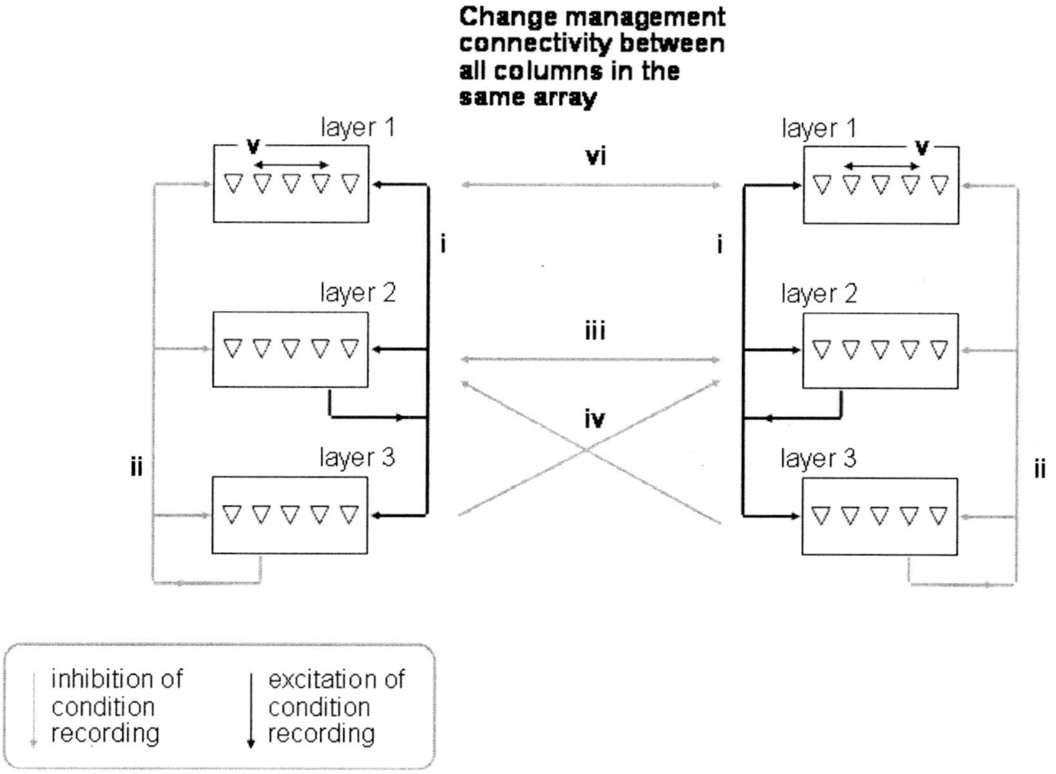

Figure 6.7. Change management connectivity within and between two columns in the same array. The roles of the different connectivities are as follows: i. active layer 2 devices in one column excite condition recording in all the layers of their column; ii. active layer 3 devices in one column inhibit condition recording in all the layers of their column; iii. active layer 2 devices in one column inhibit condition recording in all other columns; iv. active layer 3 devices in one column inhibit condition recording in all other columns; v. a high level of provisional input activity excites condition recording throughout a column which have never before recorded conditions; vi. active layer 1 devices inhibit initial recording in any column which has never before recorded conditions.

The effect of this initial provisional and change management connectivity is that if a new column has a high level of input activity, and if no other existing column has strong activity in its first layer, conditions will be recorded on devices in every layer of the new column until there is enough activity in its third layer to inhibit further recording. A high level of input activity will tend to occur when many of the favoured group of inputs to level 1 are present. Columns will therefore tend to be initiated around a group of similar input states which are different from groups to which other columns already respond, and which have often occurred in the past. After this initiation, the column will record additional conditions whenever there is no activity in the third layer of any column, if there is strong activity in its second layer, and if the proportion of active devices in its third layer is higher than the proportion in any other column. This additional recording uses provisional connectivity assigned during the same off-line sleep process.

The process flow which generates an output from an array of columns in response to an input state is illustrated in figure 6.8. Conditions will tend to be recorded if no column is producing an output. Recording will occur in the column with the strongest activity in layer 2. If no column has an adequate level of layer 2 activity, a new column will tend to be initiated, provided that no existing column has strong activity in its layer 1, and provided also that the current input state is similar to the type for which the new column has been biased. Even recording within an existing column is dependent upon the presence of adequate provisional input configurations. Hence early in the learning of a system, clustering does not generate an output in response to every input state.

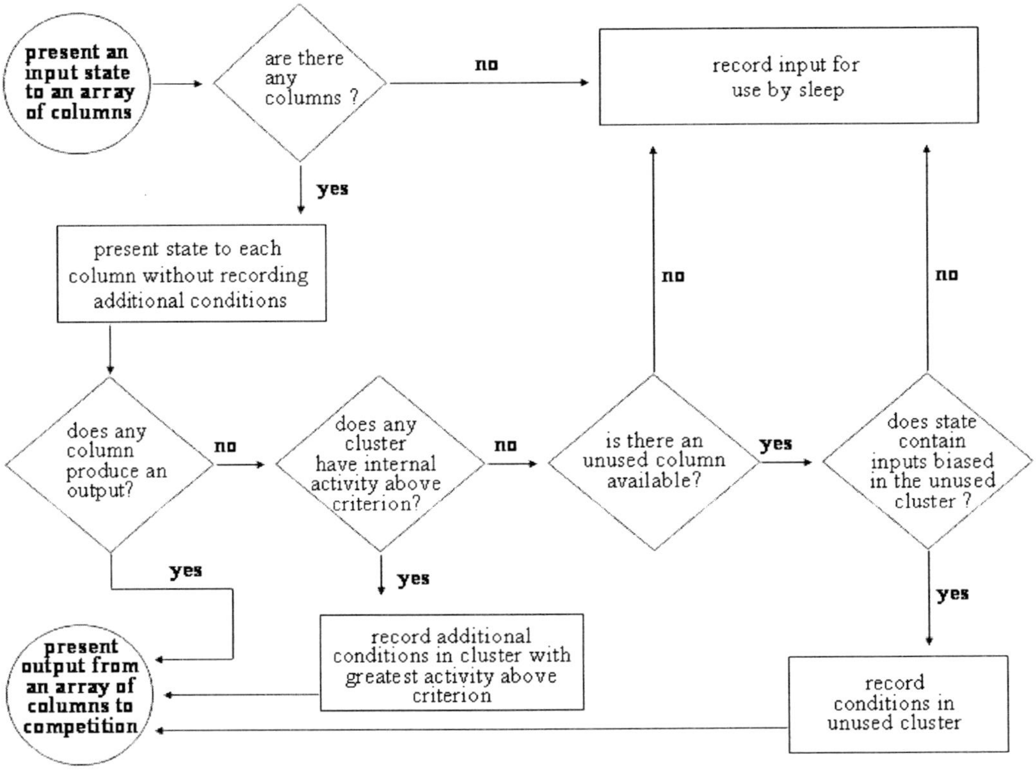

Figure 6.8. Process flow by which an output from an array of columns is generated in response to an input state.

After one new column is initiated, generation of another new column is delayed until the initiated column is generating outputs in response to a significant proportion of input states. This delay reduces the probability of two columns generating outputs in response to a very similar range of input states.

Results of these learning processes are conceptually illustrated in figure 6.9. The behaviourally relevant input space is a volume within a phase space with dimensions equal to the number of inputs. Four different columns were initiated at different points in the input space. The locations at which columns will be initiated and their subsequent expansion depend upon the specific sequence of input states making up the experience of the system. If much of the experience of the system is of behaviourally relevant input states, the volumes within which columns respond will expand to fill the input space and probably expand somewhat beyond it as illustrated in figure 6.9. Figure 6.10 illustrates how the behaviourally relevant input space is divided into fourteen different volumes, corresponding with different combinations of columns which will generate outputs when the input is within those volumes.

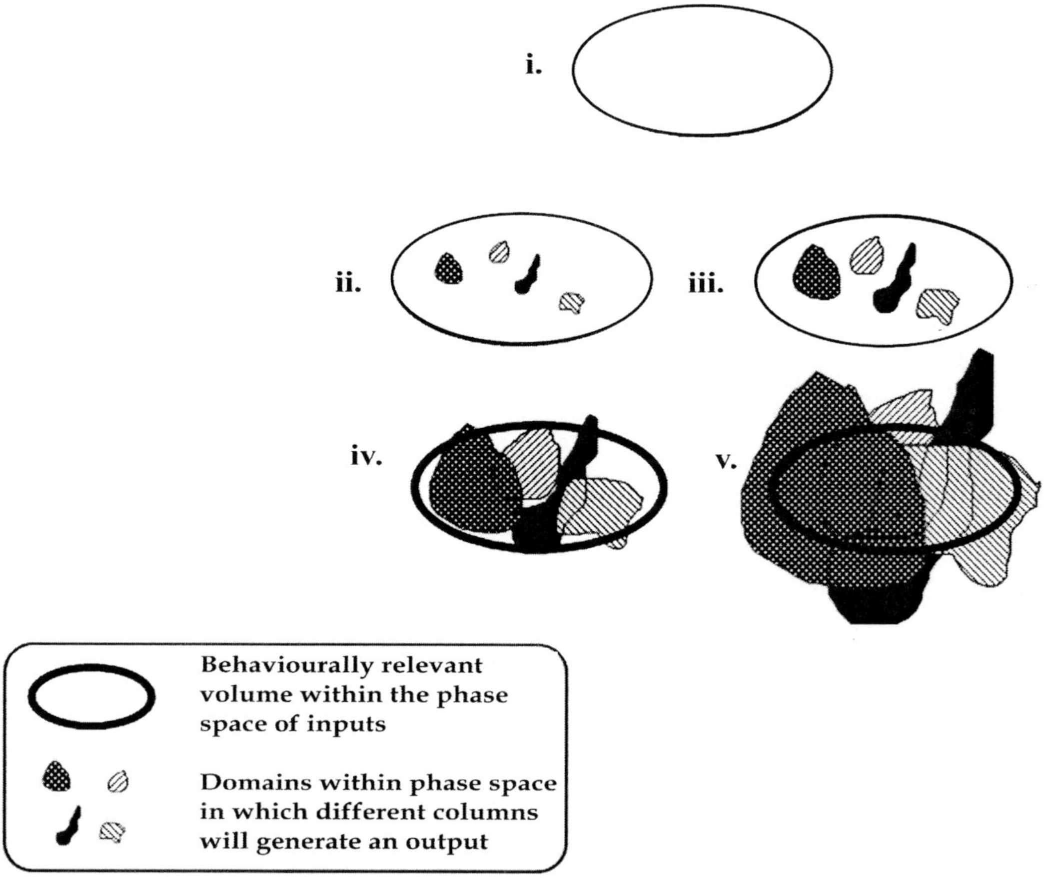

Figure 6.9. Growth of column response domains with experience, to eventually cover a behaviourally relevant input space. i. A representation of a multidimensional input space. ii. Initial volumes of input space within which each of four columns will generate an output. iii. to iv. Growth of volumes within which each column will generate an output. This growth results from experience. v. Input space almost completely covered by the four columns.

If these fourteen volumes corresponded exactly with the behaviourally significant categories of input, then the appropriate behaviour in response to any input could be managed by associating each combination with a different behaviour. As discussed below, such associations can be created by a competition subsystem. The random element in the definition of columns makes such an exact correspondence unlikely. However, there are several factors which improve the effectiveness of columns for this behavioural purpose.

The first factor is that the selection of the numbers of provisional inputs to devices in the different layers can be used to influence the probability that columns will correspond better with relevant categories. This selection influences the complexity of detected conditions as discussed in chapter 4.

The second factor is that columns are initiated around groups of inputs which tend to be active at the same time. This initiation bias affects subsequent growth and means that columns will tend to correspond approximately with different higher level similarities in the experienced input states. This second factor means that system experience during learning can be managed to improve the results.

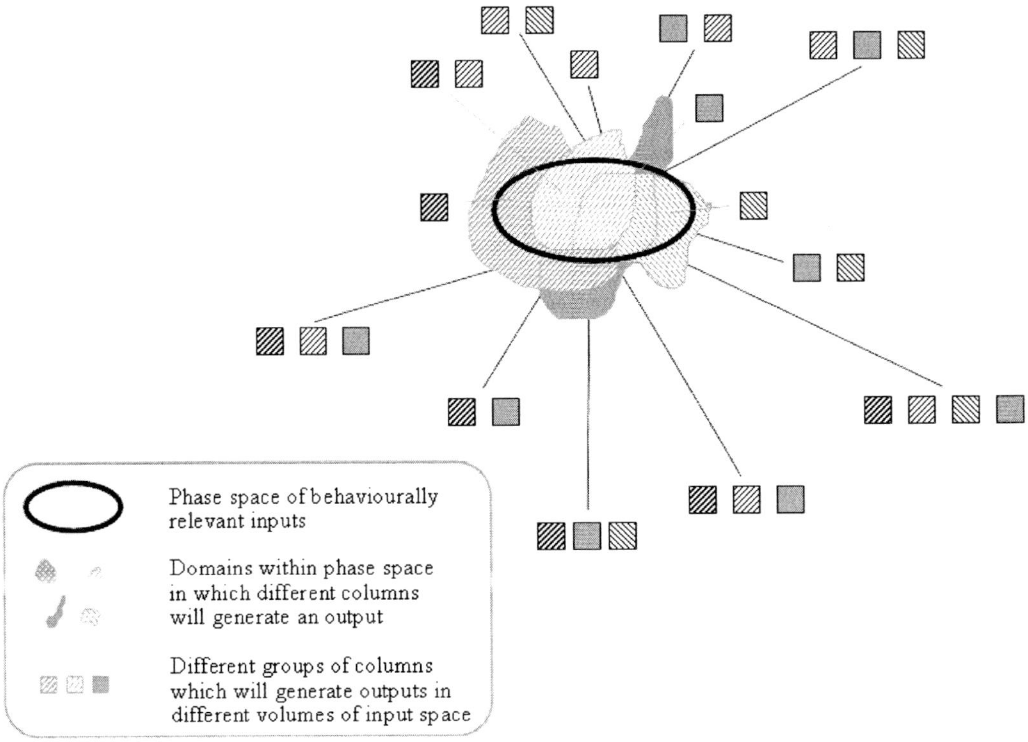

Figure 6.10. Input space divided up into fourteen different volumes within which a different combination of columns will generate an output. This discrimination permits the four columns to determine a different behaviour in response to inputs within each of the fourteen volumes.

The third factor is that the discrimination of a set of columns can be explicitly managed. For example, suppose that the volume of the input space within which two columns always respond contains two subvolumes within which two different behaviours are appropriate, as illustrated in figure 6.11. Such a situation would occur if there were two different types of sensory objects corresponding with the two subvolumes. The effect would be that the simultaneous presence of the two columns in an input state would not converge on selection of a single behaviour in competition. The system would have the experience of selecting a behaviour in response to activity of the two columns, with identical selections sometimes resulting in positive consequence, sometimes negative. This contradictory consequence feedback is an indication that the existing column set has inadequate discrimination within the problem domain. As illustrated in figure 6.11, an additional column can be initiated using inputs within that domain. If the system input states within the domain tended to be either objects of one type or of the other, then if the inputs to the initial column are biased in favour of inputs which tend to occur at the same time that column would tend to be initiated and grow within the volume corresponding with one of the object types. Growth is cut off when the contradictory consequence feedback no longer occurs. Definition of additional columns can be used to evolve the array until it provides adequate discrimination within all volumes of the input space. Note that addition of output devices to a column could also be used to provide additional discrimination.

The overall effect of column creation is therefore to compress the space of inputs which have relatively poor discrimination between different behaviourally relevant categories into a smaller space of column outputs with much better discrimination.

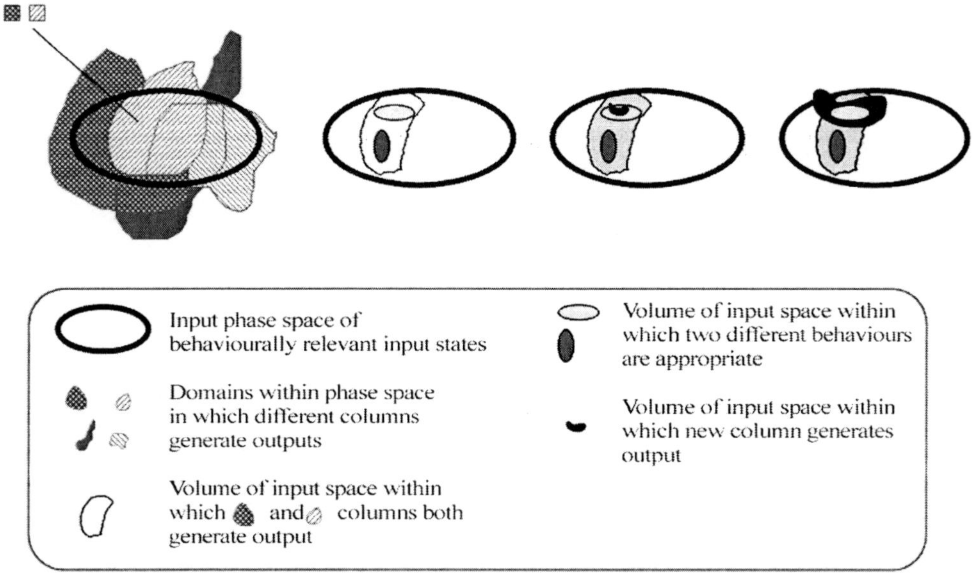

Figure 6.11. Initiation of a new column inside a volume of input space within which a specific combination of columns already generates output. Some inputs within that space indicate the appropriateness of one behaviour, others indicate a different behaviour. The existing set of columns therefore provide inadequate discrimination of the input space. The new column provides discrimination between inputs indicating the appropriateness of the two different behaviours. i. The input space divided into fourteen domains as illustrated in figure 6.9 ii. Subdomains within the domain in which the red and green column produce outputs. These subdomains correspond with input states indicating the appropriateness of different behaviours. As a result, the consequences of behavioural responses guided by the presence of outputs from the two columns will sometimes be positive, sometimes negative. iii. Within the domain generating inconsistent consequence feedback, a new column is initiated. iv. The new column expands until it provides adequate discrimination and the inconsistent consequence feedback is eliminated.

A Simple Categorization Problem

A simple categorization problem investigated by simulation [Coward, Gedeon and Ratnayaka 2004a, 2004b] has an input space of 1000 possible binary inputs or properties. Input states are therefore 1000 element binary vectors, where the presence of a property is indicated by its corresponding element being one and its absence by that element being zero. An input state is generated by random assignment of ones and zeros based on a relative probability distribution. Different probability distributions defined different types of input states. Such different types could be interpreted as different categories of input states. 60 categories were defined in the input space. The probability distribution for a category was created by random selection of a distribution shape for each block of 100 characteristics. The six possible shapes were the six segments of the curve illustrated in figure 6.12. In each selection there was an equal chance of each shape being selected plus a 40% chance of a zero probability of characteristics in the block appearing in instances of the category. An example distribution for one category is illustrated in figure 6.13. All categories contained characteristics which only appeared with very low probability and are similar to noise.

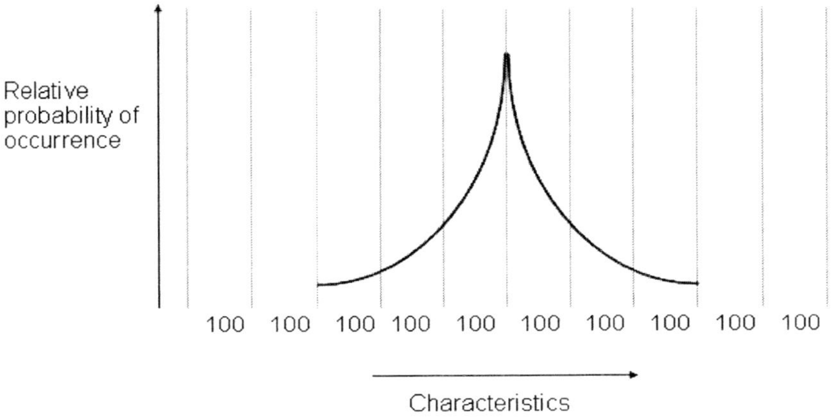

Figure 6.12. Definition of different probability distribution shapes for blocks of 100 characteristics. To construct a probability distribution for one input type with 1000 characteristics, a shape was selected randomly from the ten possibilities illustrated for characteristics 1 – 100, 101 – 200, 201 – 300 etc. A typical result is illustrated in figure 6.13.

Large numbers of different instances of each category were created. The construction process was random selection of characteristics from a set with the appropriate distribution for the category. The number of characteristics in a object varied in the range 70 – 180 and no two objects in the same category were identical. The system did not have multiple exposures to the same instance, any one instance was only presented to the system once.

Individual characteristics occurred in instances of between 15 and 33 categories out of the 60, with an average of 26.1 categories or 43.4% of the total. This percentage is a measure of the ability of individual characteristics to provide discrimination between categories of inputs.

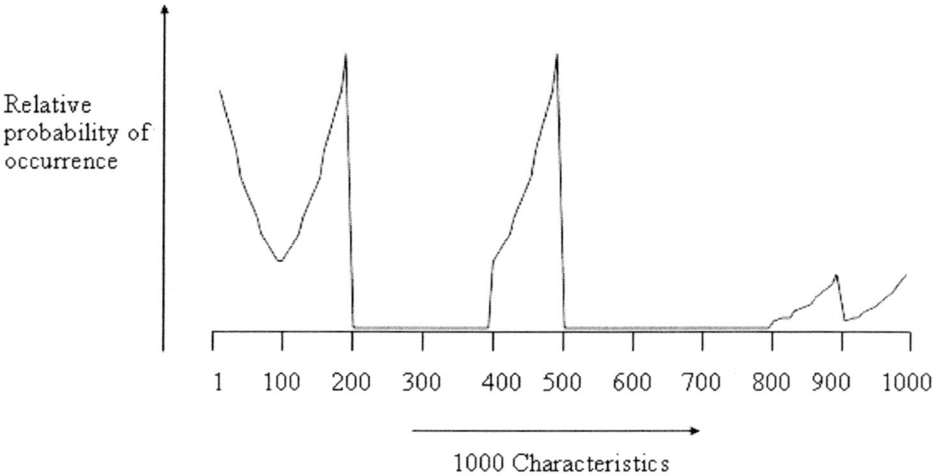

Figure 6.13. Relative probability of occurrence of 1000 different characteristics for one input type. Each input type had a different probability curve constructed by the random selection process illustrated in figure 6.12.

A clustering array was presented with sequence of different input instances of a number of the different types, alternating with periods of sleep to manage the creation of new provisional connectivity. A typical wake period was 10 sequences of one each of 50 or 60 types. This wake period was followed by a period of sleep, and then another wake period. Because no instance was presented more than once, overall array experience could include of the order of 1000 different instances of each type.

Although this scenario is much simpler that a real biological scenario it shares some key qualitative characteristics with the biological problem. Because a binary equivalent can be constructed to any form of input, sufficiently large binary conditions would be equivalent to biological inputs in information terms. The input space is fairly large, and the results scale to much larger input spaces up to 10,000 element input vectors [Coward 2001]. The relevant conditions for determining the category of an object are complex combinations of inputs, and the nature of the relevant combinations are unknown a priori to the system. All overall input conditions are different, and any one overall condition is only experienced once. The simulated problem is thus similar to the biological problem in these information terms.

Sleep Process

The process followed in sleep is illustrated in figure 6.14. In all existing clusters, three processes occur. Firstly, in all three layers, devices which have not recorded any conditions are deleted and new devices configured. The inputs to these new devices are biased in favour of inputs which frequently contributed to firing of devices which have recorded conditions in the same layer of the same cluster in the recent past. In addition, new provisional inputs are added to devices which have recorded conditions, with the same selection bias. This bias can be viewed as reducing the amount of information exchange required between devices, or equivalently as increasing the probable usefulness of the provisional connectivity for the location in which it is provided.

Secondly, if a cluster has responded to less that a minimum proportion of presentations in the preceding wake period, regular first layer device thresholds are reduced. This has the effect of roughly equalizing the proportion of input conditions to which each cluster responds.

Thirdly, if the cluster has inhibited the creation of a new unused cluster without producing output in the preceding wake period, thresholds of its second layer devices are reduced.

In addition, a new unused cluster is configured if required. The inputs to second and third layer devices in the new cluster are selected randomly, but for inputs to first layer devices the random selection is biased in favour of inputs which frequently occurred together in input presentations which did not result in an output in the preceding wake period. This cluster will not be imprinted unless the input condition contains a high proportion of these favoured characteristics.

This sleep process has been demonstrated to improve the correlation of clusters with input condition categories, and to reduce the total device and connectivity resources required [Coward 2000].

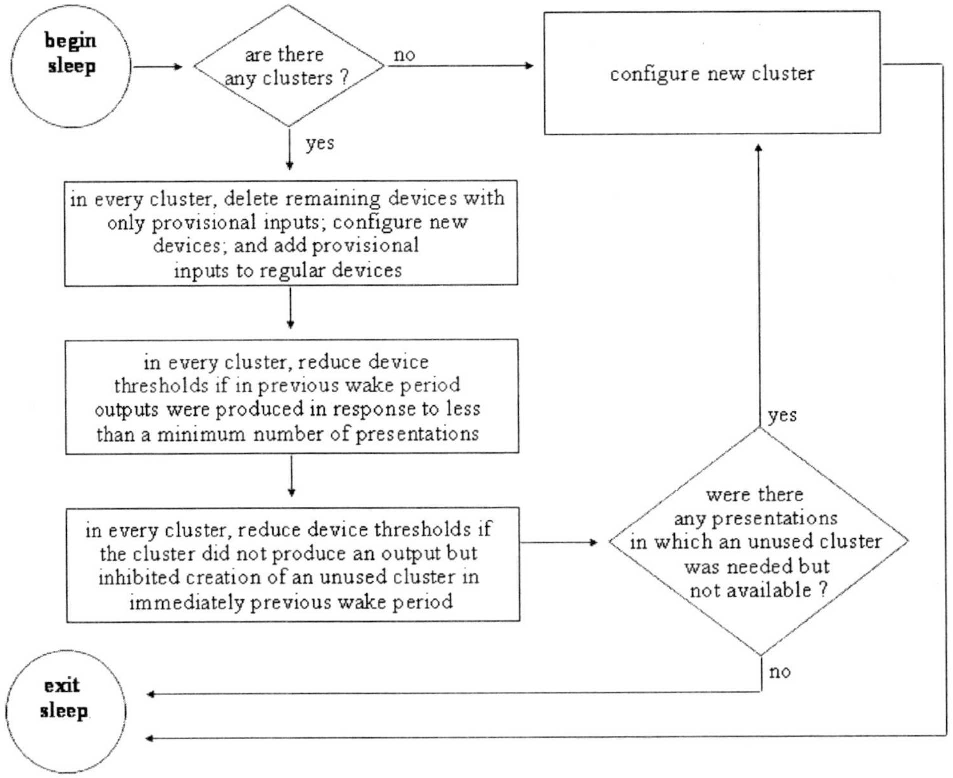

Figure 6.14. Sleep process for assignment of new provisional connectivity.

Adjustable Parameters

There are a significant number of adjustable learning parameters within an array, a list is provided in figure 6.15. Although learning of a specific type of input will be improved by tuning of these parameters to optimum values, adequate learning of a wide range of different types of input occurs for the same set of values. The actual values used for learning of such a range of different types of input are given in Coward [2001]. Learning of a wide range of different types of inputs with these parameters is described in Gedeon, Coward and Zhang, 1999; Coward, 2000; Coward, Gedeon and Kenworthy, 2001; Coward, 2001; Ratnayake and Gedeon, 2002; and Coward, Gedeon and Ratanayake, 2004.

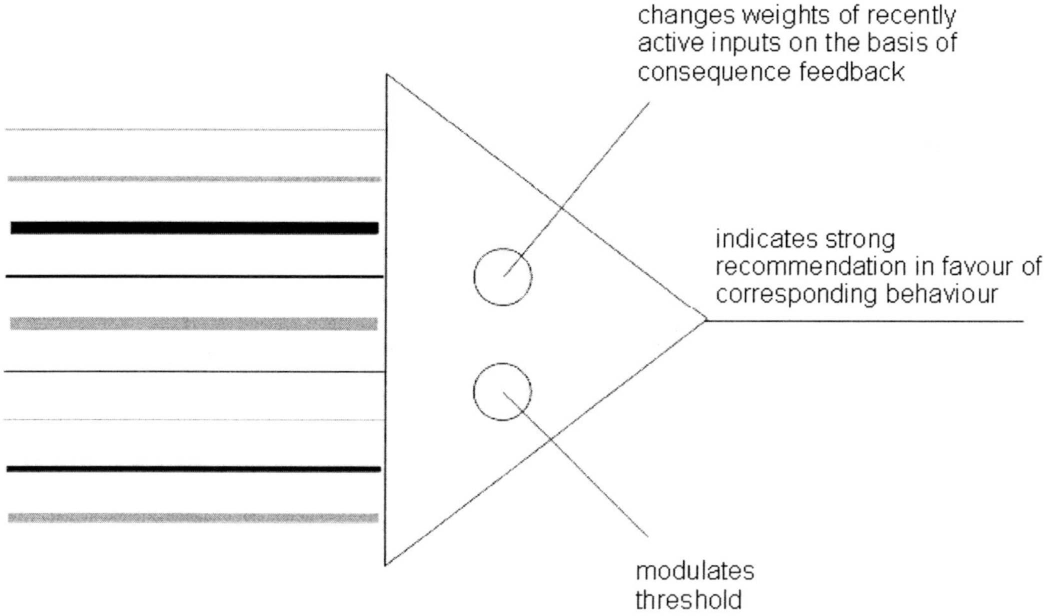

Figure 6.15. Device used in competitive subsystems. This device is generally similar to algorithms used in artificial neural networks. Inputs from clustering have continuously adjustable weights. Weight adjustments occur in response to supervised or consequence guidance. Selection of a general type of behaviour by competition at higher level is communicated to the device by adjustments to its current threshold.

Competition Organization and Algorithms

In competition there is one component for each possible behaviour. The structure of components varies with the complexity of the behavioural problem. In the simple categorization problem, competition devices have the structure illustrated in figure 6.16 and each category has a component made up of one such device. Devices have inputs from layer 4 devices in clustering. These inputs have continuously variable weights, and a device produces an output proportional to the sum of the weights of all active inputs. The category corresponding with the competition device with the largest output is selected. The selection of the largest output was by direct comparison. It is also possible to use components in which there is a more complex arrangement of devices and additional inhibitive connectivity between components to reduce the output to zero in all but one component.

Initial configuration of a cluster	**Similarity criteria**	**Learning conditions**	**Resource addition to existing clusters**	**Other**
Number of initial inputs to neurons in different layers	Minimum number of third layer neurons firing to indicate recognition	Limits on numbers of neurons which record conditions at initial learning	Number of virgin neurons added to layers in sleep	Limits on frequency of creation of new columns
Statistical bias on selection of inputs to first layer neurons	Minimum number of second layer neurons firing to indicate similarity	Limits on numbers of neurons recording conditions during regular learning	Number of provisional inputs added to neurons during sleep	Rate of reduction of neuron thresholds in the absence of cluster output
Numbers of neurons in each level	Minimum number of first layer neurons to indicate cluster familiarity	Minimum number of active inputs for condition recording	Statistical bias on selection of inputs	
	Minimum number of statistically favoured inputs present in an object to initiate new cluster	Neuron threshold after initial recording	Number of inputs added to regular neurons during sleep	
		Magnitude and timing of changes to neuron thresholds when cluster is generally unresponsive	Statistical biases on selection of inputs	

Weights are adjusted by feedback of two types. In supervised feedback, the component corresponding with the correct behaviour is indicated. A connection is established from any active layer 4 devices to that component if no connection already exists, and given a standard initial weight. If the total of the weights of all active inputs to the device corresponding with the target behaviour is less than the total for any other device, those weights are all increased by the same proportion until the output from the target device is the largest.

Consequence feedback simply indicates whether the selected behaviour was correct or incorrect. If correct, the weights of active inputs to the competition device with the largest output are increased by 5% of the standard weight. If incorrect, the weights of those active inputs are decreased by 10% of the standard weight. In addition, if consequence feedback is correct, connectivity with the standard initial weight is established to the device from any previously unconnected active clustering layer 4 devices.

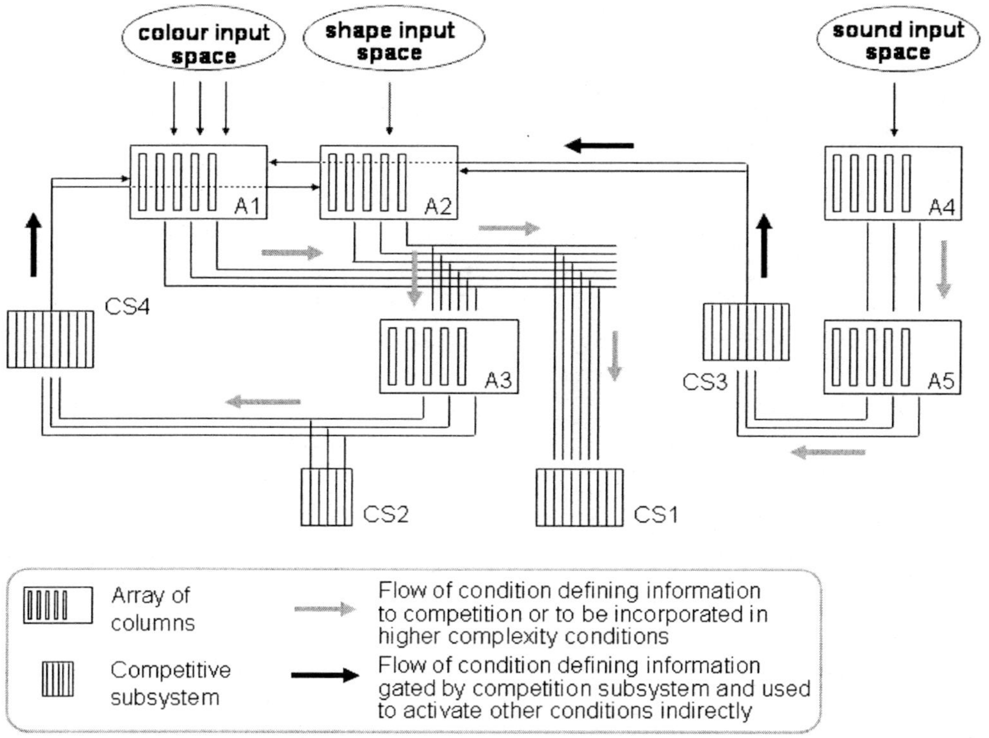

Figure 6.16. Architecture of a system which can learn to associate words with visual objects, and imagine objects which are not present.

Learning Processes and Results

Three types of learning process A, B, and C were used for the heuristic definition of columns. These processes are summarized in table 6.2. In all types, periods of time in which sequences of input objects were presented to the system (wake periods) were separated by periods (sleep periods) in which provisional connectivity was configured in new and existing columns in preparation for the next wake period. This provisional connectivity was selected randomly but with appropriate statistical biases on the selection process as described earlier.

In learning process type A, columns were defined in response to instances of the first 50 categories, then supervised feedback (i.e. the identity of the competition component corresponding with the presented category) was provided for a number of different instances of each of these categories. Recognition accuracy was then tested with five further instances of each category.

In learning process type B, columns were defined in response to instances of 50 categories, and supervised feedback then provided for a number of instances those categories. Instances of ten more categories, added in increments of two and mixed with continuing presentation of objects from the original 50 categories, were then provided. Columns were extended and added by these presentations. Supervised learning was then provided for the ten new categories and no feedback on the previous 50 categories. Recognition accuracy was then tested with five further instances of each category.

Learning process type C was the same as the second, except that consequence feedback (i.e. correct/incorrect indication for the generated category identification) was provided for the instances of the original 50 categories during the wake periods in which supervised feedback was provided for instances of the new categories.

In order to investigate whether addition of sets of columns could be a strategy for improving the overall operational discrimination of clustering outputs, four sets of columns were generated in parallel by the same object presentations. These sets each divided up the input similarity space in somewhat different ways.

Learning Process		Process steps
A	i	Clustering in response to 50 categories
	ii	Supervised learning of 50 categories
B	i	Clustering in response to 50 categories
	ii	Supervised learning of 50 categories
	iii	Clustering in response to extra 10 categories
	iv	Supervised learning of extra 10 categories
C	i	Clustering in response to 50 categories
	ii	Supervised learning of 50 categories
	iii	Clustering in response to extra 10 categories
	iv	Supervised learning of extra 10 categoriesin parallel with consequence feedback in response to original 50 categories

The differences are in the experience profiles and the type of behavioural feedback provided.

For all the learning process types, supervised learning was performed on different numbers of instances of each category. The number of instances of each category varied from 1 to 32.

The number of columns and operational discriminations were comparable for all three learning processes. An array of columns contained on average about 23 columns, of which only one new column was added during learning of the extra 10 categories. Any individual column did not in general correlate unambiguously with one category. However, on average a column generated an output in response to some instances of 23.6% of categories, an improvement in operational discrimination on the 43.4% for the average characteristic. The association between column outputs and categories learning process P1 was around 45% accurate. This accuracy improved to 75% with two sets, 85% with three sets, and over 90% with four sets. Learning capability can thus be improved by additional sets of columns in the same input space. The rest of the results reported here are for four sets of columns. The number of columns in four sets averaged 93, a compression by a factor of 10 from the 1000 characteristic input space. Clustering thus achieved both compression and higher operational discrimination.

Number of instances of each category used during supervised training	**Final recognition correctness measured on 5 instances of each of 50 original categories at end of process:**		
	A	**B**	**C**
1	67.6%	26.4%	38.0%
2	72.4%	31.2%	51.6%
4	75.6%	32.8%	70.8%
8	84.4%	40.4%	77.6%
16	88.4%	44.4%	81.6%
32	91.6%	44.8%	79.6%

In process A, columns were defined in response to experiences of instances of the 50 categories, then supervised training was provided for a limited number of additional instances. Process B was process A followed by extension of the columns by experience of the additional 10 categories, then followed by supervised training on further instances of the 10 categories. Process C was the same as process B, except that consequence feedback (correct/incorrect) was provided on the original 50 categories during process B supervised training.

Table 6.3 shows the accuracy with which instances of the first 50 categories were identified following the different learning processes. The recognition accuracy for the extra ten categories was greater than 90% at the end of processes B and C. For many of the errors in identification, the second choice identification was correct. In other words, the competition device corresponding with the correct identification had the second highest output activity.

For example, in figure 6.18 for 16 supervised instances, 88.4% of the instances of the first 50 categories were correctly identified before learning the additional 10 categories (process A), but only 44.4% of instances after the additional learning (process B). However, for 32.4% of the instances in B, the first identification was incorrect but the second choice was correct. This type of information robustness accounts for the way in which identification accuracy was restored to 81.6% if simple consequence feedback on identifications of the original 50 categories was provided during supervised learning of the new 10 categories (process P3).

Figure 6.18 demonstrates that subsequent learning of an extra 10 categories has an effect on prior learning of the 50 categories, but the effect is a gradual degradation not catastrophic destruction. The gradual degradation is the result of slight broadening of the similarity definition for some columns and of clusters with weights in competition indicating some of the original categories gaining large weights in favour of new categories. Both of these effects can be limited by providing simple consequence feedback for identifications of the original categories during the period in which supervised feedback is provided for the new categories.

Simulation of a More Complex Problem

A more complex problem modeled on cognitive processing has also been simulated [Coward 2001]. In this problem the system was presented with sequences of binary vector inputs simulating sensory inputs resulting from visual objects and speech. Six categories simulating colour (labeled red, green, brown, black, white, and clear) were constructed in a 600 bit binary space. Eleven categories simulating shape (labeled cup, coffee, tree, grass, book, pen, paper, water, chair, apple, and car) were constructed in a 1100 bit space. The visual objects simulated were shapes with a specific range of possible colours for each shape. Thus cups could be red or white; coffee could be brown or black; trees were always brown; grass was always green; books could be red or white; pens brown or black; paper always white; chairs red, white or green; apples red or green; and cars could be brown or black. Twenty-one categories simulating sounds (labeled R, G, B, W, C, T, P, H, S, L, D, E, A, I, O, U, F, K, N, CH, and M) were constructed in a 2100 bit input space. Words were simulated as sequences of three sounds, and the system was given inputs corresponding with the words for the colours and shapes.

All inputs were probabilistically defined binary vectors constructed by the same method as described earlier. Although the inputs are only simulated, they are similar in information terms to sensory inputs in the sense that a binary equivalent can be constructed to any form of input, the input space is fairly large, the relevant conditions for determining the category of an object are complex combinations of inputs, and all input conditions are different so that any one condition is only experienced once.

The system was given a series of inputs which simulated different combinations of visual and speech conditions and was required to find solutions to the following tasks. Firstly, given only the series of visual and speech inputs, define column portfolios within those inputs. Secondly, given only correct/incorrect feedback, associate combinations of repetitions found in the visual inputs with actions which include speaking the name of a visual condition (which includes both colour and shape) and taking an appropriate action. The appropriate action was determined by the sequential presence of two objects. For example, red-cup followed by black-coffee was the appropriate condition for a drink action. The condition, action pairs which received correct feedback are listed in table 1.

cup + coffee	drink
cup + water	drink
book + anything	read
pen + paper	write
apple + anything	eat
chair + anything	sit
car + anything	drive
tree + grass	walk
anything else	no action

The third task was to associate combinations of repetitions found in speech inputs with appropriate visual conditions by activating the appropriate visual condition when the word was heard. The objective was to generate the appropriate action when a visual object was followed by a pair of words corresponding with another visual object.

The fourth task was to associate combinations of repetitions currently present in visual inputs with visual conditions often present in the past when repetitions similar to the current set were also active, and to activate those past repetitions. This activation would be similar to the activation actually generated by the earlier visual object. The objective was to generate the actions of speaking the words corresponding with objects often present in the past at the same time as the current object. The functional role of such pseudo-visual activations is to expand the range of information available to determine the currently appropriate behaviour.

The architecture to solve this cognitive problem is illustrated in figure 6.19. Separate arrays of columns A1, A2, and A4 were heuristically defined in response to the colour, shape and sound sensory input spaces respectively. The outputs of arrays A1 and A2 (colour and shape) generated by one object were used to heuristically define a further array of columns A3. Outputs of array A4 generated by three successive sound inputs were used to heuristically define a further array A5.

Outputs of arrays A1 and A2 also targeted a competitive subsystem CS1 which used consequence feedback to associate different colour and shape related outputs with feature identifications (such as speaking the name of the feature). Outputs of array A3 targeted a competitive subsystem CS2 which associated the outputs with different the behaviours of table 1. The use of consequence feedback and no supervised leaning means that initially behaviours must be selected at random and the learning process is slower. As the number of different categories to be leaned increases, such unguided learning will converge more and more slowly on an appropriate combination of behaviours. However, for the problem as defined, learning speed was adequate.

Outputs of array A5 targeted another competitive subsystem CS3. This competitive subsystem interpreted these outputs as recommendations to activate devices in the first layer of arrays A1 and A2 which had often been active in the past at the same time as the currently active A5 outputs. Outputs of array A3 also targeted another competitive subsystem CS4. This competitive subsystem interpreted these outputs as recommendations to activate devices in the second layer of arrays A1 and A2 which had often been active in the past just before or just after the currently active A3 outputs.

Results with the More Complex Problem

The experience of the system illustrated in figure 6.18 consisted of the presentation of 6500 emulated visual objects and 9120 emulated words. The output of competitive subsystem CS2 can be interpreted as speaking the words corresponding with currently present conditions. The accuracy achieved by competitive subsystem CS1 in a typical run was generally well over 90%. Competitive subsystem CS2 used consequence feedback as shown in table 1 to associate combinations of array A3 outputs with actions. Correct responses climbed to about 80% within the first few hundred object pair presentations and remained at that level. The system thus demonstrates the capability learn behaviours appropriate to the presence of individual and sequential pairs of simulated objects.

Competitive function CS3 manages the generation of appropriate pseudovisual activations in arrays A1 and A2 corresponding with array A5 outputs generated in response to words. To demonstrate the effectiveness of the pseudovisual activations, outputs from array A3 were generated by a colour word and a shape to A4, resulting in indirect pseudovisual activations in A1 and A2. The outputs from A3 were used by CS2 to determine behaviour. In many cases the result was the appropriate behaviour in response to the corresponding visual object, in most other cases the system selected no behaviour. In other words, the word outputs from A5 often generated actions similar to those generated by the corresponding visual object.

Competitive function CS4 manages the generation of pseudovisual activations of objects often present in the past at similar times as currently seen objects. A series of inputs were provided to arrays A1 and A2 which included presentations of red-cup followed by black-coffee. The association between A3 outputs and indirect activations managed by CS4 in arrays A1 and A2 were established in this period. Inputs corresponding with red cup were then presented to arrays A1 and A2, and the outputs from cluster set A3 used to activate devices in arrays A1 and A2. This process generated activations corresponding with black-coffee in all such presentations. In other words, red-cup "reminded" the system of black-coffee on the basis of recent experience.

This system thus demonstrated the ability to heuristically organize inputs simulating visual and sound sensory inputs into a hierarchy of repetition of considerable functional complexity, in such a way that it can determine appropriate responses to its inputs, including actions in response to objects and sequences of objects, generation of appropriate pseudo-visual activations in response to words, and generation of pseudo-visual activations corresponding with objects which have often been present in the past at the same time as the currently perceived object.

References

Coward, L. A. (2000). A Functional Architecture Approach to Neural Systems. , 9, 69-120, 2000.

Coward, L. A., Gedeon, T. and Kenworthy, W. (2001). Application of the Recommendation Architecture to Telecommunications Network Management. 11(4) 323-327.

Coward, L. A. (2001). The Recommendation Architecture: lessons from the design of large scale electronic systems for cognitive science. 2(2), 111-156

Coward, L. A., Gedeon, T. D. and Ratanayake, U. (2004). . Proceedings of the IEEE International Conference on Fuzzy Systems, Budapest.

Coward, L. A., Gedeon, T. D. and Ratanayake, U. (2004). . Proceedings of the IEEE International Conference on Fuzzy Systems. Budapest.

Gedeon, T. D., Coward, L. A., and Zhang, B. (1999). Results of Simulations of a System with the Recommendation Architecture, , Volume I, pages 78-84.

Ratnayake, U. and Gedeon, T. D. (2002). Application of the Recommendation Architecture for Discovering Associative Similarities in Documents. .

Chapter VII

Comparison with Other Approaches

There have been a wide range of approaches to investigating the cognitive processes of the brain. One general approach is a body of thinking which has accumulated around the label of complexity theory. Another general approach is the theory of computation, ultimately based on the work of Turing who developed a conceptual machine to model the process of human calculation. The Turing machine is widely viewed as one of the roots of modern computer science.

More specific approaches include dynamical systems theory, connectionist architectures or artificial neural networks, and computational or symbolic modeling. All of these approaches, or combinations of the approaches, have been used to model different aspects of cognition.

Dynamical systems theory attempts to apply to cognition a methodology which has been very successful for modeling physical processes. In dynamical systems theory, physical processes are modeled as sequences of states in a state space, with equations of motion used to derive each succeeding future state from each current state.

In connectionist approaches, artificial neural networks (ANNs) are used to model cognitive behaviour. These networks are assemblies of devices in which each device has a number of inputs from other devices or from senses. Each inputs has a weights, and learning is modeled by changes to input weights. Memories are instantiated in the values of weights. A number of learning techniques such as statistical clustering and reinforcement learning have been implemented using ANNs. As an example of an architectural application, neural networks have been used in attempts to model the concepts of Marr.

ACT and EPAM are examples of quantitative computer modeling of psychological processes as formal manipulations of cognitive symbols. These approaches aim to model cognitive phenomena relatively unconstrained by physiological plausibility. There has also been extensive investigation into the nature of knowledge, and the degree to which the manipulation of symbolic representations is consistent with cognitive processing.

These approaches are not all fully independent, and many of them have also been investigated with a primary interest in information processing or control applications.

The operational systems approach incorporates some elements from many of the approaches, but also has some major, qualitative differences from any of these alternatives.

The similarities to and differences between the operational systems approach and these other approaches form the topic of this chapter.

Complex Systems Theory

A wide range of different systems have been regarded as "complex", including physical systems, electronic systems, economic systems, and biological systems from brains to ecosystems. Under the label of "Complex Systems Theory" there have been attempts to find commonalities in the theories applicable to these different domains. For example, the algorithmic information complexity of a string of symbols is defined as the length of the shortest program which can generate the string as an output [Solomonoff 1964; Kolmogorov 1965; Chaitin 1966] with the program generally being viewed as running on a Turing Machine (see below). This measure has been applied to estimating system complexity in systems from biological [Gusev al 1991] through cognitive [Sambrook et al 1977] to economic [De Vany 1993].

An issue with the algorithmic information complexity measure is that it has no sense of the compromises which a practical system must make between efficient use of resources and ease of change. This point is further discussed in the context of the theory of computation later in the chapter.

Physical systems can give rise to an immense variety of behaviours, with the classical example being the weather generated by the atmosphere. In a general sense, brains are physical systems. The view that both physical systems and brains share some property called "complexity" has been one source of motivation for attempts to apply to brains some of the theoretical approaches which are effective for physical systems. An example is dynamical systems theory discussed later in this chapter.

However, in a physical system the way in which the phenomenon is viewed affects whether or not it is considered complex. As Edmonds [1999] has pointed out, a brick may be regarded as simple if it is a component in a wall, but complex if it is considered at the level of its constituent particles. This leads Edmonds to conclude that it is not possible to attribute a useful complexity property to physical systems, because "estimates of ..... complexity can always be increased by the inclusion of more aspects of the system". Edmonds therefore proposes that complexity is always a property of a model, not of a real physical system. There is a close relationship between this position and the requirement for a hierarchy of models at different levels of description as discussed in chapter 2.

The issue with applying concepts derived from physical theories to understanding cognition is thus that the concepts of the physical theory may not be at a level of detail that makes human modeling of the cognitive phenomena practical. This issue is no different from the lack of usefulness of dynamical systems theory to understanding, for example, continental drift. Although in principle continental drift could probably be modeled by dynamical systems theory, the appropriate concepts are continental plates, convection currents in the mantle etc. As discussed in chapter 2, the appropriate model is determined by the boundary conditions on the domain being modeled.

Complexity theory thus has a misleading premise that the very wide range of different forms of system complexity have enough in common that similar modeling is possible. Such commonality of modeling will only be possible if there are enough similarities in the constraints within which the complexity is generated. For example, Coward and Salingaros [2004] have argued that the need for the functions of cities to be defined heuristically with efficient use of resources results in some limited similarities between the architecture of effective cities and human brain architecture.

Theories of Computation

The theory of computation [for an introduction see Lovasz and Gacs 1994] begins with the definition of the concept of "algorithm". The theory had its origins in the 1930s with the work of Church and Turing. These mathematicians were independently working on the definition of processes which it is possible for a human being to perform. Church made a hypothesis, which can be paraphrased that it is possible to refer to computable and uncomputable mathematical functions independent of the human or machine computational resources which are available. Church [1936] also formulated a family of problems which he could prove were not decidable by any algorithm, a result closely related to the Godel result that there are theorems which are undecidable in any given axiom system. To achieve this he created a definition for an algorithm. Turing [1936] working independently defined a conceptual Turing machine as a definition of processes which could be performed by a human being "working mechanically", hypothesizing that such a machine could do anything that could be described as purely mechanical [Turing 1948]. Turing's hypothesis was therefore that an "algorithm" could be defined as any process which could be performed on a Turing machine. The Turing machine definition turned out to be equivalent to the Church definition for an algorithm.

The Turing machine was found to be a more useful starting point for further investigations of the properties of algorithms, and also led to concepts important in computer science. A Turing machine has a tape player with an unlimited tape. The tape is separated into sections, each section containing a 1, 0 or blank. The tape head views one section at a time. The Turing machine also has a number of states, and at a particular moment in time is in one of those states. The machine operates by reading one section of the tape, and on the basis of the contents of that section and its current state can write on the current tape section, move the tape one section left or right, and/or change state.

A separate Turing machine must be constructed for each algorithm. However, it is possible to construct a machine such that, using suitable programs, everything can be computed which is computable on any Turing machine. Such a machine is called a universal Turing machine.

The Turing machine requires unbounded memory to perform its functions. Its memory is not accessible immediately: in order to read a far memory cell, all intermediate cells must be read. The limitation on immediate access is overcome by the random access machine which can reach an arbitrary cell in a single step. The random access machine and the Turing machine can compute exactly the same algorithms. However, in order to reach a cell immediately, that cell must have an address, and that address must be stored. Since the number of cells is unbounded, the number of addresses is also unbounded, and the registers which contains addresses must be able to contain arbitrarily large numbers.

Logic circuits are also used as a model for computation, where circuits implement Boolean functions. A given logic circuit allows only a given size of input, but a logic circuit exists to solve every function, input size combination. Finite state machines are logic circuits with the addition of timing and the additional ability to store intermediate results. Unbounded arrays of interacting finite state machines are called cellular automata. A Turing machine can be simulated by a cellular automaton, and tasks requiring a lot of computation can be performed faster on a cellular automaton than on a Turing machine.

Investigation of a range of conceptual machines of these types has indicated that all functions computable on any such machine are also computable on a Turing machine.

A number of problems have been demonstrated to be algorithmically undecidable, including whether a Diophantine equation is solvable, and whether a plane surface can be tiled for certain specifications of tile and tiling conditions.

The difficulty of algorithmically decidable problems has been investigated, where difficulty is defined as the rate of increase of the resources required as the size of the input space increases. There are two types of resource considered, the number of steps to complete the algorithm (viewed as equivalent to time), and the amount of storage required. For example, the number of steps could increase linearly with input space n (e.g. in the case of addition this resource increases as n), or increase polynomially (e.g. in the case of multiplication, as n^2), or increase exponentially (e.g. in the case of determining whether the vertices of a graph can be coloured with three colours and no adjacent vertices the same colour, as 3^n).

There is a tradeoff between storage and time. Suppose for example that in the course of a computation the values of a small but complex Boolean function were used repeatedly. Then on a random access machine it would be worth computing the values once for all inputs and using a table look up.

The Turing machine is deterministic. If random functions (e.g. coin tossing) are allowed as part of algorithms, then problems can be solved which a deterministic Turing machine cannot solve, and in other cases random choices can significantly accelerate the solution. For example, testing whether a large number is prime can be carried out using Fermat's little theorem that if p is prime then $a^{p-1} = 1 \bmod p$ for all $a < p$. The converse is not true in that for a composite number c, $a^{c-1} = 1 \bmod c$ is sometimes true. However, it can be demonstrated that at most half of all $a < c$ meet the Fermat condition. If the test is performed for a randomly selected series of values of a, the primeness of p can be demonstrated to a very high degree of probability by a much shorter process than the deterministic process of testing the divisibility of p by all primes $< \sqrt{p}$.

The Turing machine can be modified to allow random processes, such a machine is known as a randomized Turing machine. For example, for any given tape section content/machine state combination there could be several possible actions, with the selection between them being probabilistic.

As discussed earlier, the algorithmic information complexity of a description or set of symbols is defined as the length of the shortest program which will cause the printing of the description on a universal Turing machine. There are some restrictions on the universal Turing machine to ensure that complexities are consistently defined. A completely random sequence can then be defined as one for which the algorithmic information complexity is the same length as the description. The problem of constructing a sequence of length n with a algorithmic information complexity > n/2 is algorithmically unsolvable, but can be solved with high probability on a randomized Turing machine.

Despite the practical demand for parallel computing, there is no canonical model for parallel computation. The main problem lies in modeling the communication between different processors and subprograms. Approaches tend to investigate how two processors can solve their tasks with minimum information exchange, and assume that each processor "knows" the meaning of any information received from the other processor.

In summary of this discussion of the theory of computation, the starting point is the definition of an algorithm as any process which can be performed on a Turing machine. The focus of the theory is on problems which can be solved with this definition, and on the resources required to solve different types of solvable problem. The introduction of a random process makes it possible to solve problems which are unsolvable by deterministic algorithms. Finally, modeling of parallel computation is unsatisfactory owing to the difficulty of modeling communication between processors.

Physical Systems and the Turing Machine

In a physical system, the enormous range of inputs which affect behaviour mean that in general the system exhibits what is called chaotic behaviour. In particular, slight differences in inputs may lead to large differences in system state. The classical illustration is the butterfly flapping its wings in Brazil which many weeks later may cause a tornado in Kansas. In living and man made systems there are severe limits on both the information space from which inputs can be taken and the behaviour space of possible behaviours. Although these spaces may be numerically large, they are very small compared with the analogous spaces for purely physical systems.

Furthermore, in contrast with the chaotic behaviour of physical systems, it is important that in cognitive systems (and electronic systems), similar input states often lead to similar behaviours.

The Turing machine assumes access to unlimited storage resources. In addition, it implicitly assumes extremely large processing resources if it is emulating a system in real time. It has even been suggested that the Turing machine could emulate any natural physical system. This suggestion is sometimes known as the Church-Turing hypothesis, but there appears to be no evidence that either of these mathematicians made such a suggestion. The Turing machine in computational terms is therefore analogous with a theory like quantum mechanics in physical terms. Although any conceivable process, including cognitive processes, might in principle be modeled by quantum mechanics or by the Turing Machine, the level of detail would make such descriptions incomprehensible to a human mind.

Thus although the Turing machine provides important insights into the nature of computing, it is insufficiently constrained to model operationally complex systems, whether natural or artificial.

Computing and Cognition, and the Deterministic and Non-Deterministic Turing Machines

The distinction between the deterministic Turing machine and a Turing machine with access to a random process was discussed earlier. Because most electronic systems are deterministic, their processes can ultimately be modeled by a deterministic Turing machine. However, because the recommendation architecture requires random processes for selection of conditions, ultimate modeling of its processes would require a non-deterministic Turing machine.

An implementation of some form of random process makes it possible for recommendation architecture systems to be implemented on electronic hardware. The nature of random has been the topic of intense discussion, but for the purposes of this discussion can be regarded as a selection based on a criterion which employs very little information relevant to the objective of the selection. It is probably impossible in principle to reduce "very little" to complete zero whether in electronic or biological systems. However, the results of a deterministic Turing machine are themselves only probabilistic, for example in testing the prime-ness of a number as discussed earlier. The requirement is therefore to make "very little" small enough that the result is valid at a high enough probability level.

Dynamical Systems Theory

Dynamical systems theory attempts to apply techniques from the physical sciences, and in particular classical mechanics, to modeling the processes of the brain [Beer, 2000; Van Gelder, 1998]. A distinction can be made between dynamical modeling and dynamical systems theory. Dynamical modeling provides dynamical models for natural phenomena. Dynamical systems theory models systems geometrically, in terms of positions, distances, regions, and system trajectories over time in a conceptual phase space of all the possible states which the system could enter. The mathematician Henri Poincare realized that even if there is no known way to model system behaviours in detail as functions of time, this geometrical approach could extract significant information about system behaviour. For many mechanical systems, the system state tends over time towards a small subset of all the possible system states. This subset is called a limit set. For some limit sets, all system trajectories which pass close to the set tend to converge on the set. Such limit sets are called attractors because small perturbations away from the attractor will tend to return to the attractor. Dynamical systems theory can sometimes construct maps of the system phase space showing the attractors and the regions around them in within which trajectories will tend to enter the attractor.

Dynamical systems theories of cognition make two assumptions. One is that cognitive agents are dynamical systems, the other is that significant information on the behaviour of cognitive agents can be extracted by this geometric approach. The theory is sometimes positioned as an alternative to the hypothesis that computer science is a valid framework for understanding cognition. In other words, dynamic systems theory plays much the same role in dynamical cognitive science as computer science plays in traditional cognitive science. Neither computer science nor dynamic systems theory are theories of cognitive processes. Rather, they provide a set of tools for use in developing accounts of particular aspects of cognition.

Attempts to model the EEG using an ensemble of neurons [Wright 2000] form one example of dynamical systems theory applied to the brain. Effort is made to model synaptic dynamics with some degree of accuracy. In dynamic systems theory, it is argued that cognitive concepts can be modeled by concepts like attractors and trajectories.

Although brains are dynamical systems, the issue is whether entities like attractors and trajectories etc. are concepts which can be useful for creating causal descriptions of cognitive phenomena as discussed in chapter 2. As Crutchfield [1998] has pointed out, it is not clear how dynamical systems would support the type of information processing required for a cognitive system. There are some considerable differences between the mechanical systems, for which dynamical systems theory creates useful conclusions, and cognitive systems. These differences are of the boundary condition type discussed in chapter 2 and therefore tend to suggest that in most cases entities appropriate for modeling one domain will not be appropriate for the other.

A mechanical system has large numbers of relatively simple and almost identical components. In a cognitive system there are significantly greater differences between individual neurons. A cognitive system must consistently act upon itself and its environment in order to achieve objectives. In contrast, mechanical systems have no objectives, only rules governing component level behaviour. Mechanical systems typically exhibit chaotic behaviour: slight differences in input conditions will in general result in radically different later states. In a cognitive system the rules governing component level behaviour must be carefully defined to ensure that in many cases similar input conditions generate similar output states.

Thus the relationship between cognitive modeling and dynamical systems theory is analogous with the relationship between quantum mechanics and macro chemistry discussed in chapter 1. All the phenomena of macro chemistry could in principle be described in terms of quantum mechanical concepts, but comprehensible descriptions generally require concepts on a much more general level. It is of course possible that occasionally a reference to quantum mechanics may provide some interesting macro chemical insight. Dynamical systems theory is unlikely to provide more than the occasional insight to the modeling of cognitive systems.

Connectionist Approaches

The recommendation architecture is in some ways a connectionist architecture in the sense that it is based upon large numbers of devices which interact by interconnectivity. However, generally the device learning algorithms used in connectionist approaches are limited to adjustment of input weights. The permanent recording of conditions in the recommendation architecture makes the approach qualitatively different.

Statistical clustering is a mathematical approach to organizing large bodies of data, and because it can be applied to organizing sensory inputs is a type of learning algorithm potentially relevant to cognitive systems. Many statistical clustering techniques can be implemented as connectionist architectures. Statistical clustering will first be discussed separately, then various ANN approaches including statistical clustering described. A number of problems with application of these approaches to cognition will be identified, and the way in which the recommendation architecture overcomes these problems described. Finally, the extensively investigated adaptive resonance architecture will be compared in more detail with the recommendation architecture.

Statistical Clustering

Statistical clustering attempts to organize unlabeled input vectors into clusters or "natural groups" in such a way that vectors within a cluster are "more similar" to each other than to vectors assigned to other clusters [Gokcay et al 2002]. In such clustering there may be no a priori information indicating natural groups or similarity measures. Statistical clustering techniques thus divide a large number of sets of observational data into meaningful groups (or clusters) based on some similarity criterion. In statistical clustering analysis the number of clusters and other information about their composition may be unknown [Kaufman and Rousseeuw 1990]. Some statistical clustering methods are hierarchical and proceed by producing a sequence of partitions of the data space corresponding with different numbers of clusters. The process can be one of merges among a relatively large initial set of clusters or splits among a relatively small initial set. Other clustering methods involve relocation of data sets between a fixed number of initially specified clusters [Fraley and Raftery 1998].

Statistical clustering is therefore one approach to heuristically defining operationally useful conditions, and various statistical clustering algorithms are instantiated by different unsupervised ANNs (see next section). To divide vectors into groups which possess strong internal similarities, a criterion function is used and a grouping is sought which optimizes the value of the function (i.e. maximizes or minimizes). For example, self organizing maps [Kohonen 1995] assign each member of a large population of possible input vectors to one or more of a fixed small number of outputs. Assignment is on the basis of the similarity between the input vector and the reference vector for each output. An input is assigned to the output with the greatest similarity, and the reference vector for that output is changed slightly in the direction of the input.

There are four major differences between statistical clustering analysis and clustering as defined in the recommendation architecture. These differences are in the areas of data availability, indication of data set membership in a cluster, changes to the indications of data set membership, and membership of multiple clusters.

In statistical clustering analysis, the same data sets are available for repeated comparison with the developing cluster set. In the recommendation architecture, each individual data set is available only once and only information recorded about that data set at the time is in any way available for later processing. In statistical clustering analysis, membership of a data set in a cluster is binary (true or false), although a numeric indication of membership probability may be available in some approaches. In the recommendation architecture the indication of membership is the particular combination of outputs generated which can contain considerable information about where within the similarity domain defined by the cluster the current input is located.

In statistical clustering analysis, the binary or numeric indication of cluster membership can change as clusters are reorganized and redefined. In the recommendation architecture, once a given subset of a data set has generated an indication of membership in a cluster, the same indication will always result if the same subset occurs in any subsequent data set. Finally, in statistical clustering analysis an individual data set is assigned to only one cluster, while in the recommendation architecture data sets can generate outputs from multiple clusters.

The significance of these differences is that while statistical clustering analysis is very effective for the problem domains at which it is targeted, it is not effective for heuristic real-time management of a complex combination of control features. In a system which learns to perform a complex combination of features, components must determine the portfolio of input information conditions to which they will respond on the basis of a sequence of input conditions which never repeats exactly. The presence of a programmed information condition must be indicated with enough information richness that other components can use the output information effectively, and because any component could choose to use an output at any time, once an output has been generated, the same output must always result from the same input condition.

A range of statistical clustering approaches have been implemented by ANNs, and some useful applications have resulted. However, as discussed in the next section, these applications tend to be in the domains of pattern recognition or simple control problems, and severe difficulties are encountered when they are applied to more operationally complex problems.

Neural Networks

Artificial neural networks (ANN)s achieve learning using layers of interconnected devices modeled to some degree on biological neurons [see e.g. Jain et al, 1996; Ballard, 1997]. In ANNs, device inputs have weights which generally vary continuously, although weights are binary in "weightless" neurons [Ludermir et al, 1999]. Device learning algorithms repeatedly adjust individual input weights in the network learning process.

In ANN unsupervised learning, the network finds some way to associate inputs into classes using no information over and above the inputs themselves. Device input weight adjustment is typically determined by a competitive algorithm. In such algorithms the network is organized into clusters which indicate the presence of different types of input by producing an output from the cluster corresponding with the input type. If a new input is sufficiently similar to an existing type, device weights characterizing the type are adjusted to shift the type towards the new input. If the new input is not similar to any existing type, a new cluster is defined by adjustments to network structure and device input weights. Learning algorithms include adaptive resonance theory (ART) [Carpenter et al, 1988] and Kohonen mapping [Kohonen, 1995]. There are two key parameters in competitive learning algorithms. One parameter is a vigilance parameter which defines "sufficiently similar". The second is a learning rate parameter which defines the degree of adjustment to an existing cluster which can occur in response to a new but sufficiently similar input. The learning algorithm used in recommendation architecture clustering has a general similarity to ANN unsupervised competitive algorithms. The algorithm has a vigilance parameter determining when a cluster can change, but because of the use of permanent condition recording there is no equivalent to the learning rate parameter. There is some similarity between the vigilance parameter dependence on "resonance" between higher and lower level devices in ART and the feedback between higher and lower arrays in the recommendation architecture required to achieve activation consistency when multiple independent conditions are recorded on individual devices. However, this array feedback is based on consistency between the timing of past changes to lower level devices and past activity of currently active higher level devices, not on consistency between current activity of lower and higher level devices as in ART.

In reinforcement learning in ANNs, training only provides feedback on the general correctness of an output in response to an input, and device input weight adjustments are determined by this correctness measure. A correctness measure could be an indication of the consequences of an output given the input. Learning algorithms include Q-learning [Watkins et al, 1992], which has been implemented by an artificial neural network [Sun et al, 2001]. The recommendation architecture competition subsystem uses a form of reinforcement learning, but differs from ANN algorithms in that it is organized as a modular hierarchy with simple but well defined operational meanings to all exchanges of information.

In Hebbian learning [Hebb 1949] the ability of one neuron to activate another increases when the two neurons are active at the same time. However, this mechanism has not been successfully used to model complex cognition. In the recommendation architecture clustering subsystem, a range of mechanisms based on simultaneous activity are used to support learning. One of these has some resemblances with Hebbian learning, but the strength of the activation ability decays after the simultaneous activity unless it subsequently results in a behaviourally useful activation. Furthermore, such activations cannot occur unless other active substrates also encourage secondary activations of the type.

The recommendation architecture thus uses unsupervised device algorithms in the clustering subsystem, but introduces a qualitatively different type of device algorithm in which learning is managed by a range of mechanisms using information on past and current simultaneous activity of different populations of devices. In this algorithm, once a device has generated an output in response to a specific combination of inputs, it will always produce an output in response to any exact repetition of that combination. With this algorithm a greater degree of meaningful context can be sustained throughout learning. The algorithm also means that the ranges of complexity with respect to system inputs of the conditions detected in each device layer are more tightly specified. A modular hierarchy overlaid on the layers can therefore use specific ranges of complexity for different behavioural and change management purposes.

The recommendation architecture uses reinforcement type algorithms in the competition subsystem, the major differences being the organization of competition into a behavioural hierarchy, and the lack of direct influence by reinforcement feedback on the conditions detected by the system.

Although the distinction between unsupervised and supervised learning (see e.g. Card 1998) is thus analogous with the separation between clustering and competition, the use of consequence information (e.g. category identification) in both types of ANNs means that the essential recommendation architecture separation is not maintained. In ANNs the term "competition" is used to indicate any process of resolution between alternative activations [e.g. Taylor and Alavi 1993]. Competition in this sense occurs in both separations of the recommendation architecture, but the competition function of that architecture, competition refers strictly to the use of consequence information to resolve between alternative behavioural recommendations.

A further key concept in the recommendation architecture is the indirect use of reinforcement type information to refine the clustering hierarchy without changing existing cluster outputs, by causing specific clustering modules to produce additional outputs under some already learned input conditions.

There are thus a number of major differences between ANNs and the recommendation architecture. In general in ANNs no approach to partitioning of a complex set of operations is discussed. There is little concept of a hierarchy of operational management signals. The only exception is in the case of Adaptive Resonance [Carpenter and Grossberg 1988] which makes use of an indication of input similarity as a precondition for learning. This, however, is just a single higher level management signal.

There are four major problems with ANNs as cognitive models. One is the lack of attention to the maintenance of an adequate context for information exchanges, which results in the second problem of a tendency for later learning to completely overwrite earlier learning, a problem known as catastrophic interference [French]. An issue closely related to these first two problems is the lack of capability to perform complex combinations of interacting features. The fourth problem is the physiological plausibility of ANNs.

Information Exchange Context

The context for information exchange between modules is minimally addressed in ANNs. The typical ANN approach tries to target module outputs on specific cognitive features. For example, a typical problem attempted by Kohonen Nets is self organization of natural language information [Honkela 1995] in which map outputs indicate individual words and relative position on the map indicates the syntax of the word.

In a system which learns a complex combination of operations with limited resources and limited a priori knowledge, there must be some means to ensure that the conditions which are defined and not excessive but are adequate to distinguish between behaviourally significant differences in input states. Once an association between a group of conditions and a behaviour has been defined, any change to one of the conditions in the group may reduce the validity of the association. In an operationally complex system with limited condition recording resources, the detection of any one condition will acquire multiple meanings, and such meanings must be adequately preserved. The critical problem is therefore the type and degree of changes to conditions which can be consistent with preservation of adequate operational meanings.

The output of a device in response to the presence of a specific combination of active inputs indicates that the condition specified by that combination of active inputs is present. Networks synthesize the outputs of many devices into a network output indicating the presence of some more operationally significant condition. The network output is itself the outputs(s) of one or more devices assigned to provide that output. In an operationally complex system a device output could acquire a wide range of different meanings. The alternative, that no condition have multiple operational meanings, would require an impractical level of resources.

In ANNs, the condition detected by a device is defined by the weights assigned to its inputs. These weights are evolved independently. In supervised learning, such evolution is on the basis of feedback of the consistency between network output in response to an input and some known target output. In unsupervised learning, evolution is on the basis of some criterion of similarity between the network input and network prototype inputs. A device will not necessarily produce an output in response to an exact repetition of an earlier input condition, and the relationship between different conditions generating an output from the same device is complex. Weight evolution algorithms are therefore ineffective for maintaining complex combinations of operational meanings

The Learning Issue

Ineffective maintenance of the operational meanings for information exchanges is at the root of the tendency for later learning to interfere with or completely delete earlier learning in ANNs. French [18] has reviewed this problem and the various approaches to avoiding it. One approach is sparse coding of internal representations implemented in a range of ways, but such sparse coding creates problems when a large number of patterns must be learned. A second approach is to limit learning to domains which have a high degree of internal structure, and early on present examples of all possible pattern types. However, this approach limits learning to the pattern types presented early on. A third approach is to restrict the receptive fields of inputs, which in the extreme makes the network localist. However, as receptive fields expand to allow more distributed representations, interference increases.

Another approach discussed by French [18] is to have new learning always include some reiteration of old learning. One way to achieve this reiteration is by direct inclusion of a selection of past inputs. Another way is to separate two subsystems for new learning and permanent storage, with new learning occurring first in the new learning subsystem, and learning in the permanent storage subsystem resulting from a combination learning in the new learning subsystem and pseudopatterns corresponding with existing learning generated from permanent storage [e.g. Ratcliff 1990]. However, presenting a mix of old and new inputs again mitigates but does not directly address the issue that any change in the meaning of a device output in favour of some system objectives will inevitably result in loss of meaning for other system objectives. Robins [1995] introduced the concept of repeating pseudo-inputs or approximations of the original inputs. However, as operational complexity increases, achieving an adequate balance between old and prior learning will be increasingly difficult.

Yet another approach is to place additional constraints on the degree to which the conditions programmed on devices can change. An extensively investigated example of this approach is adaptive resonance theory (ART) [Carpenter et al 1988]. In ART, as in a self-organizing map, an input vector is assigned to an output condition if it is sufficiently similar to the reference vector for that condition, and such an assignment results in a modification of the reference vector in the direction of the input. In contrast with regular self organizing maps, the number of output conditions is not predefined, and if an input vector is not sufficiently similar to any existing output vector, a new output is defined with the input as its initial reference vector. Although this approach reduces the degree of change, an exact repetition of an input condition will still not necessarily generate an output and the risk of catastrophic interference remains.

Capability to Perform Complex Combinations of Interacting Features

Problems solved by ANNs include pattern recognition and control problems. However, in terms of the definition of operational complexity developed earlier, pattern recognition problems are operationally simple in having a fairly small number of alternative behaviours (i.e. different categories which can be identified) and no conflicting internal objectives. The control problems tend to have relatively small input spaces and few possible behaviours, or to have different behaviour types which are fully determined by independent input spaces, or to have little learning after an original non-operational training period. Such control problems are therefore operationally relatively simple or not required to learn without interference with prior learning in their operational mode. Such networks are therefore not as subject to the constraints which give rise to the recommendation architecture. There is a general problem of catastrophic forgetting in ANNs, and attempts to avoid this problem tend to break down as the magnitude and complexity of the learning task increases [French, 1999].

There have been attempts to define modular hierarchies of ANNs to handle more complex problems. One approach uses modules detecting unambiguous features, objects etc. Such approaches can be effective for pattern recognition (e.g. recognition of features in MRI brain scans [Sabisch et al, 1997]) but do not scale to manage a complex set of operations. Another approach is based on networks of networks in which the meaning of information exchange is not defined, and coordination between networks is dependent upon a process analogous with convergence on attractor states in physical dynamical systems [e.g. Anderson et al, 1997]. However, how dynamical systems would support information processing is not clear [Crutchfield, 1998].

Another attempt to define an ANN modular hierarchy is the expert architecture developed by Jordan and Jacobs [1994]. In this approach it is assumed that the system input space can be separated into segments, and appropriate supervised learning can occur in independent expert networks receiving inputs from each segment. A gating network can then determine appropriate system behaviour by taking a weighted average of all the expert network outputs. However, the assumptions made in this architecture, including the existence of a clean separation of the input space into segments and the absence of information exchange between different expert networks, mean that the target problem space is operationally relatively simple. The approach has only been applied to relatively operationally simple problems [e.g. Anderson et al, 1994].

Physiological Plausibility

A difficulty in mapping ANNs into physiology is the contrast between weight modification by mathematical manipulation in ANNs and synaptic strength modification by synaptic activity in biological neurons. In particular, the widely used backpropagation algorithm is regarded as implausible because it requires a supervisor providing quantitative targets for network behaviour.

The Recommendation Architecture Compared with the Marr Architecture

It is of interest to compare the recommendation architecture to the work of Marr on cognitive architectures [Marr]. In his model of the neocortex, Marr supposed that there is a potentially infinite number of input patterns which can be grouped into a number of different classes. He used a class of type "poodle" as an example. As discussed in work to simulate Marr's model [Willshaw et al 1997]:

"Each single event that represents a different occurrence of a poodle contains certain, but not all, of the features of the poodle class......Two events that represent different poodles have many features in common - many more than the number of features shared by an event representing a poodle with one that does not. Marr referred to such clusters of events over the set of input fibres as . The simplest problem that can be considered is where a network has a single output cell, and it has to learn to distinguish between the inputs from two mountains, occurring equiprobably.....The problem is to find values for the weights and the threshold such that the outputs cell responds to events from one mountain only."

In Marr's approach the implicit objective is to link cell outputs with functionally unambiguous "mountains" of similarity in a sequence of input conditions, and indicate the presence of such mountains by an unambiguous output (in the example, the output from one cell). In the recommendation architecture, the system defines clusters of repetition analogous with Marr's mountains, but these clusters do not correlate unambiguously with functionally relevant categories of input conditions. Instead, the outputs indicating the presence of the ambiguous "mountains" have enough information richness that they can be combined with outputs from other mountains to generate high integrity behaviours (simple examples would be category identifications). Once the objective becomes "to find values for the weights and the threshold such that the outputs cell responds to events from one mountain only" the system is being implicitly designed using unambiguous information contexts, and for a sufficiently complex functionality will either be unrepairable and impossible to modify or will be forced into the von Neumann architecture.

The Recommendation Architecture Compared with ART

Grossberg [1995; 1999] has attempted to use ART mechanisms as a basis for understanding large scale brain architecture, with a particular focus on the visual cortex. In this approach the layers of the visual cortex are required to address the partially conflicting requirements between the need to learn without interference with prior learning and the two processes of perceptual grouping and attention. Perceptual grouping is the process of finding boundaries that can segregate inputs derived from different visual objects. Attention is the process which enables selective processing of higher priority information.

Grossberg argues on the basis of visual illusions that perceptual grouping can activate detailed visual information for boundaries which are incomplete. In other words the presence of higher level information indicating that a boundary is present can activate detailed visual information creating the illusion of boundary segments not actually visible. In Grossberg's definition, attention sensitizes (or primes) an observer to expect an object to occur at a given location, or with particular properties, but cannot actually activate visual information at a detailed level unless that information is actually present [Grossberg, 1999].

In ART, limiting the interference between new and prior learning is achieved by only allowing learning if there is a consistency (or resonance) between information detected at detailed and higher levels. Grossberg identifies the presence of such resonances with the process of attention, arguing that attention favours detailed information which is consistent with higher level "expectations" [Grossberg, 1995]. However, it is not clear how different behavioural priorities could be used to select perceptual information in this model.

In perceptual grouping, higher level information can activate simple visual information which is not present, but such information cannot be activated in ART or the attention process as defined. Therefore Grossberg [1999] identifies two problems with the overall model. Firstly, because perceptual problems can form preattentively, it is hard to see how attention-like feedback can be present to manage the stability of learning. Secondly, visual illusions indicate that detailed visual information can be activated and remembered in the absence of real visual input. This appears inconsistent with the ART mechanism which requires consistent information at detailed and higher levels for learning to occur.

Grossberg's solution to these problems is a multilayer device architecture to generate perceptual groups in a way that allows stable learning using the ART process which he labels "attention". This multilayer structure provides a functional role for the LGN and layers 2/3, 4 and 6 in visual areas V1 and V2.

There are some major differences between this multilayer architecture and the recommendation architecture. In the recommendation architecture it is the imprinting algorithm rather than consistency between activations at higher and lower levels which ensures learning in a way that limits interference with prior learning. Because ART depends upon changes to relative weights of device inputs, maintenance of adequate information contexts will become more difficult as operational complexity increases. Another major difference is that in the recommendation architecture, attention is a behaviour which is managed by detection of conditions in clustering and interpretation and selection by competition. The behaviour type is the gating of information flow into clustering or between clustering modules and is unrelated to managing learning without interference with prior learning.

Symbol Manipulation and Theories of Knowledge

The third approach frequently used in the modeling of cognitive systems is information processing of cognitive symbols. An issue with such an approach is establishing a genuine theory of content, in other words what makes a particular symbol "about" some external object [Millikan 1990]. This issue, known as the binding problem, arises because any symbol must be instantiated in some way (for example by specific device connectivity or state) within the architecture. The issue is how to establish one to one correspondence between a set of cognitively useful symbols and a set of corresponding instantiations.

In the recommendation architecture approach, the information elements used in cognitive processing are conditions which actually occur in system experience. The way these conditions are defined therefore makes them "about" cognitive objects, and the binding problem is avoided. The cost of this "about-ness" is of course the cognitive ambiguity of the conditions. However, any attempt to make the information processing elements unambiguous will make learning of a complex combination of cognitive capabilities impractical as discussed in chapter 3. This impracticality will be illustrated by discussion of two symbolic architectures: EPAM and ACT-R.

The EPAM Model

It is of interest to consider a cognitive model such as EPAM III [Feigenbaum 1984] from the point of view of the issues which arise in learning complex real-time behaviours. EPAM is an attempt to model performance and learning in a recognition system. The input to EPAM is a set of features provided by an (unspecified) sensory-perceptual system. Its output is a symbol that accesses an image stored in semantic long term memory (also unspecified). An image in EPAM is defined as a feature list, and recognition is a tree search for an image which contains a large set of the features in the currently perceived object and no features that are not in the object. Learning is of two types. In familiarization either a new image is defined because no image corresponding with the current object could be found, or an image containing a subset of the object features is modified by addition of some of the missing features from the object. In the second type of learning, discrimination, the tree used in the search is updated because it generated an image which did not correspond with the object.

The first question is, how could EPAM model the management of complex behaviour? The only apparent way would be for sets of images retrieved from semantic long term memory to become associated with behaviours. Consider the problem which develops when there is a need to learn. Suppose that the system is faced with a situation in which two objects with identical feature lists require different behavioural responses. For example, a familiar person might develop a subtle expression which is not one of the existing features but which indicates that they did not regard the behaviour of the agent as appropriate in current circumstances. The sensory-processing subsystem would somehow have to be told to add new features, but the recognition system has no knowledge of the type of features required, because all its information is derived from the set currently delivered. Hence the sensory-processing system would have to add features at random until one that could be used was identified.

In fact, EPAM offers no mechanism for such feature addition. The implication is that features must be sufficiently simple that every feature that could conceivably be relevant to the recognition system is provided a priori. The level of complexity of such features would be similar to that of shape primitives (e.g. boundary elements) of which there would be a very large number. Furthermore on this model there must be an object or small set of objects for every behaviour, again a very large number. Hence the implied EPAM model for handling functionally complex behaviour is a tree search from a very large number of primitive features to a very large number of object sets. Such a model would experience severe real-time problems.

This discussion of EPAM illustrates a number of points. The first is that the capability to map high level recognition of a behaviour problem to a detailed level at which the problem can be solved is essential. The second is that the boundary between two subsystems must be made in such a way that problems of this type can be resolved. The third is that when information is exchanged between two modules, if that information is a predefined unambiguous set there will be problems with learning.

Production Systems

Production systems as proposed by Newell and Simon [1972] model cognition as an ordered list of productions, where each production is the combination of a condition and an implied action. At each point in time a set of conditions are detected within sensory inputs, and an internally defined priority list selects the combination of actions to be implemented. These conditions and actions are unambiguous cognitive knowledge.

ACT-R [Anderson and Lebiere 1998] is an extensively investigated implementation of the production system type. In ACT-R there are two ways in which cognitive knowledge is represented. Declarative knowledge is represented as , and models the type of knowledge a human might have when they attempt to solve a problem. A chunk is defined by its and its . Types correspond with categories or facts and slots correspond with the attributes of those categories or facts. For example, a chunk with a type dog might have slots indicating colour or size. A chunk for the fact that a dog chased a cat would have type chase and slots corresponding with the agent (dog) and object (cat). A chunk for the fact that 4 + 3 = 7 would have type addition and slots corresponding with two addends (4 and 3) and one sum (7).

Procedural knowledge is represented as production rules, where a production rule specifies a goal, a list of conditions or chunks, and the definition of the actions to be taken to achieve the goal if the chunks are present. An example might be a goal of adding two digits and , a chunk that + = , and an action to write in a specific location.

ACT-R has been very successful in modeling a range of cognitive processes, in some cases quantitatively. However, it is not specified how ACT-R could bootstrap its symbolic knowledge from sensory experience. It is assumed that the sensory inputs are preprocessed in some way into useful cognitive categories. ACT-R therefore fails to address resolution of the binding problem. It also does not address the issue of finding a compromise between information storage resources and ease of learning. In other words, it does not have a way to map the operations of the architecture into plausible physiology.

It has been demonstrated in the history of science that even quantitative descriptions of a phenomenology is not a guarantee that a theory can be mapped into deeper levels of physical description. As discussed in chapter 2, at the time of the Copernican revolution in astronomical thinking, the model of the solar system in which the sun and planets moved in circular "epicycles" around notional points which moved in circular orbits around the earth matched observations more accurately than the new model in which the earth and planets moved in orbits around the sun. However, the Copernican model had far more potential to be consistent with physical descriptions on a deeper level of detail.

Conclusions

A range of approaches to understanding human cognition have been investigated, including analogies with other types of systems labeled "complex"; fundamental theories of computation; dynamic systems theory; connectionist architectures; and symbolic modeling.

The key requirements for an adequate theory of cognition are firstly a consistent hierarchy of causal descriptions on many levels of detail, including both physiological and psychological, and a capability to bootstrap cognitive capabilities in such a way that learning of a complex combination of cognitive capabilities can occur with limited resources and minimal interference between early and later learning. None of the alternatives adequately address these requirements.

References

Anderson, J. A. and Sutton, J. P. (1997). If we compute faster, do we understand better? 29 (1), pages 67-77, 1997.

Anderson, C. W. and Hong, Z. (1994). . Proceedings of the IEEE International Workshop on Neural Networks Applied to Control and Image Processing.

Anderson, J. R. and Lebiere, C. (1998). . NJ: Erlbaum.

Ballard, D. H. (1997). . MIT Press.

Beer, R. D. (2000). 4(3), 91 – 99.

Carpenter, G. A. and Grossberg, S. (1988). The ART of Adaptive Pattern Recognition by a Self-Organizing Neural Network, , 3, pages 77-88, 1988.

Chaitin, G. J. (1966). On the Length of Programs for Computing Finite Binary Sequences. 13, 547-569.

Church, A. (1936). 'An Unsolvable Problem of Elementary Number Theory'. 58, 345-363.

Coward, L. A. and Salingaros, N. A. (2004). An Information Architecture Approach to Understanding Cities. 30(2), 107 - 118.

Crutchfield, J.P. (1998). Dynamical Embodiments of Computation in Cognitive Processes. 21, 635-637.

De Vany, A. Information, Bounded Rationality, and The Complexity of Economic Organization. Discussion paper, Dept. of mathematics, University. of California, 1993.

Edmonds, B. (1999). . Ph.D. Thesis. University of Manchester.

Feigenbaum, E. A. and Simon, H. A. (1984). EPAM-like models of recognition and learning. 8, 305-336.

Fraley, C. and Raftery, A. E. (1998). How Many Clusters? Which Clustering Method? Answers Via Model-Based Cluster Analysis. , University of Washington.

French, R. M. (1999). Catastrophic Forgetting in Connectionist Networks. 3(4), 128-135.

Gokcay, E. and Principe, J.C. (2002). Information Theoretic Clustering. 24(2) 158-169.

Grossberg, S. (1995). The Attentive Brain. American Scientist 83, 438 - 449.

(1999). How Does the Cerebral Cortex Work ? Spatial Vision 12, 163 - 186.

Gusev, V. D., Kulichkov, V. A. and Chupkhina, O. M. (1991).Genome Complexity Analysis 1: Complexity Measures and the Classification of Structural Features. 25, 669-677.

Hebb, D. O. (1949). . New York: Wiley.

Honkela, T, Pulkki, V. and Kohonen, T. (1995). Contextual Relations of Words in Grimm Tales, Analyzed by Self-Organizing Map, in Fogelman-Soulie, F. and Gallinari, P. (eds.), , 2, 3-7. Paris: EC2 et Cie.

Jain, A. K., Mao, J. and Mohiuddin, K. (1996). Artificial Neural Networks: A Tutorial. 29(3), 56 - 63.

Jordan, M. I. and Jacobs, R. A. (1994). Hierarchical Mixtures of Experts and the EM Algorithm. Neural Computation 6, 181 - 214.

Kaufman, L. and Rousseeuw, P. J. (1990). . New York: Wiley.

Kohonen, T. (1995). . Springer Series in Information Sciences 30.

Kolmogorov, AN. Three Approaches to the quantitative definition of Information. , 1965, 1, 1-17.

Lovasz, L. and Gacs, P. (1999) . http://www.cs.yale.edu/HTML/YALE/CS/HyPlans/lovasz/complex.ps

Ludermir, T., Carvalho, A., and de Souto, M. (1999). Weightless Neural Models: A Review of Current and Past Works. 2, 41 - 61.

Marr, D. (1991). . Edited by L. M. Vaina. Boston: Birkhauser.

Millikan, R. (1990). Compare and Contrast Dretske, Fodor, and Millikan on Teleosemantics. 18, 2, 151-160.

Newell, A. and Simon, H. A. (1972). . Prentice-Hall: New Jersey.

Petitot, J. (1985) . Paris: Maloine.

Ratcliff, R. (1990). Connectionist Models of Recognition Memory: Constraints imposed by learning and forgetting functions. 97, 285 - 308.

Robins, A. (1995). Catastrophic Forgetting, rehearsal, and pseudorehearsal. 7, 123 -

Sabisch, T., Ferguson, A. and Bolouri, H. (1997). Automatic registration of complex images using a self organizing neural system. , Anchorage, Alaska.

Sambrook, T and Whiten, A. On the Nature of Complexity in Cognitive and Behavioural Science. , 1977, 7, 191-213.

Solomonoff, RJ. A Formal theory of Inductive Inference. , 1964, 7, 1-22,224-54.

Sun, R., Merrill, E. and Peterson, T. (2001). From implicit skills to explicit knowledge: a bottom-up model of skill learning. 25(2), 203 -244.

Taylor, J. G. and Alavi, F. N. (1993). Mathematical Analysis of a Competitive Network for Attention, in Taylor, J. G. (ed.), , Elsevier.

Turing, A.M. (1936). 'On computable numbers, with an application to the Entscheidungsproblem'. , 42, 230-265.

Turing, A.M. (1948). Intelligent Machinery. In Meltzer, B. and Michie, D. (eds) 1969. . Edinburgh University Press.

Van Gelder, T. (1998). 'The Dynamical Hypothesis in Cognitive Science'. 21, 1-14.

Watkins, C. J. C. H. and Dayan, P. (1992). Q-learning. 8(3), 279 - 292.

Willshaw, D., Hallam, J., Gingell. S., and Lau, Soo Leng (1997). Marr's Theory of the Neocortex as a Self-Organizing Neural Network. 9, pages 911-936, 1997.

Wright, J. (2000). Brain Dynamics: Modeling the Whole Brain in Action. In
, ed. E. Gordon. The Netherlands: Harwood.

Chapter VIII

Neurophysiological Evidence for the Recommendation Architecture

The argument made to this point is that as the operational complexity of the tasks performed by a system increases, practical considerations including the needs to make the best use of available resources and to make changes to some operations without undesirable side effects on other operations will severely constrain the architectural form of the system. If operational changes must be learned through experience, these practical considerations constrain the system into the recommendation architecture form, and some of the detailed structures and processes likely to be present in any system with that architecture have been described.

The mammal brain must perform a complex combination of interacting behaviours, and learns a significant proportion of these behaviours through experience. It is therefore of interest to compare the forms and processes of the mammal brain with those expected in a system with the recommendation architecture. Figure 1 illustrates the major subsystems of the recommendation architecture discussed in chapter 4, mapped to the corresponding physiological structures in the mammal brain.

In this mapping, the cortex corresponds with the clustering subsystem, managing the definition, permanent recording and detection of information conditions within sensory inputs. The thalamus, basal ganglia and cerebellum correspond with the competitive subsystems respectively selecting attention behaviours, general behavior types and specific behaviours within an already selected general type. As discussed in chapter 4, conditions detected in the associative cortex may influence attention behaviours, either by targeting the thalamus or by the basal ganglia managing some more complex cognitive attention behaviours. This topic will be discussed in more detail in chapter 9. The hippocampus system including both cortical and subcortical components corresponds with the clustering and competition subsystems managing the assignment of condition recording resources, and maintaining the connectivity which supports activation on the basis I past temporally correlated condition recording. The amygdala corresponds with the structure which manages the level of activity which devices within one area generate in response to detection of a given proportion of their programmed conditions, and therefore manages the relative probability of a behaviour of the area type in response to current conditions.

The types of conditions detected at different points in clustering differ in information complexity as discussed in chapters 3 and 4. The typical complexity ranges of conditions detected at different points can be labeled by the type of unambiguous cognitive condition which would be in a similar range of complexity (see for example figure 9.1 in the next chapter). Such unambiguous conditions would be cognitive features, objects, groups of objects etc. However, this labeling is conceptual and only intended as an aid to explanation. In practice, conditions will be defined within many different ranges of complexity, and different ranges will correlate to different degrees with different types of unambiguous cognitive conditions. Conditions within a number of different ranges of complexity may be useful for guiding any one behaviour type.

The experimental evidence in support of this mapping is the primary topic of this chapter. Six areas of experimental evidence will be considered: physical structure; pattern of connectivity; context for information exchange; module management; device algorithms; and general effects of physical damage and deficits. More specific discussion of memory deficits resulting from physical damage will take place in chapter 10.

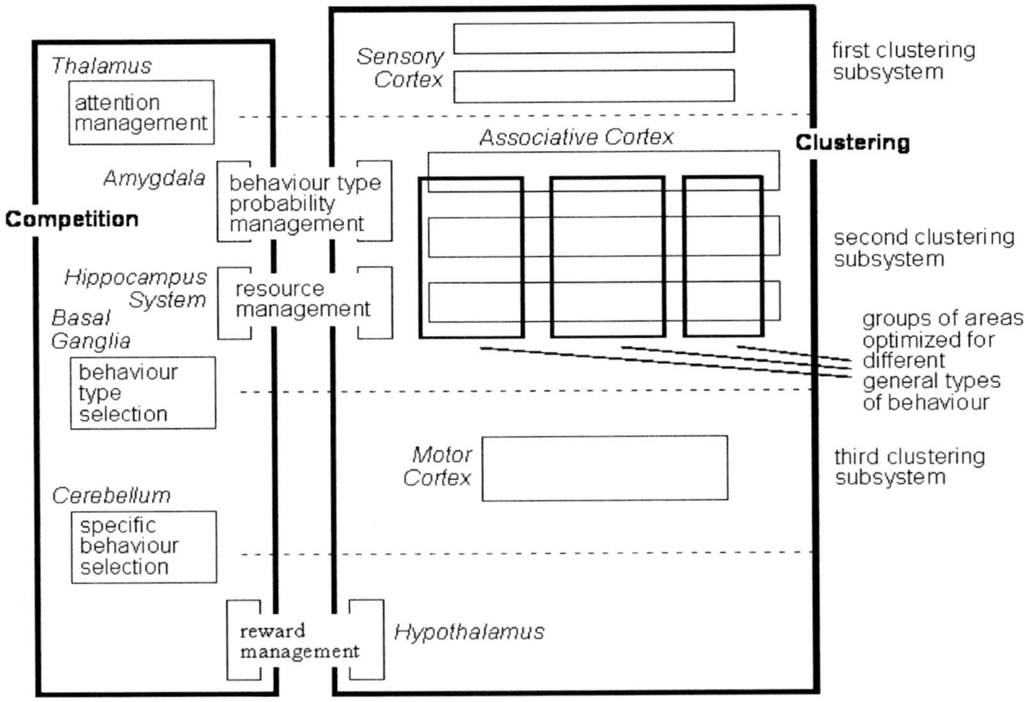

Figure 8.1. The highest level architectural separations of the recommendation architecture from figure 11 in chapter 4, mapped into major physiological structures of the mammal brain. The experimental evidence for this mapping is the topic of this chapter.

Physical Structure

The separation of the mammal brain between cortex and various subcortical structures including thalamus, basal ganglia and cerebellum resembles the recommendation architecture separation between clustering and various competitive subsystems in a number of ways. The resemblances include physical structure, types of connectivity between structures, and operational roles of different structures.

As discussed in chapter 3, the required information synchronicity can be achieved in clustering by the organization of into layers. A modular hierarchy must be overlaid on the layers, with a module defined as a group of devices with more connectivity within the group than externally with peer groups. Modules on one level will be generally similar, but not identical.

The cortex is organized into layers of neurons, with a pronounces modular hierarchy of columns and areas overlaid on the layers. A primary means for identifying these modules is greater internal connectivity relative to external connectivity [Van Essen et al, 1995; Crick et al, 1986; Tanaka, 1993; Calloway, 1998]. The observed connectivity is thus consistent with minimization of information exchange between modules as far as possible, as required by the recommendation architecture.

Also as required by the recommendation architecture, the observed modules (columns and areas) are physically similar in a general sense, but heterogeneous at all levels, as demonstrated for example by the patchiness of connectivity between areas, with one site within an area projecting to or receiving inputs from multiple discrete loci within another area [Van Essen et al, 1995; Barone et al, 2000]. As would be expected for a heuristically defined modular hierarchy, there is variation between individuals of the same species (e.g. in the location of cortical areas [Van Essen et al, 1995]).

As expected for a modular hierarchy, modules appear to be roughly orthogonal, for example neurons located nearby in the cortex respond to similar but slightly different conditions [Tanaka, 1993] and neighbouring neurons seldom fire at the same time [Crick et al, 1986]. Consistent with a partial operational ambiguity of information, the conditions detected by neurons do not appear to bear any simple relationship with cognitively useful features [Tanaka, 1993].

In the recommendation architecture, the majority of the connectivity in clustering is excitatory and indicates the detection of conditions, while a much higher proportion of connectivity in competition subsystems is inhibitory. In the mammal brain the majority of the connectivity in the cortex is excitatory, while a much higher proportion of connectivity in the thalamus, basal ganglia and cerebellum is inhibitory [Crick et al, 1986].

In the recommendation architecture there is a general requirement for multiple levels of competition. The first level selects input access and information flows within clustering from a range of alternatives provided by clustering. The second level makes selections between general types of behaviour, and the third level makes selections, within the general type already selected, between different combinations of simple movements. The competitive subsystems gate but do not change clustering outputs, because such change would be operationally complex. The requirement that almost all sensory information must pass through the thalamus before reaching the cortex, combined with strong projections from the cortex back to the thalamus, resembles the connectivity for first level competition management of input access and information flow. The association of the thalamus with the attention function is also consistent with this interpretation. The arrangement of the motor cortex as a control map of the physical body, but with cortex output needing to pass through the basal ganglia and cerebellum before being released to control [Kingsley, 2000] resembles the two level selection between clustering recommendations.

The basal ganglia provides the crucial physiological link between the idea of movement and the motor expression of that idea, with most neuronal activity in the basal ganglia occurring before any movement begins [Kingsley, 2000]. The functional anatomy of the basal ganglia, with strong excitatory projection from the cortex to the striatum, strong inhibitory projection from the striatum to the substantia nigra and globus pallidus [Kingsley, 2000], in qualitative connectivity terms resembles the structure of the simple competitive system described above. The basal ganglia in connectivity and functional terms thus resembles a second level competitive subsystem in the recommendation architecture which selects general type of behaviour. This view of basal ganglia functionality is consistent with the evidence that it has a general cognitive role and is not just involved in the control of movement [Middleton, 2000]. The observation that cerebellar lesions do not produce weakness or paralysis or affect the ability to initiate or terminate voluntary movement, but rather prevent precise coordination of movements [Kingsley, 2000] is consistent with a role of selecting detailed combinations of movements from alternatives provided by the cortex.

Pattern of Connectivity

A major forward flow of information is observed in the cortex from raw sensory input to steadily increasing information complexity. An example is the ventral pathway V1 → V2 → V4 → posterior inferotemporal cortex → anterior inferotemporal cortex → prefrontal cortex [Tanaka, 1993]. There is also a predominant vertical flow of information through cortex columns [Kingsley, 2000]. These flows are excitatory and resemble the flow of condition defining information in the direction of increasing information complexity in a clustering system. This forward flow is overlaid with a complex combination of other projections which can be either excitatory or inhibitory, as for the management connectivity in clustering. As expected if it is necessary to achieve a practical minimum of information exchange, projections are from a limited number of specific layers to a limited number of other specific layers within a column [Lund et al, 1988; Lund et al, 1991; Van Essen et al, 1995], from one column to a limited number of other columns [Saleem et al, 1993], and from one area to only a handful of other areas [Saleem et al, 1993].

In a number of cases detailed cortex connectivity can be interpreted as the information exchange required to achieve specific management functions in a recommendation architecture. For example, it can be argued that there is a primary excitatory information flow in the visual cortex of the macaque monkey: input from the thalamus → layer IV → layers II/III → layer V → layer VI [Calloway, 1998; Buonomano et al, 1998]. There are excitatory projections from V to II/III, from VI to IV, and within layer V [Calloway, 1998]. There are also inhibitory projections from V to II/III and from VI to IV [Lund et al, 1988], inhibitory projections within V and within VI [Lund et al, 1988]. This connectivity is almost exactly the connectivity required to implement the management of the conditions under which new information conditions are recorded while limiting additions to input and output populations as described in section 6.3.2. The mapping between functional layers and cortex layers is provided in figure 2. The missing connectivity is excitatory between neurons in layer VI. An experimental prediction is therefore that excitatory connectivity of this type should be present.

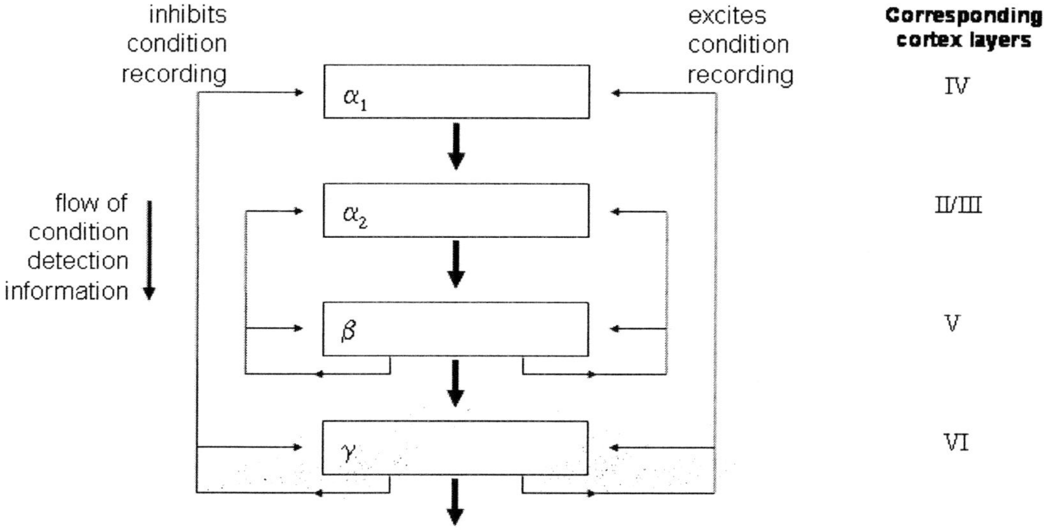

Figure 8.2. Column structure to reduce information exchange as discussed in chapter 3, mapped into cortex layers.

A second connectivity example is the observation that "As a general rule, each of the cortical sensory systems sends "forward" projections from primary areas to the middle layers of higher order cortical areas (e.g. V1 to V2) which are reciprocated by "backward" projections that terminate in layer I of primary areas (e.g. V2 to V1).

A sensory evoked potential in the primary somatosensory cortex that predicts touch discrimination behaviour in awake monkeys is generated by excitation of layer I, apparently as a result of backward activation from higher cortical processes. Reversible lesions in V2 of squirrel monkeys shows that these corticocortical projections modulate the overall responsiveness of V1 neurons to visual stimuli without changing their directional or orientation preferences.

Layer I is primarily a synaptic zone filled with mostly asymmetric round, presumably excitatory synapses upon the dendritic spines of pyramidal neurons whose cell bodies lie deeper in layers II, III, and V.

We found [in the rat somatosensory system] that backward cortical projections were a major source of extrinsic inputs to layer I [and] fibres extended horizontally within layer I for at least 1 mm in any direction" [Cauller et al, 1994].

These characteristics are exactly the properties required for the process to maintain information context when multiple semi-independent conditions are recorded on individual devices: backward excitatory connectivity from higher complexity information conditions which does not change the conditions programmed at low complexity, with the backward projection distributed to allow connectivity when determined to be relevant; and the connectivity targeted at a level early enough to prevent recording of additional information conditions, such as level II/III.

Context for Information Exchange

There are several ways in which cortex phenomena resemble the maintenance of partially ambiguous information contexts within the clustering subsystem of the recommendation architecture. One is the operational ambiguity of information, a second is the presence of mechanisms that appear to support context maintenance, a third is the existence of critical periods for learning, and a fourth is the permanence with which apparently irrelevant information is recorded.

As discussed earlier, cells and columns in the cortex correlate with visual conditions that have no obvious connection with behaviourally useful features [Tanaka, 1993]. This is consistent with the requirement that clustering modules detect repetitions of input information conditions which cannot be evolved towards patterns relevant for specific behaviours.

There are a number of mechanisms in the cortex which resemble mechanisms required to maintain context in a recommendation architecture. For example, a specific temporal pattern of activity (post synaptic bursting) accompanies synaptic potentiation [Paulsen et al, 2000]. Such a pattern of activity could be used to communicate when an additional condition is being recorded, for example in order to establish the feedback connectivity required to discriminate between multiple conditions on the same device. The observation of synchronous gamma band activity in the cortex [Singer et al, 1995] and the observation that neuron firing rates and thresholds are linked to high frequency fluctuations in membrane potential [Azouz et al, 1999] support the view that synchronicity of neuron outputs communicate additional context information, for example that a particular population of neuron outputs are derived from a single visual object.

As a general principle, the requirement to maintain context means that the more widely an output from a clustering module is distributed, the less that output can be permitted to change. Therefore if a domain of sensory experience is widely used in a complex fashion for higher behavioural purposes, there will be a tendency for the domain to be organized into an array of columns in an early experience period and only widely distributed after that period. Changes after the early period will be extremely limited or not allowed at all. The requirement for specific early learning periods after which no further learning is possible resembles the existence of critical periods for learning in the early development of mammals in which basic verbal and visual sensory types are learned, and cannot be learned if absent from the experience during the appropriate period [e.g. Berardi et al, 2000].

It is also of interest that an early learning period in which experimental column structures are extensively modified before being brought into use for behavioural purposes is consistent with the absence of effective episodic memory from the very early stages of human development.

Cortex Maps

A ubiquitous feature of early sensory processing in the visual cortex is the existence of maps. The visual cortex appears to contain a topographic map of retinal inputs, plus superimposed maps of orientation, colour, direction of motion etc. Up to three maps have been observed in the same patch of cortex [Hubener, Shoham, Grinvald and Bonhoeffer, 1997]. There is evidence that these maps are relatively uniform with respect to different input conditions [Swindale, Shoham, Grinvald, Bonhoeffer and Hubener]. In other words, for any given combination of features such as orientation, spatial frequency and retinal location on the same map, the total neural response across the map is independent of the specific combination chosen.

However, maps are nor consistent across different individuals of the same species, and individual neuron responses do not correlate with one specific combination of feature values, but with bands of such values.

This mapping capability is as expected if cortex areas heuristically organize their inputs into groups or portfolios of conditions which tend to occur at the same time, with some general a priori guidance on input sources but not specific guidance on the definition of individual portfolios.

However, it is also possible that for early visual processing in which portfolios can be fairly completely defined before widespread release of outputs to higher cognitive processing, the requirement that conditions not be changed after initial definition could be relaxed. In this case, a different portfolio organization algorithm would be feasible, perhaps along the lines of Kohonen maps. However, such relaxation would not be possible for portfolios with outputs which are widely distributed while learning is still under way.

Module Management and Sleep with Dreaming

A major requirement of a system with the recommendation architecture is for a process to heuristically minimize information exchange between modules at all levels. This process must find a compromise between the operational advantages and the increased architectural complexity resulting from additional information exchanges between modules. As discussed earlier, the indicator of the probable operational advantage of information exchange between two unconnected modules is frequent simultaneous activity of the two modules in the past. One way to determine the degree of such correlated activity which avoids the need to collect and store large amounts of information over a long period of experience is to perform an approximate rerun of experience with a bias towards the more recent.

As discussed in chapter 11 and in more detail in Coward [1990], sleep with dreaming in mammals closely matches this information exchange minimization process. Sleep with dreaming is an off-line process which includes an apparent rerun of experience, with a bias towards the more recent [Wilson at al, 1994]. Note that this process is radically different from the memory consolidation process proposed as the functional role of sleep with dreaming originally by Marr [1969 - 1971 (1991)]. No memory consolidation is required in the information minimization process: the model for short term memory in the recommendation architecture is the ability for higher complexity information conditions acting through a competitive subsystem to reactivate simpler, recently active information conditions (see chapter 10). Provided that provisional connectivity assignment continued on a random basis without the bias, the major effects of deprivation of REM sleep would be a long term increase in the difficulty of learning new behaviour without interfering with existing behaviour and in the difficulty of recovering functions after physical damage. In other words, declarative memory would still operate, but would be somewhat less efficient than with REM sleep. This is consistent with the absence of a major effect on memory following the reduction or elimination of REM sleep by antidepressant drugs [Vertes et al, 2000].

A second global process required for clustering is resource management. A plausible implementation of this process requires an explicit map of resources, a clustering subsystem which receives input from all clustering resources and generates resource assignment recommendations, and a competitive subsystem which selects the most appropriate actions. The hippocampus structure [see e.g. Amaral et al, 1990] includes Ammon's horn with a regular, repetitive structure consistent with the map subsystem, cortex type resources (including the perirhinal, parahippocampal and entorhinal cortex) with extensive connectivity to and from all other cortex areas consistent with the required specialized clustering subsystem, and connectivity to and from a proposed competitive structure (the thalamus) for example via the cingulate gyrus [Kingsley, 2000]. The similarities between the effects of damage to the resource management system and of damage to the hippocampus structure are discussed below.

Neuron Algorithms

In the recommendation architecture, the algorithms by which devices determine their outputs from their inputs are the result of the requirement to support the partially ambiguous but meaningful information processing functions of the device. There are many possible algorithms, but any algorithm must be able to discharge these information processing needs, and will therefore be constrained to certain general forms.

The condition recording device in clustering has a large number of inputs from other condition recording devices in previous layers. These inputs are separated into semi-independent groups corresponding with different conditions. Any one such input indicates the presence of a component in a condition programmed on the target device. Such inputs are therefore excitatory. The condition recording device has a number of other types of inputs from various sources, these inputs do not form part of groups. Inputs from activity detection devices manage changes to groups and may be excitatory or inhibitory. Inputs from subsequent layers excite prolongation or initiation of device activation in various ways, and some of these inputs may be segregated into other groups. Inputs from competitive subsystems manage access of inputs to and release of outputs from devices in clustering modules.

The cortex contains a number of neuron types including pyramidal, chandelier and basket cells. The pyramidal neuron in the cortex has a large number of inputs from pyramidal neurons in preceding layers, and these inputs are separated into groups by the dendritic tree. The pyramidal neuron also has a range of inputs from chandelier and basket cells and from higher level pyramidal neurons. Pyramidal neuron outputs are excitatory. Cells like chandelier and basket cells have much less extensive dendritic trees and outputs can be excitatory or inhibitory [Crick at al, 1986]. In general terms therefore, the physical form of pyramidal neurons is consistent with the function of the condition recording device and chandelier and basket cells are consistent with the activity detection devices.

A condition recording device generates an output in response to the activity of its inputs from other condition recording devices. Such activity could be the activity of a large subset of one of its groups or smaller subsets of a number of such groups. In information terms the output indicates the presence of an information condition within the portfolio of the device. This output can have a structure which communicates information in addition to the presence of a condition. Such information could include the size of the current active subset, the input population from which the active subset is derived, and the consistency of the output with higher level conditions.

It was suggested in chapter 5 that the rate of generation of voltage spikes of pyramidal neurons could indicate the size of the current active subset, the frequency modulation of the generation rate could indicate the input domain from which the active inputs were derived, and the length of time voltage spike generation persisted could indicate consistency with the activity of the targets of the output.

In chapter 5 the suggestion that the 40 Hz gamma band frequency within the EEG could reflect this frequency modulation of firing rates was discussed. This model would need an integration period significantly less than 25 milliseconds, with the optimum period depending on the detailed shape and amplitude of the modulation. Integration periods are generally believed to be of the order of the passive membrane constant (i.e. the time for a synaptic potential to decay) which is in the observed range 8 - 15 milliseconds [Konig et al, 1996]. The effect of such frequency modulation will be the observed irregularity in the output timing of cortex neurons.

What is known of pyramidal neuron change management algorithms also appears consistent with the recommendation architecture model. Increases to excitatory post synaptic potentials (EPSPs) take place if the EPSP occurs slightly before a postsynaptic action potential (AP) [Markram et al, 1997; Paulsen et al, 2000]. In other words, changes occur if there are enough EPSPs from different synapses to generate a post synaptic AP. As expected for the change management process in a recommendation architecture, such changes appear to be reduced by the presence of inhibitory signals [Tsumoto 1992, Buonomano et al, 1998].

A simple model for the permanent condition recording process using such mechanisms models a provisional condition structure as a group of inputs of three types on to one dendrite of a pyramidal neuron. Type one inputs would be excitatory and would come from the preceding layer of the column. These inputs define the provisional condition. Type two inputs would also be excitatory and would come from a specific module of neurons. This module would be a particular layer of the column in which the neuron was located. Type three would be inhibitory, and again come from one specific module of the column. The synaptic strengths of the type one inputs would be inadequate to generate an AP, but if the synaptic strengths of the types two and three inputs would be such that if most of the type one and type two were strongly active and the type three relatively inactive, an AP could be generated. If an AP resulted, the synaptic strengths of the type one inputs would increase and that of types two and three decrease. The functional effect would be that a set of active inputs corresponding with the provisional condition would not cause the neuron to fire unless the appropriate combination of change management inputs were also active, but if they caused the neuron to fire, the result would be a permanent increase in the tendency for an exact repetion of the condition to fire the neuron in the future, independent of the states of the types two and thre inputs.

Deficits and the Effects of Physical Damage

The types of high level deficits which can occur as a result of physical damage or other single causes reveal considerable information about system architecture. The observation that failure or damage affects one type of system behaviour and not another implies that an architectural separation exists between the two types and that the system architecture allows such a separation.

Damage to the Cortex

Consider first the effect of physical damage on clustering separated into area modules. Local damage to early areas would result in perception deficits. Local damage to the final areas would result in paralysis deficits. However, local damage to the middle, behavioural areas would result in changes to the relative probability of different types of behaviour in response to a given type of object. In other words, such damage would result in personality changes. Information recorded in the past in response to a given object will exist in all behavioural areas as a result of the generation of recommendations of the different types with respect to that object. Local damage to one area will therefore have little effect on, for example, the ability to discriminate between familiar and unfamiliar objects. In other words, declarative memory will be unaffected.

These expectations correspond well with observation of the effects of cortex damage. Damage to visual or somatosensory areas produces sensory deficits, and damage to motor cortex produces paralysis deficits [Kingsley, 2000]. As illustrated by the classic Gage case [Harlow, 1868], damage to the associative cortex results in personality changes but no apparent loss of declarative memory.

The effects of cortex damage on memory are discussed in much more detail in chapters 10 and 11.

Damage to the Hippocampus System

Physical damage to the system managing assignment of resources to clustering can be expected to result in very specific memory and other deficits as discussed in chapter 10. Damage preventing future assignment of resources will result in loss of the ability to record new conditions, and therefore the ability to create new declarative memories. However, skill learning based on adjustment of relative behavioural weights assigned to detections of previously recorded conditions will be largely unaffected.

Condition recording resources assigned at the same time have a high probability of recording conditions at similar times. Provisional connectivity to support indirect activation on the basis of temporally correlated past condition recording is therefore most efficiently managed by the same resource management function, resulting in the permanent connectivity being routed through that function. Damage could therefore also result in loss of the ability to access all episodic memories from a specific period of time.

A more subtle effect derives from the fact that because no two sensory experiences are identical, generation of behavioural recommendations will in general always require a small amount of recording of new information conditions. The absence of the recording capability will therefore tend to result in passivity, and only strong arousal allowing generation of recommendations using only lower proportions of previously recorded conditions will produce behaviour.

Loss of ability to create new declarative memories, loss of time slices of past declarative memories, unaffected skill learning, and passivity are exactly the symptoms exhibited by patients with damage to the hippocampus system, for example that caused by Korsakov's Syndrome [Butters, 1984]. It is also of interest that such patients also exhibit a surge of REM sleep in the early stages of the deficit [Greenberg et al, 1968]. Given the role of REM sleep in the assignment and configuration of resources, the surge can be interpreted as a system response to the absence of resources.

Basal Ganglia Deficits

Deficits in the competitive subsystem which selects general types of behaviour could be expected to manifest itself in a number of ways. This competitive subsystem requires a global activation process which ensures that one and only one type of behaviour is selected in response to a sensory condition. Problems with this activation process in extreme cases would result either in slow or no selection or multiple selections.

Slow or no selection would result in behavioural slowness and in the extreme case apparent paralysis. Multiple selections cannot lead to multiple different simultaneous inconsistent behaviours because different physical movements cannot occur at the same time. Multiple selections at the general level would therefore be forced into a single, coordinated behaviour by the second competitive subsystem. However, this forcing will be on the basis of physical constraints rather than behavioural selection and will therefore tend to be less stable. Hence the multiple selection deficiency will result in rapid behavioural shifts with no apparent consistency.

In chapter 4 it was demonstrated that a general behavioural selection subsystem with a global process to limit behaviour selection to one type was required. Identification of the basal ganglia and the dopamine cycle within the basal ganglia with these subsystems is consistent with functional and deficit observations. On this model, if some problem caused the dopamine level to be too low, no behaviour type would be selected by the basal ganglia and the result would be behaviour slowness and paralysis. These symptoms resemble Parkinson's disease [Kingsley, 2000]. If on the other hand a problem caused the dopamine level to be too high, the result would be failure to limit behaviour selection to just one type. Multiple types would then be presented to the cerebellum for specific behaviour selection. A single behaviour would generally be selected by the cerebellum, but the selection would be unstable, resulting in rapid behavioural shifts. These symptoms resemble those of Tourette's syndrome [van Woerk, 1983]. For more detailed discussion see Coward [1990].

Experimental Verification

There are a number of areas in which experimental verification of the recommendation architecture model for the mammal brain is possible. However, within the architectural framework a number of different implementations could achieve a given function. The two different implementations of a column module discussed earlier are one example. Hence experimental tests require investigation of whether actual processes are consistent with a possible implementation.

One fundamental requirement is that cortex neurons under appropriate conditions permanently and instantaneously record new combinations of their inputs. For example, on the basis of the column model illustrated in figure 6, a pyramidal neuron in layers II or III if presented with a combination of inputs on the same dendrite from layer IV which did not cause it to produce an output should be caused to fire by excitatory input received from layer V provided no inhibitory input from layer V was present. Once it had been caused to fire, presentation of the same combination of inputs from layer IV should cause it to fire independent of layer V inputs.

The requirement at the cortex column level is thus for a process in which specific cortex layers inhibit and excite recording of new information conditions in target layers of the same column. Again starting from the model illustrated in figure 6, the prediction is that neurons in layer V projecting excitatory synapses to layers II/III will be activated at lower levels of overall layer V activity than neurons projecting inhibitory synapses to layers II/III. Increases in EPSPs should occur in layers II/III pyramidal neurons provided there is excitatory synaptic input from layer V neurons, but will be cut off if there is inhibitory input from (different) layer V neurons. Excitatory connectivity should exist between neurons in layer VI.

Another prediction concerns the device algorithms to support the maintenance of information context if multiple relatively independent conditions are programmed on each device. Again on the basis of the column functionality in figure 6, excitatory projections from higher areas to layer II and III neurons should cause those neurons to produce output for longer periods than with the same inputs from layer IV in the absence of the back projections. The excitatory synapses to layer II and III pyramidal neurons should be strengthened by the post synaptic bursts which indicate increased EPSPs in the source neuron, provided the target neuron is active.

A further experimental test is suggested by the identification of different areas of the associative cortex with different general types of behaviour. The responses of all neurons within one such area to their operational inputs should be modulated in a similar fashion by the presence of neurohormones associated with anger, fear, or hunger etc. and/or the probability of an output from different areas proceeding to the motor cortex should be affected in different ways by such neurohormones.

The recommendation architecture model for the hippocampus system as a resource manager for the cortex is significantly different from the temporary memory store model [Marr, 1991(1969-1971)], and the differences could be tested experimentally. For example, suppose that the pattern of hippocampus and cortex activity was measured when a radically novel object was presented for the first time, when it was presented on two separate subsequent occasions before sleep and two more occasions after sleep. There is considerable variability in hippocampus and cortex response to the same stimulus, but the temporary memory store model would predict systematic differences based on the update of the cortex by the hippocampus between the two repeats before sleep and the two repeats after sleep. The resource management model would predict much less systematic variability before and after sleep, such variability would only result from a generally very small employment of the new resources assigned in the intervening sleep period.

Other tests are suggested by the proposed role of sleep with dreaming. The process involves creation of new provisional synapses in appropriate locations determined by recent correlated activity. Direct observation of the creation of provisional connectivity may be experimentally difficult, but on a behavioural level, the deficits resulting from dream sleep deprivation are expected to be dependent on waking experience during the deprivation period. Because resources are depleted and not replaced efficiently, some behavioural shift away from behaviours frequently used in waking could occur.

Conclusions

Systems which perform sufficiently complex combinations of operations tend to be constrained within specific architectural forms by practical considerations including limits to information recording and processing resources and needs to change some operational features without side effects on other features. If changes to features must be learned, the architectural forms are called the recommendation architecture.

The mammal brain learns to perform a complex combination of operations, and there are some strong resemblances between the forms of the recommendation architecture and the physical structures and processes of the mammal brain. The layers, columns, areas, and connectivity of the cortex resemble those of the modular hierarchy in the recommendation architecture which defines and detects conditions within system inputs. The partial operational ambiguity of conditions in the recommendation architecture resembles the cognitive ambiguity of conditions detected by the cortex. The separation of functionality between the thalamus, basal ganglia and cerebellum subcortical structures and their connectivity with the cortex resembles the competitive subsystems in the recommendation architecture which use conditions detected by the modular hierarchy to select the subset of the input space which will influence current behaviour, the general type of behaviour, and the specific behaviour within the already selected general type. The process to minimize the global exchange of information within the modular hierarchy resembles REM sleep in the mammal brain.

The recommendation architecture provides an approach to understanding the mammal brain as a system, relating information recording and processing at the neuron level through various intermediate structures to cognitive phenomena.

Predictions of the type of neuron algorithms to be expected in the cortex and the functional role of different types of connectivity within the cortex are made.

References

Amaral, D. G. and Insausti, R. (1990). Hippocampal Formation. In . ed. G Paxinos. Academic Press.

Azouz, R. and Gray, C. M. (1999). Cellular mechanisms contributing to response variability of cortical neurons in vivo. 19(6), 2209-2223.

Barone, P. and Kennedy, H. (2000). Non-uniformity of neocortex: areal heterogeneity of NADPH-diaphorase reactive neurons in adult macaque monkeys. 10, 160-174.

Berardi, N., Pizzorusso, T. and Maffei, L. (2000). Critical periods during sensory development. 10, 138-145.

Buonomano, D. V. and Merzenich, M. M. (1998). Cortical plasticity: from synapses to maps. 21, 149-186.

Butters, N. (1984). Alcoholic Korsakov's Syndrome: An Update. 4(2), 226-244.

Calloway, E. M. (1998). Local circuits in primary visual cortex of the macaque monkey. 21, 47-74.

Cauller, L. J. and Connors, B. W. (1994). Synaptic physiology of horizontal afferents to layer I in slices of rat SI cortex. 14(2), 751-762.

Coward, L. A. (1990). . Praeger: New York.

Crick, F., and Asanuma, C. (1986). Certain aspects of the anatomy and physiology of he cerebral cortex. In volume 2, editors McClelland, J. L., Rumelhart, D. E. and the PDP Research Group. Cambridge, MA: MIT Press.

Greenberg, R., Pearlman, C., Brooks, R., Mayer, R. & Hartmann, E. (1968). Dreaming and Korsakov's Psychosis. 18, 203-209.

Harlow, J. M. (1868). Recovery from Passage of an Iron Bar through the Head. 2, 327-346.

Hubener, M., Shoham, D., Grinvald, A. and Bonhoeffer, T. (1997). Spatial relationships among three columnar systems in cat area 17. 17, 9270 – 9283.

Kingsley, R. E. (2000). . Baltimore MA: Lippincott, Williams and Wilkins.

Konig, P., Engel, A. K., and Singer, W. (1996). Integrator or Coincidence Detector ? The role of the cortical neuron revisited. 19(4), 130 - 137.

Lund, J. S., Hawken, M. J. and Parker, A. J. (1988). Local circuit neurons of the macaque monkey striate cortex: II. Neurons of laminae 5B, and 6. 276, 1-29.

Lund, J. S. and Yoshioka, T. (1991). Local circuit neurons of the macaque monkey striate cortex: III. Neurons of laminae 4B, 4A, and 3B. 311, 234-258.

Markram, H., Lubke, J., Frotscher, M. and Sakmann, B. (1997). Regulation of synaptic efficacy by coincidence of postsynaptic APs and EPSPs. 275, 213-215.

Marr, D. (1991). . Edited by L. M. Vaina. Boston: Birkhauser.

Middleton, F. (2000). Basal Ganglia Output and Cognition: Evidence from Anatomical, Behavioural, and Clinical Studies. 42, 183-200.

Paulsen, O. and Sejnowski, T. J. (2000). Natural patterns of activity and long-term synaptic plasticity. 10, 172-179.

Saleem, K. S., Tanaka, K. and Rockland, K. S. (1993). Specific and columnar projection from area TEO to TE in the macaque inferotemporal cortex. 3, 454 - 464.

Singer, W. and Gray, C. M. (1995). Visual feature integration and the temporal correlation hypothesis. 18, 555-586.

Swindale, N. V., Shoham, D., Grinvald, A., Bonhoeffer, T. and Hubener, M., (1997). Visual Cortex maps are optimized for uniform coverage. 3(8), 822 – 826.

Tanaka, K. (1993). Neuronal mechanisms of object recognition. 262, 685-688.

Tsumoto, T. (1992). Long-term potentiation and long-term depression in the neocortex. 39, 209-228.

Van Essen, D. C. and Andersen, C. H. (1995). Information processing strategies and pathways in the primate visual system. In An Introduction to Neural and Electronic Networks. Academic Press.

Van Woerk, M. H. (1983). Neuroleptics in Neurologic Disorders. In , 65-74. New York: Raven.

Vertes, R. P. and Eastman, E. E. (2000). The Case Against Memory Consolidation in REM Sleep. 23 (6).

Wilson, M. A., and McNaughton, B. L. (1994). Reactivation of hippocampal ensemble memories during sleep. 265, 676-679.

Chapter IX

The Recommendation Architecture Cognitive Model

In order to support the phenomenology of human memory and cognition, a cognitive architecture of the recommendation architecture type is required which incorporates all of the capabilities discussed so far. These capabilities include a range of information recording, information access, information management and support capabilities.

The purpose of this chapter is to describe such a cognitive architecture and to demonstrate how it has the general capability to carry out complex cognitive processes which it can learn with limited and plausible a priori (genetic) guidance. In particular, the system does not require a priori definition of cognitive features, categories or processes but can bootstrap appropriate behaviours in response to such features and categories.

Discussion of how this architecture supports the specific phenomena of memory, cognition and consciousness will be provided in chapters 10, 11, 12 and 13.

An Architecture to Support Complex Cognitive Processes

The architecture at the highest level is illustrated in figure 9.1, which follows figure 4.11 in chapter 4 (and the physiological mapping illustrated in figure 8.1 of chapter 8). Any sufficiently complex operational system will exhibit the primary separation between clustering and competition as illustrated in the figure, but at a more detailed level the architectural form will depend upon compromises needed to achieve a balance between effective operations and resource economy.

For example, in the figure there are three special purpose subsystems managing behaviour type priority, clustering resources, and rewards. These three subsystems exhibit the standard separation between clustering and competition, but these subsystems are also separated from the subsystems managing primary system behaviour.

The behaviour type management subsystem has special clustering and competition subsystems, and its behaviours target different areas within the primary clustering subsystem and components within competition. These different areas detect condition portfolios which have been optimized for different general types of externally directed behaviours and the components manage the selection of those behaviour types. Externally directed behaviour types could include food seeking, aggressive, avoidance and sexual behaviours. Genetic information can specify a limited number of such behaviour types by identification of competition components and of clustering resources at some levels of condition complexity for each type, combined with a bias on the outputs from these clustering resources in favour of targeting the corresponding behaviour types in competition. The optimizations of the portfolios are achieved by means of contradictory consequence feedback for the behaviour type as discussed in chapter 4.

The behaviour type management clustering subsystem detects conditions within current system information which indicate that its corresponding behaviour type is particularly appropriate under current circumstances. There is a high degree of genetic influence over the definition of these conditions. For example, conditions correlating with low blood sugar would indicate a higher priority on food seeking behaviour. The presence of such conditions is interpreted by the special competition subsystem as a recommendation to temporarily lower the thresholds of devices in the area in primary clustering optimized for that type and in the corresponding competition components. As a result, there is a higher probability that strong recommendations of the type will be generated under current sensory conditions and that those recommendations will be accepted.

The resource management subsystem again has special clustering and competition subsystems, but in this case the clustering subsystem detects conditions correlating with shortage of resources in different arrays in the primary clustering subsystem, and recommends resource assignment behaviours including device and connectivity assignment.

The reward management subsystem also has special clustering and competition subsystems. The clustering subsystem detects conditions which correlate with circumstances in which it is appropriate to increase or decrease the recommendation weights of behaviours recently implemented by the primary competition system. There will generally be a high degree of genetic programming in the definition of these conditions and their recommendation weights. Learning can result in conditions which frequently occur just before a reward acquiring recommendation strength in favour of the reward [Coward 2001].

Another compromise to note is that in figure 9.1 there are three levels of primary clustering separated by information gates. These levels correspond with sensory, associative and motor cortices. Attention management predominantly receives inputs from the first clustering level, but may receive some inputs from the second clustering level and so on. In general, relatively simple attention behaviours will be recommended by sensory cortex outputs, and only much more complex attention behaviours forming parts of higher cognitive processes will require outputs from the associative cortex. Conversely, relatively simple action behaviours may be recommended by the sensory cortex directly to behaviour type selection or even specific behaviour selection, rather than being recommended via the associative and motor cortices. For example, a behaviour such as ducking a flying object may depend only on sensory cortex conditions.

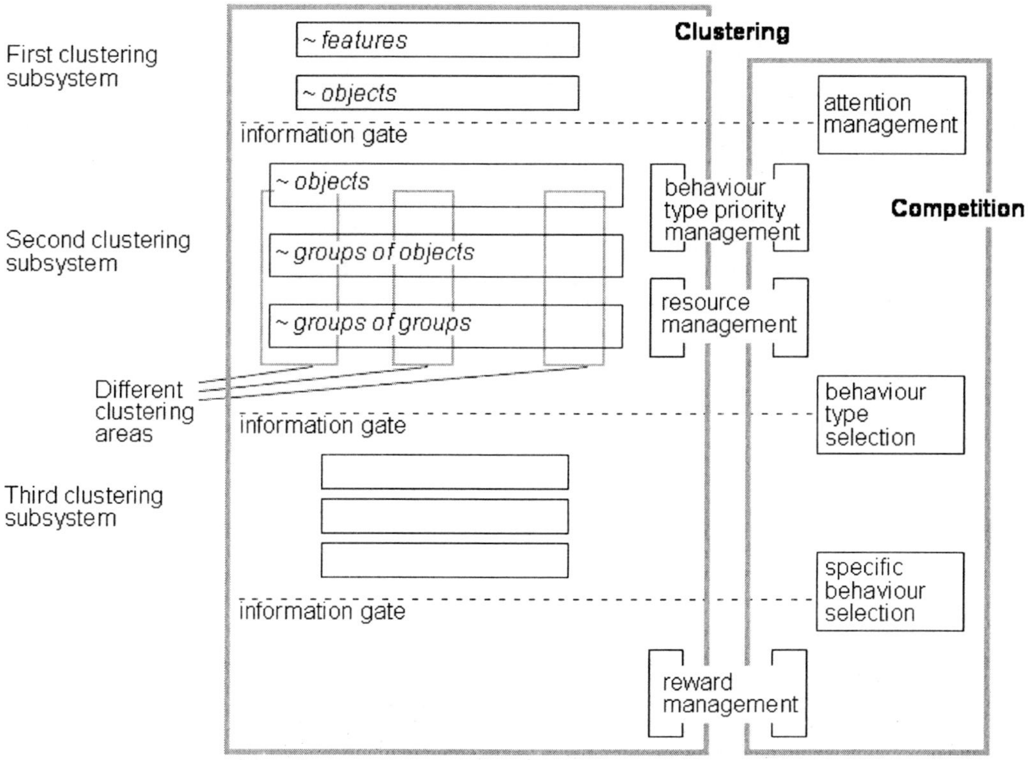

Figure 9.1. The highest level separations of a cognitive model with the recommendation architecture. There are a number of separations within the primary separation into clustering and competition. The first, second and third clustering and competition subsystems manage attention, selection of general behaviour type and selection of specific behaviours within an already selected general type. The clustering subsystems detect condition portfolios on a number of different ranges of complexity, some of which roughly correspond with the complexities of cognitive features, objects etc. However, the ranges in the different subsystems may overlap as shown. Within the second clustering subsystem there is a separation into areas corresponding with different general types of behaviour such as aggressive, food seeking, sexual etc. There are also three additional clustering, competition subsystems managing behaviour type priority, clustering resource assignment, and generation of reward signals.

Furthermore, as discussed in chapter 4, both the first and second clustering levels (i.e. the sensory and associative cortex) may detect conditions within similar ranges of information complexity. For example, conditions within ranges of complexity comparable with sensory objects may be detected in both subsystems. This will tend to occur if the optimal condition portfolios for recommending behaviours influenced by the first clustering subsystem are similar in complexity range but different in detail from those optimal for recommending behaviours influenced by the second clustering subsystem. An example could be if portfolios in the second clustering subsystem needed to provide more or different discrimination between sensory input states than portfolios in the same range of condition complexity in the first subsystem. Detection of different conditions within similar ranges of complexity implies that some inputs to the second level will be from intermediate ranges of complexity within the first clustering level.

Finally, different areas may detect conditions within different input domains (such as visual, auditory and propioceptic). Information from such different areas may be merged at some level of complexity to detect composite conditions, or condition portfolios detected within one sensory domain may indirectly activate portfolios defined within another domain.

A simplified clustering architecture processing visual, auditory and propioceptic sensory inputs and able to support complex cognitive processing using the available information handling processes is illustrated in figure 9.2. In this simplified architecture the separation between the first and second clustering levels is omitted, and it must be borne in mind that the relevant conditions may be detected in one or other of these subsystems depending on the effectiveness compromises made in the actual system. The modeling of cognitive processes is not qualitatively affected by this omission.

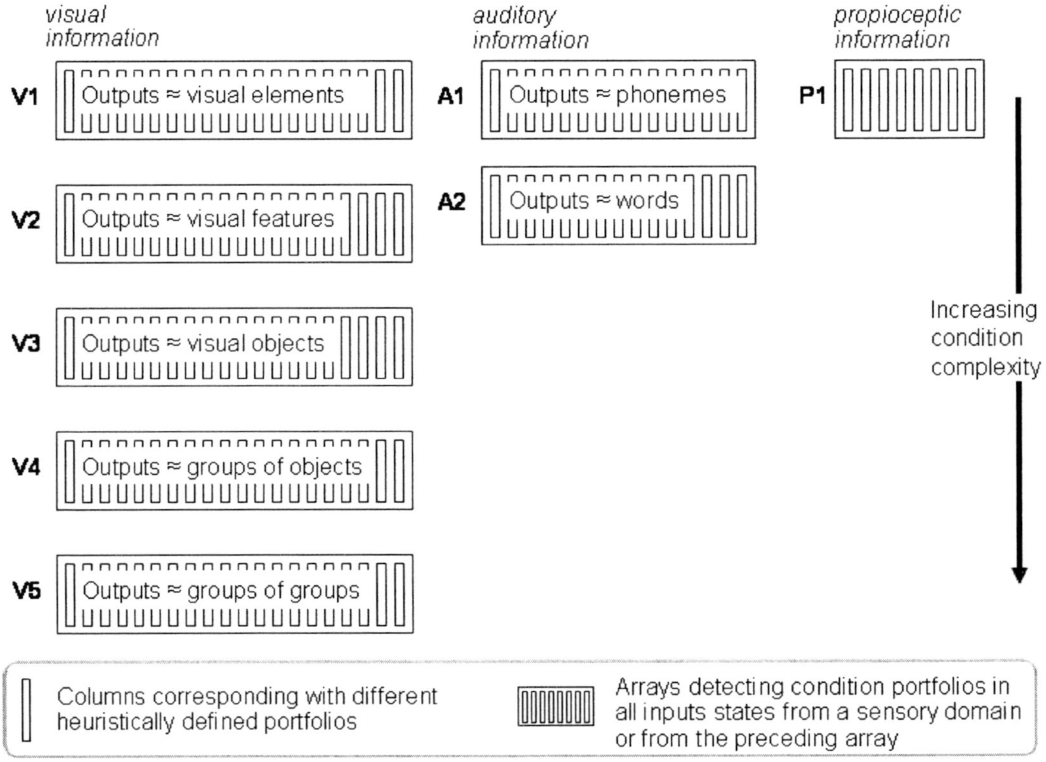

Figure 9.2. The clustering architecture used in the cognitive model. Parallel sequences of arrays of columns detect condition portfolios independently in different sensory input spaces: visual; auditory; and propioceptic. Different arrays in a sequence detect portfolios within different behaviourally useful ranges of condition complexity. In the figure, these ranges of complexity correspond roughly with the complexity of cognitively relevant circumstances such as the presence of features. Portfolios at that level will have the best discrimination between such features. In practice sequences will have many more arrays of columns at different complexities and portfolios on several levels of complexity will be relevant to discriminating between cognitive features. However, to simplify explanations the number of arrays has been reduced. Process descriptions are not significantly changed by this simplification.

The architecture has five clustering arrays detecting portfolios of conditions within visual inputs in different ranges of complexity, two arrays detecting portfolios within auditory inputs, and an array detecting portfolios within propioceptic inputs.

Each clustering array has a set of competition components which receive outputs from the level as illustrated in figure 9.3, and again only a subset of the competition components corresponding with an array are illustrated. A more extensive list of the types of behaviours which can be recommended by clustering outputs is provided in table 9.1. Behaviours include attention shifts, externally directed actions and internally directed information actions.

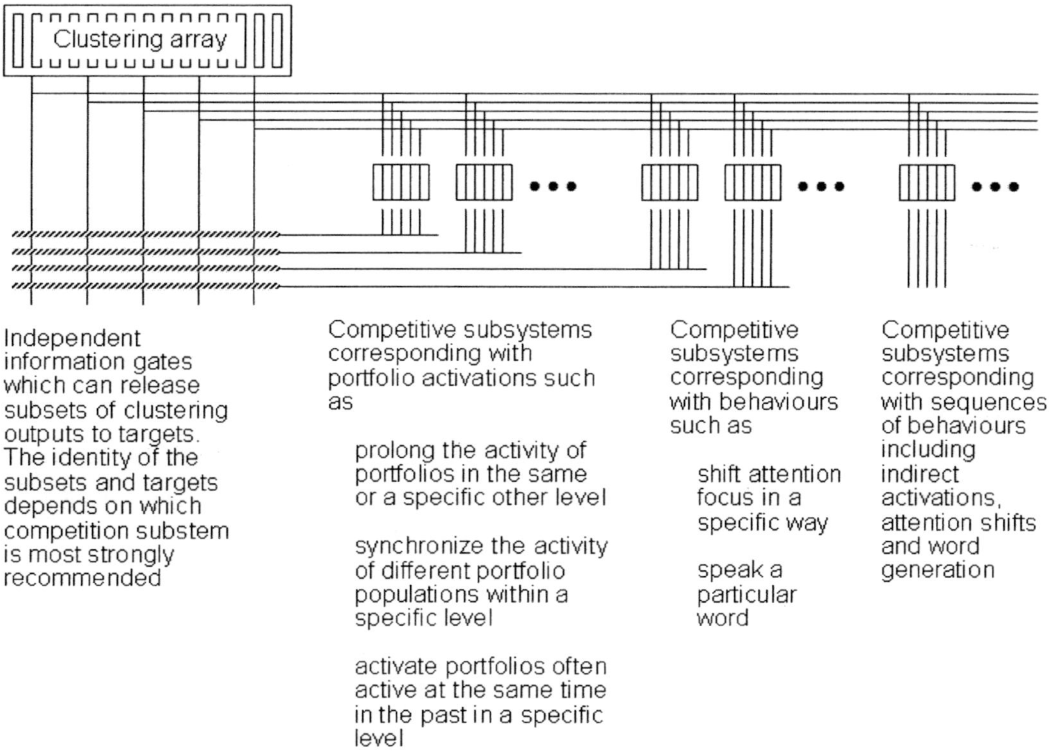

Figure 9.3. Structure of competition making use of clustering outputs from one array. Each output from clustering may have recommendation strength in favour of many different behaviours. Competition manages to selection of the most strongly recommended behaviours. In many cases a behaviour is the release of a specific subset of clustering outputs to another clustering array.

There are additional arrays of portfolios which discriminate between different types of information states in the primary sensory arrays. Examples are illustrated in figure 9.4. One type of array I1 detects conditions which discriminate between different degrees of overlap between independent populations of activated portfolios in array V3. Such populations could have been activated, for example, in response to different visual objects. In other words, some portfolios activated in response to the first object may also be activated in response to the second object on some levels of condition complexity. Array I1 detects the proportion of identical portfolios in the two populations.

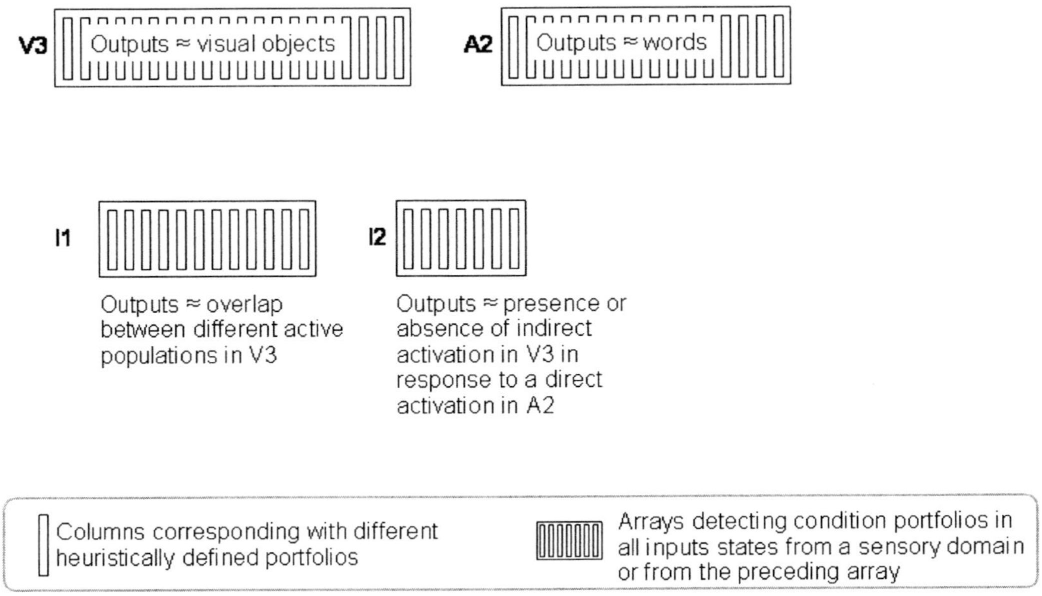

Figure 9.4. Additional arrays in clustering which detect information circumstances in clustering itself. One array detects the degree of overlap between two successive active portfolio populations in a particular other array. The other detects the presence or absence of an indirect activation in one array in response to a direct activation in another array.

A second type of array I2 detects conditions which correlate with the presence or absence of an indirect activation in array V3 in response to an activation in array A2. As will be discussed in later chapters, outputs from I2 would indicate whether or not a word which was heard corresponded with a visual circumstance learned earlier.

Such arrays use condition detection to identify information circumstances within the brain which may be relevant to determination of behaviour. The existence of such arrays and an initial connectivity bias may be specified by genetic information, but detailed definition of portfolios will be heuristic.

The clustering architecture illustrated in figure 9.2 is a simplification of a real brain architecture in a number of ways. One important simplification is that the typical complexity ranges of conditions detected within visual and auditory inputs have been labeled by the type of unambiguous cognitive condition which would be in a similar range of information complexity. Such unambiguous conditions are cognitive features, objects, groups of objects etc. However, this labeling is conceptual and only intended as an aid to explanation. In practice, many different arrays will detect conditions defined within many different ranges of complexity, and different ranges will correlate to different degrees with different types of unambiguous cognitive conditions. Conditions within a number of different ranges of complexity may be useful for guiding any one behaviour type. Furthermore, additional ranges of complexity may be required to support functions like consistency management as discussed in chapter 4. A somewhat more realistic clustering architecture is illustrated in figure 9.5. However, the simpler version is adequate in most cases to understand higher cognitive processes.

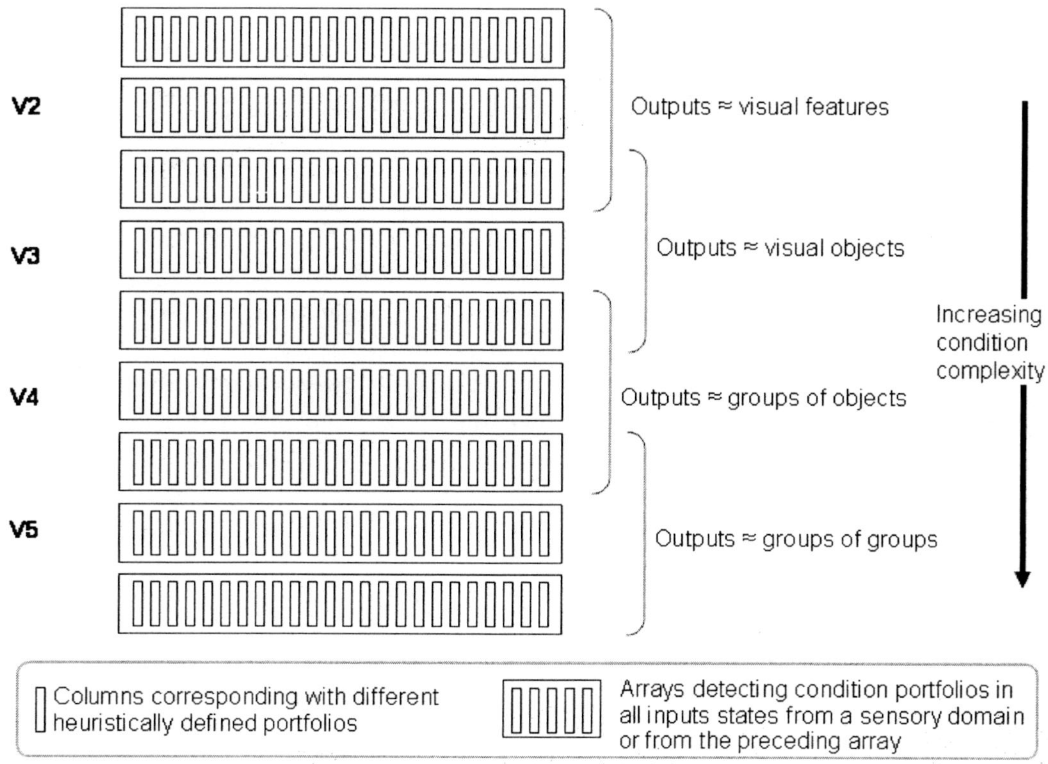

Figure 9.5. More realistic version of clustering arrays handling visual inputs. Several sequential arrays all contribute to discrimination between cognitive categories on different levels of complexity such as features, objects and groups of objects. Furthermore, the arrays providing discrimination between categories on one level may overlap with those providing discrimination on another level.

Another simplification is that the (to some degree genetically specified) preprocessing of raw sensory inputs into conditions with better but still limited discrimination between different behaviourally relevant circumstances discussed in chapter 4 is omitted.

Information Handling Processes

In the recommendation architecture, information ultimately derived from the state of the environment or from the internal state of the brain at different points in time is recorded and accessed to determine appropriate behaviour. Information can be recorded by two mechanisms. One is permanent recording of conditions in the clustering subsystems. The other is change to recommendation weights in the competition subsystems. Information is accessed in clustering by activation of conditions, and in competition by comparison of recommendation weights.

As discussed in chapter 4, past temporally correlated activity is the only source of information which can be used to guide condition definition and activation. A priori genetic information can influence initial resource allocation but cannot provide specific guidance on, for example, cognitive categories. Positive and negative consequence feedback can influence recommendation weights but cannot directly influence conditions. Supervision by some external teacher can influence the environment available to the learning system including providing models for imitation, and can provide consequence feedback but cannot directly act on the internal processes of the learning brain.

Indirect Activation of Information within Clustering

Within clustering, the primary units of information are columns of devices which record portfolios of similar conditions. As discussed in chapter 4, portfolios do not correspond exactly with cognitive categories of any type, although some portfolios can discriminate between certain types of category better than others. An activated portfolio influences behaviour through its extensive range of recommendation strengths in favour of different behaviours.

A portfolio is directly activated by the presence of any of its recorded conditions within current sensory inputs. In addition, a portfolio can be indirectly activated if there is some indication that its recommendations have a reasonable chance of being relevant under current circumstances. Such indirect activations expand the range of information available to determine the most appropriate behaviour. A critical question is what kind of indications could be used.

Temporal correlations in past activity are the only available basis for such indications. A portfolio which does not have any of its conditions present in current sensory inputs can nevertheless be activated if it has been active in the past at times which correlate in some way with the past activity of portfolios which are currently active. The implication is that correlated past activity indicates an increased chance that the indirectly activated portfolios could contribute to determining the most appropriate current behaviour. Once a portfolio has been indirectly activated, it could of course influence the activation of yet other portfolios by the same type of mechanism.

However, indiscriminate indirect activations on the basis of any past correlations would lead to excessive and largely irrelevant active populations. Indirect activations are therefore behaviours which must compete with other behaviours in competition to achieve implementation. A portfolio can have recommendation strengths in favour of indirect activation of other portfolios in addition to its recommendation strengths in favour of externally directed behaviours.

There are a number of different types of temporal correlation which could be used. These different types are indicated in table 9.1. One approach is activation on the basis of recent past activity or condition recording (table 9.1A). In this case the currently active portfolios must have adequate recommendation strength in favour of such reactivations, which are not portfolio specific. A second approach is activation on the basis of frequent past activity of an inactive portfolio which was simultaneous with, just before or just after the past activity of currently active portfolios (table 9.1B). In this case the currently active portfolios must have adequate recommendation strength in favour of activations of specific portfolios. A third approach is activation on the basis of past temporally correlated condition recording (table

9.1C). Again the currently active portfolios must have adequate recommendation strength in favour of reactivations of specific portfolios.

A. Activation on the Basis of Recent Activity
1. Reactivate portfolios which have recently been active
2. Reactivate portfolios which recently recorded conditions

B. Activation on the Basis of Frequent Past Activity
1. Activate portfolios which have often been active at the same time in the past as the currently active portfolio
2. Activate portfolios which have often been active just after past activity of the currently active portfolio
3. Activate portfolios which have often been active just before past activity of the currently active portfolio

C. Activation on the Basis of Frequent Past Condition Recording
1. Activate portfolios containing conditions recorded at the same time in the past as some conditions in the currently active portfolio
2. Activate portfolios containing conditions recorded just after some conditions in the currently active portfolio
3. Activate portfolios containing conditions recorded just before some conditions in the currently active portfolio

D. Managing Information Availability
1. Prolong the activity of currently active portfolios
2. Shorten the activity of recently activated portfolios
3. Synchronize the activity of several different populations of currently active portfolios in the same array
4. Release outputs of one array to the next array

E. Attention management
1. Perform a general sequence of attention and internal activation behaviours
2. Perform a specific sequences of attention and internal activation behaviours
3. Perform an individual attention behaviour

F. Speech Behaviour Management
1. Generate a sound
2. Speak a word
3. Say a phrase
4. Express meaning verbally by a sentence

G. Motor Behaviour Management
1. Perform a general sequence of motor behaviours
2. Perform a specific sequence of motor behaviours
3. Perform an individual motor behaviour

A key issue is how the appropriate recommendation weights can be acquired. As discussed in chapter 4, any temporally correlated activity of two portfolios creates strong recommendation weights in favour of one portfolio activating the other at a later point in time. However, these weights decay with time. If before the weight has decayed an indirect activation actually occurs, and is followed by positive consequence feedback, the decay is stopped and may be reversed somewhat. Future indirect activations followed by positive consequence feedback can increase the weight.

General Information Activation Management within Clustering

There are a number of behaviours listed in table 9.1D which influence the activation of portfolios in a more general sense. These behaviours include prolonging the activity of currently active portfolios, shortening the activity of recently activated portfolios, synchronization of the outputs of different active portfolio populations, and releasing the outputs from one array to the next array.

In order to determine the most appropriate behaviour in response to, for example, a group of visual objects, conditions incorporating information from all the objects in the group must be detected. Portfolios of relatively simple conditions can be detected within one object by focusing the attention domain on that object as discussed in chapter 5. If conditions within a group of objects are needed, these portfolios must be maintained active until portfolios in response to all the objects in the group are active. Such activity prolongation is a behaviour which must be recommended by adequate portfolio recommendation strength.

Note that relatively simple behaviours could be determined by detecting conditions in array V3 of figure 9.2 where the group of objects is processed as a single attention object. More sophisticated behaviours would require processing each object separately to activate conditions in array V3 before integrating the information into group conditions detected in V4.

In order to urgently determine appropriate behaviour in response to the sudden appearance of an object in the visual field, it may be important that the activity of portfolios activated prior to the appearance of the new object be removed as soon as possible. Recommendation strength in favour of shortening the activity period of recently activated portfolios makes such accelerated removal possible.

Separations can be maintained between populations of activated portfolios which are activated in the same array in response to different input domains corresponding with, for example, different visual objects. These separations are maintained by using different frequency modulation phases for device outputs within one clustering array as discussed in chapter 5.

Once separate portfolio populations have been activated in array V3 in response to all the objects in a group, portfolios containing information derived from all the objects can be activated by synchronization of the phases of the outputs of the different populations in array V3 and releasing those outputs to array V4 detecting higher complexity conditions. Such synchronization and release is a behaviour which must occur at the appropriate time. For example, if an externally directed behaviour must be generated in response to a particular group, conditions must be detected within all the members of the group before synchronization and release to detect conditions within the group as a whole. Synchronization and release are therefore behaviours which must be recommended by an adequate population of currently active portfolios to be implemented.

Resource Management within Clustering

Because portfolios are defined heuristically, the condition recording resources which will be required by different clustering modules cannot be exactly known a priori. It is therefore necessary to assign such resources to modules as required. Such resource assignment includes additional columns for arrays, devices for columns and additional connectivity (including provisional conditions) for devices. It is also important to be able to identify column and device resources which have already been assigned by tagging them in some way, to prevent double assignment. As discussed in chapter 4, the simplest way to achieve this management is to create a map of all clustering column and device resources, with an active boundary which gradually moves across the map. Resources corresponding with map positions on one side of the boundary are already assigned, those on the other side are still available, and resources corresponding with the current position of the boundary are assigned to modules requiring them.

The management subsystem must then detect modules which require resources. A possible indicator would be failure to record conditions when required to generate an adequate population of active portfolios. Another possible indicator could be degree of recent condition recording.

Once resources have been assigned, they require provision of appropriate connectivity. There are several types of connectivity which are necessary. One is the connectivity to define provisional conditions out of which permanent conditions can be defined later. This connectivity must be derived from the input space of the module to which the resources have been assigned. A second type of connectivity is to support change management, and must be derived from the source modules providing that connectivity to the assigned module. A third type is connectivity to support indirect activations on the basis of temporally correlated past activity. Because the map indicates the temporal order in which resources have been assigned, it is an approximate, implicit map of the time sequence of condition recording. In particular, the boundary indicates the location of resources most likely to record conditions at the same time in the immediate future, and can be used to guide the assignment of provisional indirect activation connectivity.

As discussed in chapter 4, there will always be some random element to provisional condition definition. In order to make the probability that conditions recorded in the future will be as behaviourally useful as possible to the module in which they are recorded, provisional conditions are defined within the input space to their assigned module with a bias in favour of inputs often active when the module was itself active in the past. To identify such inputs, the system performs a partial, input driven rerun of past experience, with a bias in favour of the more recent past. Inputs often active when the target module is active can then be favoured for the definition of provisional conditions.

As discussed in chapter 8, the hippocampus system appears to match the requirements for the resource management system, and REM sleep matches the partial rerun of recent experience required to support effective definition of provisional conditions.

The Behavioural Component Hierarchy

Genetic information must specify the spectrum of available behaviours at the level of "atomic" or indivisible behaviours. Higher level behaviours made up of sequences and combinations of atomic behaviours can be specified genetically or defined heuristically.

For example, atomic behaviours can be externally directed such as generating a specific set of muscle movements. An example might be a combination of muscle movements which contributes to the production of sounds relevant to human speech. Atomic behaviours can be for attention management purposes, such as the set of muscle movements and internal activation actions needed to shift the attention domain to the next object on the left or on the right. Atomic behaviours can also be to support indirect activations or general information management behaviours.

Organization of Behaviourai Components

Every behaviour has a corresponding component in competition, and these components are organized in a hierarchy. One way in which such a hierarchy could be organized is illustrated in figures 9.6 through 9.9. In figure 9.6 there are high level components corresponding with attention, indirect activation, information availability, speech and motor behaviours. Each of these high level components contains more detailed components correspond with specific behavioural types within the general types. For example, the indirect activation general component contains more detailed components corresponding with indirect activation on the basis of past temporally correlated portfolio activity, past temporally correlated condition recording, and simple recent activity. Each of these more detailed components contain subcomponents corresponding with yet more specific behaviour types as illustrated in figure 9.7. For example, the component corresponding with indirect activation on the basis of past temporally correlated activity contains subcomponents corresponding with past activity just before, at the same time as, and just after past activity of a currently active portfolio. Another level of component detail is illustrated in figure 9.8, in which, for example, activation on the basis of past activity at the same time is subdivided by the target array in which such past simultaneous activity will result in activation. A component in figure 9.8 corresponding with activation in one array is further subdivided into components corresponding with activation of individual portfolios in the target array as illustrated in figure 9.9.

General Behavioural Selection Process

The overall organization of this component hierarchy is specified by genetic information. The default (genetically defined) competition process is that outputs from clustering initially were provided to all the highest level components illustrated in figure 9.6. A competition between these components resulted in the selection and activation of one component, for example the indirect activation component. This activation excited all the subcomponents of that component illustrated in figure 9.6, corresponding with indirect activation on the basis of past activity, with indirect activation on the basis of past condition recording, and with indirect activation on the basis of recent activity etc. Clustering inputs available to these subcomponents result in one of them being activated, for example the component corresponding with indirect activation on the basis of past activity. Activation of this component results in excitation of all its subcomponents illustrated in figure 9.7, corresponding with activation on the basis of past activity at same time as currently active portfolios, with activation on the basis of past activity just after activity of currently active portfolios and with activation on the basis of past activity just before activity of currently active portfolios.

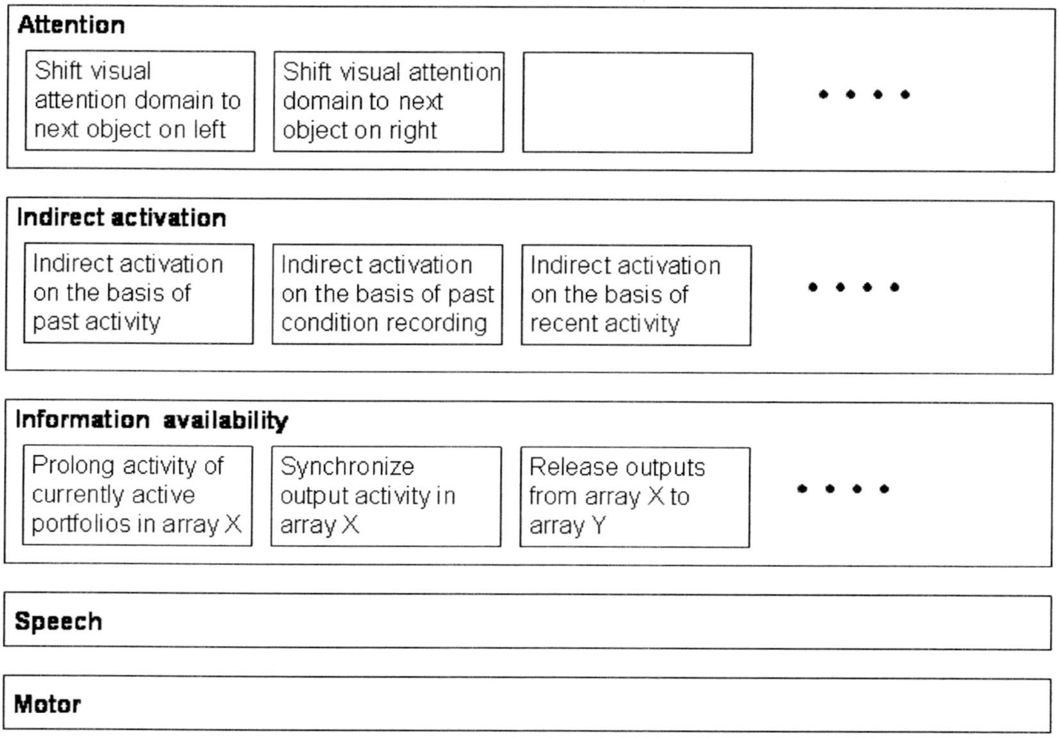

Figure 9.6. Some high level competition components corresponding with general types of behaviour. Illustrated types are attention, indirect activation, information availability, speech and motor behaviours. Each high level component can be activated by appropriate inputs from clustering, and there is inhibitive connectivity between the components so that only one behaviour type or a small consistent set is selected. If activated, a component excites a set of more detailed components corresponding with more detailed behavioural types of its general type.

Again, inputs directly to these components from clustering results in activation of just one of the subcomponents, for example the component corresponding with activation on the basis of past activity at same time as currently active portfolios. This component also has subcomponents as illustrated in figure 9.8 corresponding with indirect activation of portfolios in different arrays. One of these subcomponents is selected and excited.

Activation of this component excites all of its subcomponents illustrated in figure 9.9. These components correspond with excitation of individual portfolios or groups of portfolios in the target array. They may have inputs from a large number of different source portfolios, and if enough of these portfolios are active and if the component has been excited by the behaviour selection process described, the target portfolios will be indirectly activated.

In this behaviour selection process, a detailed behaviour can only be selected if its behaviour type has previously been selected. The competition hierarchy may only allow release of clustering outputs to detailed components if their behaviour type has been selected.

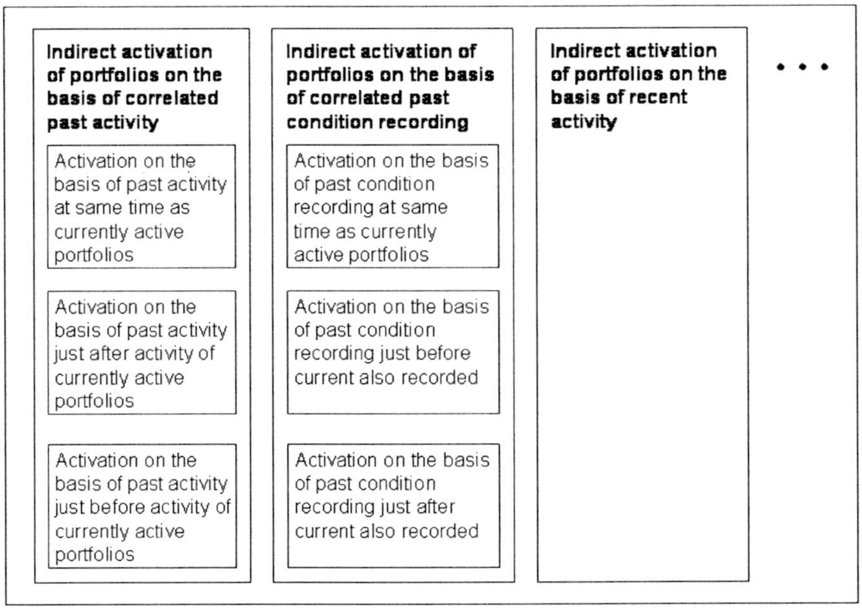

Figure 9.7. More detail of the indirect activation component. This component is one of the general behavioural components of competition illustrated in figure 9.6. The subcomponents of this component and their subcomponents are illustrated. If the indirect activation component as a whole is activated it excites its more detailed components. Each of these sub-components also receives excitatory inputs from clustering and inhibitory inputs from the other sub-components. The sub-component with the strongest net input strength is activated. If activated, the selected sub-component excites its sub-sub-components.

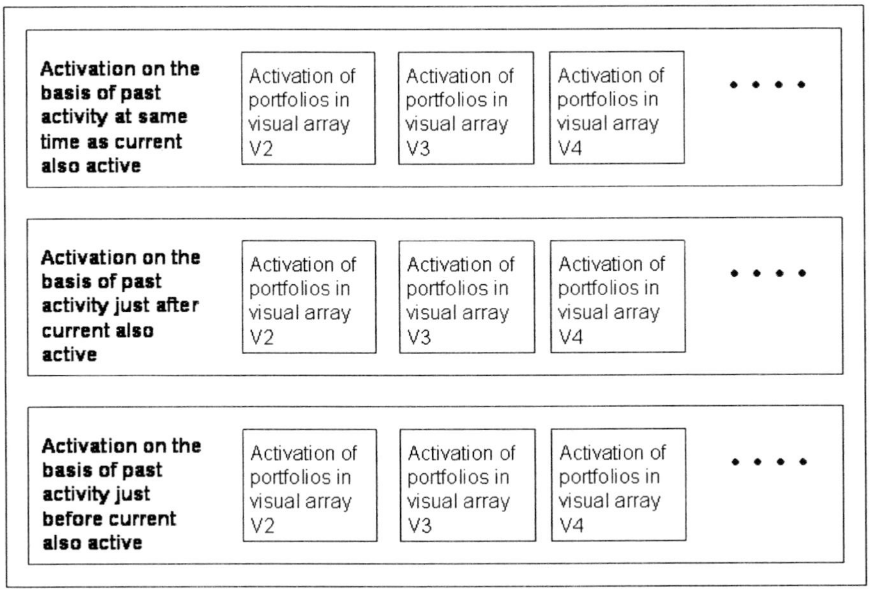

Figure 9.8. More detail of the component corresponding with indirect activation on the basis of correlated past activity. This component is one of the subcomponents of the indirect activation component illustrated in figure 9.7. If it is activated it excites its more detailed components. Each of these sub-components also receives excitatory inputs from clustering and inhibitory inputs from the other sub-components. The sub-component with the strongest net input strength is activated. If activated, the sub-component excites its sub-sub-components illustrated in figure 9.8.

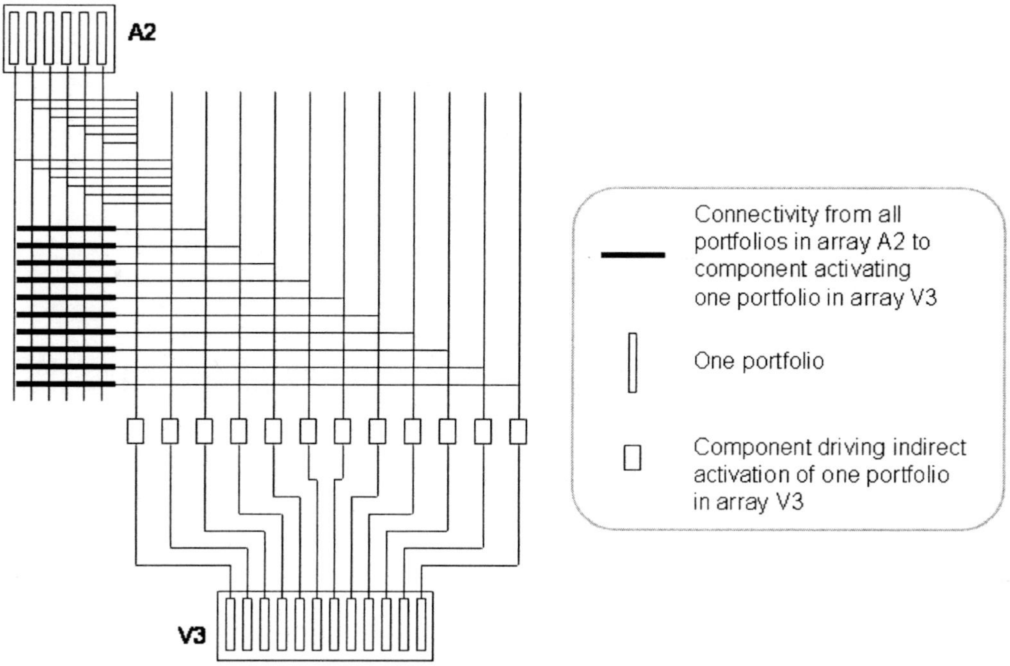

Figure 9.9. The most detailed level of the component corresponding with indirect activation of portfolios in array V3 on the basis of simultaneous past activity with other currently active portfolios. In the diagram these other portfolios are located in array A2, but that may not be the only source. Detailed competition components correspond with the activation of each individual portfolio or group of portfolios in array V3. If the higher level component corresponding with the behaviour type has been activated and if there are enough active inputs to a detailed component corresponding with a portfolio in V3, the portfolio in V3 will be indirectly activated.

Components Managing Specific Behavioural Sequences

This default selection process can be bypassed under some circumstances by special purpose behaviour sequence components. Such a component is illustrated in figure 9.10. This component receives inputs from clustering and can be activated by adequate recommendation strength. If activated, it first activates a more specific target component at a fairly detailed level. This component, along with many others, will be receiving inputs from clustering. However, the activation lowers the thresholds of devices within just the target component, and the effective recommendation strengths of its inputs will therefore be increased. A behaviour of the target component type will therefore tend to be selected, although the specific behaviour will be determined by its current clustering inputs. When the special purpose component receives an indication that a behaviour of that target component type has been implemented, it activates a second target component and so on. The activation of the special purpose component thus encourages a standard series of behaviour types, although the specific behaviours within the types are determined by (perhaps different) outputs from clustering.

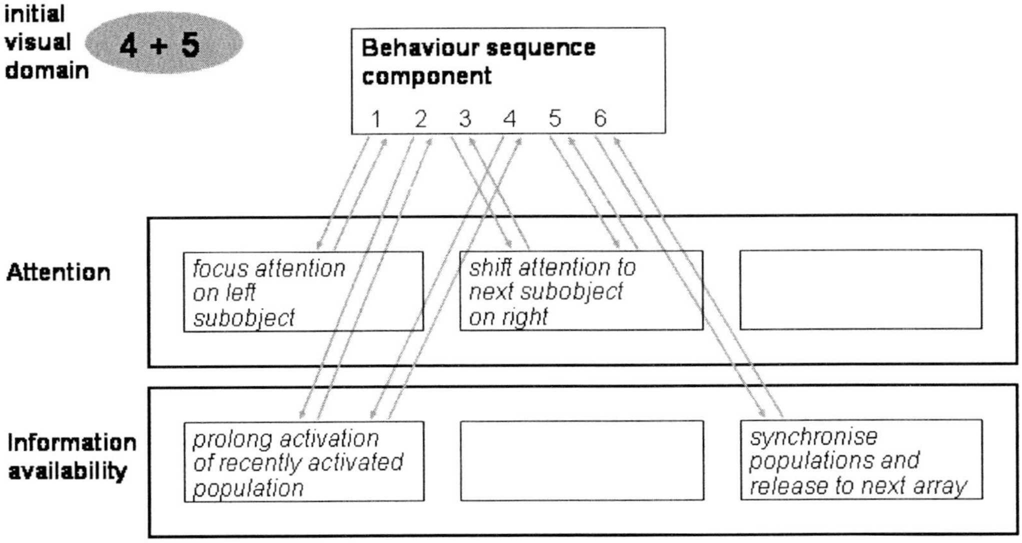

Figure 9.10. Example of a behaviour sequence component. Portfolios activated when the attention domain covers the visual object "4 + 5" have recommendation strength in favour of activating the sequence component. This component activates a sequence of more specific behavioural components, each component being activated when a behaviour corresponding with the preceding component has been accepted. To be accepted, a more specific component must have recommendation strength from portfolios in addition to those activated by the original object.

As will be discussed in much more detail in later chapters, an example of a behaviour sequence might be generating a response to the equation 4 + 5 as illustrated in figure 9.10. Portfolios activated initially in response to (4 + 5) viewed as a single visual object activate the behaviour sequence component. This component activates a series of six components in sequence: firstly the behaviour of focusing the attention domain on the left sub-object, in this case (4); secondly of prolonging the activity of the resultant activated portfolio population; thirdly of shifting the attention domain to the next object on the right, in the example to (+); fourthly of again prolonging the resultant population; fifthly of again shifting the attention domain to the next object on the right, in the example to (5); sixthly of synchronizing the outputs of the three populations and releasing the outputs to the next array.

The effect of this sequence is that the perception of the equation activates three separate portfolio populations in response to the three arithmetic objects making up the equation, then synchronizes the three populations and detects portfolios incorporating information from the three objects. Such portfolios would have recommendation strength in favour of saying "nine" as discussed in detail in chapter 12.

Such behaviour sequence components could be defined by genetic information or could be defined heuristically. Heuristic definition could occur as a result of a genetically defined tendency to create a component to correspond with frequently occurring sequences, and preserving the component if its sequence was often followed by positive consequences. As discussed below, imitation is one way in which components could be defined to implement behaviourally useful sequences.

The behaviour sequence component example was for a cognitive sequence, but similar components could be defined genetically or heuristically for motor action sequences, for example sequences of muscle movements involved in playing the piano.

Acquisition of Recommendation Strengths

Twenty-two types of competition component corresponding with twenty-two types of behaviour which could be recommended by the outputs from a clustering array are listed in table 9.1. Any one portfolio could have recommendation strengths in favour of many or all of the different behavioural types, and many different more specific behaviours within a type. The selected behaviour depends upon the total relative recommendation strengths across the currently active population, including the modulation of recommendation strengths resulting from one set of portfolios activating a component receiving inputs from a different set.

Learning must result in each portfolio acquiring an appropriate combination of recommendation strengths. This learning must be effective using only the information which is actually available to the brain. Genetic information can influence the initial assignment of resources and connectivity in and between clustering and competition, and some of the general circumstances which will result in positive and negative consequences, but cannot specify cognitive features and categories in detail. A specific example of the influence and limits of genetic information discussed later in this chapter is imitation. A genetic tendency to generate positive consequences when portfolios detected within perceptions of an external source overlap on some levels of complexity with portfolios detected shortly afterwards within perceptions of self will create a tendency to imitate.

A further genetically defined tendency results in the interpretation of other portfolios as recommending positive consequence feedback. These portfolios have genetically defined bias on connectivity which results in their correlation with, for example, perceptions of smiling or certain voice tones. Such a genetically defined tendency means that if the initial genetically defined recommendation strengths in support of the imitation capability are limited and only approximate, they nevertheless provide a good starting point which can be refined by feedback from an external teacher.

Dynamic Portfolio Populations in the Cognitive Architecture

In order to illustrate the dynamics of portfolio activation during cognition in the recommendation architecture model, a very simplified description of how reading could be performed in the cognitive architecture illustrated in figures 9.2 and 9.3 will be outlined. Discussion of how syntax information could be reflected in portfolios will be excluded. The description will be of the processes for responding to a specific sentence "The neighbour's black dog chased a little cat up the tree".

At the start of the reading process, there is a behaviour sequence component which is activated by the portfolios activated in response to conditions detected within the perception of a group of words with a capitalized word on the left and a period on the right (although other cues could be used). This sequence of behaviours moves the attention domain from word to word as illustrated in figure 9.11, and includes specific information availability and indirect activation behaviours at various points. The behavioural sequence component excites a sequence of more detailed behaviour type components. The next component in the sequence is excited once a behaviour of the preceding type has been performed. The detailed nature of the behaviour at each point is determined by the outputs from clustering at that point. The first ten steps in its sequence are illustrated in figure 9.12. The steps in this behaviour sequence (following the step numbering in figure 9.12) are as follows:

1. The attention domain is focused on the first word on the left in figure 9.11, and portfolios are detected in clustering level V2 of figure 9.2 within the direct sensory inputs which result. Such portfolios would correlate with different shapes characteristic of different words.
2. At this point the sequence component has activated the indirect activation subcomponent corresponding with activation of portfolios in level V3 on the basis of past simultaneous activity. The portfolios activated in V2 have recommendation strengths of this type, in favour of portfolios which have often been active in the past at the same time as the portfolios within the visual perception of the word. These recommendations are therefore accepted. The portfolios activated as a result in V3 will tend to be those which have often been active in response to direct perception of the object corresponding with the word, since the words and the corresponding objects were often present in the past at the same time during a learning period. A pseudovisual activation is therefore generated which approximates to the activation which would have been generated by direct perception of the object, but excluding portfolios in V1 and V2 close to sensory inputs.
3. The activation of the component recommending activity prolongation in V3 of the sequence component means that self prolongation recommendations by the pseudovisual activation are accepted.
4. The activation of the shift attention component means that the attention domain is then moved to the next word.
5. Secondary portfolios in level V3 are activated which have often been active in the past at the same time as the portfolios within the visual perception of the second word. These secondary portfolios are activated at a different frequency modulation phase from the secondary portfolios corresponding with the first word.
6. The activity of the populations in V3 is prolonged by acceptance of the self prolongation recommendations of those populations.
7. At this point the sequence component activates two subcomponents. One is the component managing attention shifts to the next object on the right. The other is the component managing synchronization of outputs from V3 and release to V4. Outputs from V3 are available to both of these components. If the words so far make up a phrase, the synchronization recommendation will be stronger and will be accepted. Otherwise attention will shift to the next word.
8. The process of indirect activation is repeated for the third word.

9. The recommendation to prolong activation is again accepted.
10. Again at this point two subcomponents are activated, one to synchronize activity in V3 and release to V4, the other to move attention to the next word without release.

Figure 9.11. Positions of the attention domain as the sentence "The neigbour's black dog chased a little cat up the tree" is processed.

Once the active populations in V3 correspond with the words of a phrase, the effective recommendation strengths to synchronize the population outputs and release to V4 will be stronger than the recommendation strengths to shift to the next word without synchronization. At this point an active portfolio population will be generated in V4 corresponding with the information content of the phrase. The activity of this population will be prolonged as a result of the self prolongation recommendation strengths of the population and the activation of the corresponding subcomponent by the sequence component. The activity in V3 is extinguished and the process repeats for the next phrase. At the end of this repeat, two populations will be active in V4 corresponding with the first two phrases.

Phrase populations are added in V4 until the effective recommendation weight of the overall activity in favour of synchronization of outputs and release to V5 exceeds that in favour of moving to the next phrase. Portfolios activated in response to visual inputs derived from a period will contribute to synchronization recommendation strengths. The population activated in V5 will approximate to the population which would be activated if the objects corresponding with the words of the sentence were visually experienced, again without the portfolios close to actual sensory input. The population in level V5 could generate, for example, speech recommendations to say “The neighbour should not have allowed his dog to be loose”.

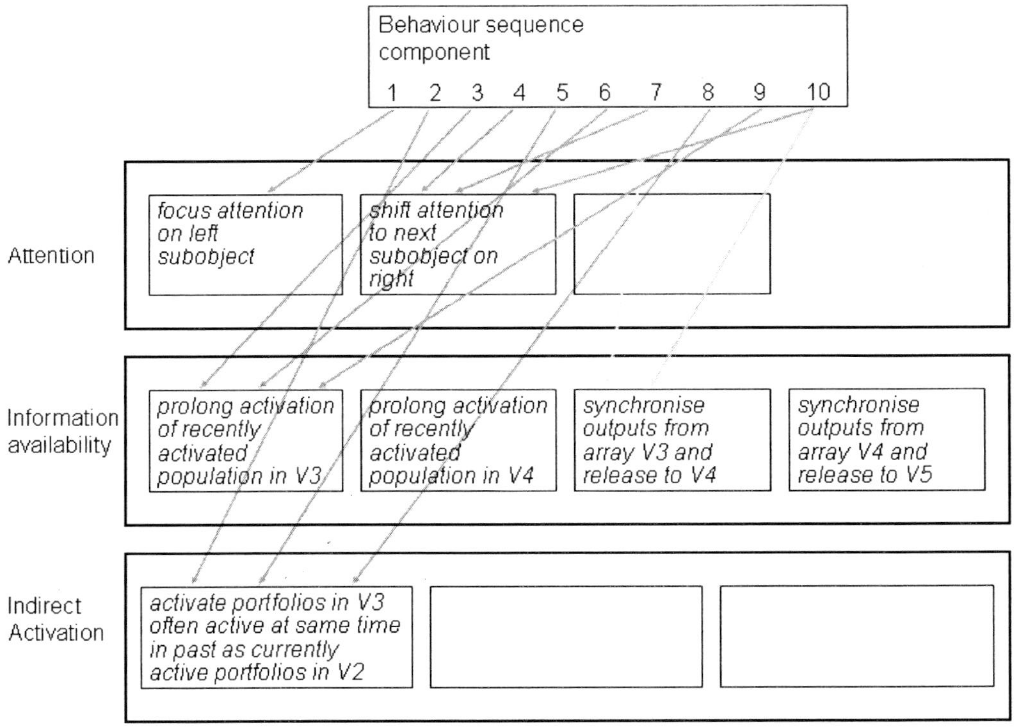

Figure 9.12. Behaviour sequence component for generating behaviours appropriate for processing a sentence. Only the first ten steps in the sequence are illustrated. The behaviour sequence component is activated by portfolios such as those which correlate with the presence of a period followed by a capital letter. The behaviour sequence component drives the attention domain and an associated group of internal activation behaviours.

In the case of words for features or objects like “dog” or “black”, this process activates a series of populations of portfolios in level V3 (or perhaps V2) which are subsets of the portfolios which could be activated in that level by direct visual perception of the actual visual object or feature. Some other words, like “chased”, “up”, or “little” will activate populations of portfolios recorded in a number of circumstances where the corresponding relationships between objects were present.

For example, consider how this could happen for the word “chase”. Over a period of experience, the word might be heard at the same time as a range of visual perceptions of different subjects chasing different objects. Examples might include a cat chasing a squirrel, a dog chasing a ball, a lion chasing an antelope, or a policeman chasing a suspect. In each of these experiences a population of portfolios is activated in level V3 in response to visual inputs when attention is focused on the subject, a separate population in level V3 in response to the object, and a population in level V4 with conditions combining information from subject and object. Because of the cognitive ambiguity of individual portfolios, there will be some overlaps between the active populations in level V3 in response to different subjects and between those in response to different objects. For example, portfolios which contain conditions characteristic of body positions chasing and chased will occur frequently in level V3 in response to perceptions of pursuits. Some portfolios in level V4 will contain information derived from these subject and object chase-biased portfolios, and these portfolios will also be activated in response to many chase perceptions. The portfolios most often active at the same time as the word "chase" will therefore be these more frequently occurring portfolios in level V3 (subject), level V3 (object) and level V4. The word "chase" will therefore tend to activate these portfolios, and such an activation will be experienced as if a perception of a fuzzy "subject" chasing a fuzzy "object" was present, but without the corresponding more specific sensory conditions in levels V1 and V2 or more specific object conditions in level V3.

If the subject/object combination in a chase relationship being read has been actually perceived directly at an earlier point in time, the activated population will include a substantial proportion of the portfolios active at that time in levels V3 and V4 and little condition recording will be required. If the subject/object combination has not been perceived in a chase relationship before, the level V4 active population will initially be weak, being only the relatively small group of frequently active portfolios derived from chase. Some almost active portfolios (i.e. internally active but not producing output) will be triggered to record conditions until they activate. The expanded portfolios would be the basis for later recall memory of the reading as discussed below, and the expansion would also tend to add them to the general "chase" portfolio population.

Other words like "the" or "a" influence the identity of the indirectly activated subset corresponding with the word they qualify, so the combination of the two words may be the primary population which generates a secondary population in level V3.

The activated population in level V4 in response to a phrase will then be a subset of the same population within which activated subsets would be generated in response to direct perception of the objects and relationships within the phrase. Similarly the activation in level V5 in response to a sentence will be a subset of the same population within which activated subsets would be generated in response to direct perception of the objects and relationships within the sentence.

As in the case of direct visual perception, most of the activated conditions will have been recorded in the past, and only novelty with respect to past experience will result in recording of additional conditions. For example, suppose that the dog had been described as pink. Some conditions would be activated in levels V2 and V3 corresponding with dog and with pink with little recording required, but in level V3 and perhaps V4 there will be recording of conditions incorporating information derived from both dog and pink as this combination had not been encountered before. These conditions are the basis for imagining the experience of pink dogs.

The degree of recording is also influenced by the activity of other portfolio populations. For example, the presence of neurochemical concentrations corresponding with a strong emotion will activate portfolios in clustering. These portfolios may have recommendation strengths in favour of recording a higher level of conditions in portfolios which might contribute to determination of the behaviour appropriate to the presence of the emotion.

The absence of meaning in a sequence of words, indicated by strong activation in level V3 but minimal activity in level V4 may result in little condition recording in levels V4 and V5.

Note that at any point in the reading process the current activated population is recommending many other behaviours in addition to the reading process. For example, the population of portfolios activated in response to "neighbour's" would also be recommending activation of portfolios recently active in the past when they were active. If such a recommendation were accepted in preference to the reading related recommendation, portfolios in which conditions were recorded during a recent discussion of the fence at the boundary with the neighbour might be activated, generating a recommendation to think about fixing the fence rather than reading. The recommendation accepted will therefore be determined by a complex relationship with recently active portfolio populations.

Genetic Information

As briefly discussed earlier, a requirement for an effective cognitive architecture is to make use of biologically plausible a priori (i.e. genetic) information, without requiring implausibly detailed such information. In the recommendation architecture, a priori information can bias the initial resource and connectivity assignments in support of general behaviour types without specifying these assignments in detail.

Thus natural selection could ensure that substantial and appropriately connected resources are assigned to a behaviour type with high survival value. Note that such assignments do not require an abstract "knowledge" of the behaviour type. The only requirement is that assignment of such resources has survival value.

Appropriate connectivity means three things. Firstly, the inputs to the condition recording resources are biased in favour of the type of inputs more likely to be relevant for the behaviour type, and provisional conditions are defined on levels of complexity which are similarly more likely to be relevant. Secondly, combinations of primitive actions likely to be relevant to the behaviour type are define as behaviours. Similar such behaviours can be grouped into types. Thirdly, outputs from the clustering resources are biased in favour of the competition modules managing the behaviour types, and significant provisional recommendation weights are assigned to such outputs.

Experience will change both input and output connectivity, but the initial biases and resources ensure that the speed of convergence on appropriate behaviour will be much faster.

For example, consider the resources which could be assigned to support imitation behaviours. Such resources would be biased in favour of conditions likely to be present in perceptions of the arms, hands, or fingers of other people, and have outputs biased in favour of competitive modules controlling arm, hand and finger movements. Note that this is only a bias: many conditions eventually recorded in the resources may be irrelevant, and conditions recorded elsewhere may be relevant. However, given the biases, there will be a tendency towards imitation behaviour, and bootstrapping of appropriate behaviours will be more rapid and efficient as discussed in the next two sections.

Another possible example of resource biasing is the existence of deficits in which subjects can draw inanimate objects with reasonable accuracy, but when asked to draw animate objects are only able to produce a very generic sketch [Satori, Job, & Coltheart 1993]. This example is discussed more fully in section 4.4. A further possible example of resource biasing is regional specialization for major types of behaviour. Such regional biasing could occur if different major types of behaviour (for example, aggressive, fearful, food seeking, sexual etc.) would be optimized better with different portfolios. Such regional specialization would be a compromise between the behavioural advantages and resource costs of duplicate portfolio definition in the same input space.

Imitation

Portfolios are defined heuristically, and there are limits to the degree to which genetic information can define the associations between the resultant portfolios and appropriate behaviours. Supervision by some external teacher can provide feedback on the appropriateness or otherwise of a behaviour, but cannot directly cause the learner to perform an appropriate behaviour.

In the absence of guidance towards appropriate behaviours, genetic or from supervision, initial selection of behaviour in unknown circumstances could only be random, resulting in a slow and inefficient learning process. Imitation provides a way for a supervisor to give specific guidance in the absence of direct internal access to the learning brain.

The term "imitation" covers a wide range of phenomena which require different degrees of a priori genetic information to support. For the imitation of speech, the imitator has fairly similar sensory input from both the target sounds and the attempted imitation. The portfolios activated on some levels of complexity will therefore be similar. An architecture to support imitation under these circumstances in illustrated in figure 9.13. Genetic information specifies definition of a set of competition components corresponding with generation of verbal sounds. Genetic information biases connectivity within clustering to favour the creation of portfolios within auditory inputs on a level of complexity which provides good discrimination between different verbal sounds, and favours targeting of the outputs of the auditory array on to these components with significant initial recommendation strength. Genetic information also specifies the existence of a reward system. This reward system compares the groups of portfolios in the auditory array which are active before and after a verbal behaviour. If there is strong similarity between the groups, the recommendation strengths of members of the group into the behaviour are increased, otherwise decreased. The effect of this reward system is to encourage imitation of sounds which are heard. The contribution of such a system to speech learning is discussed in the next section.

In the case of imitation of physical behaviours such as object manipulations, the sensory inputs will be somewhat less similar, for example being derived from perceptions of the hand of the supervisor for the target and from the hand of the learner for the imitation. However, there will still be some degree of similarity and an analogous system could support imitation.

This system could be refined by support from an additional reward system as illustrated in figure 9.14. This system depends upon a genetically programmed rewarding behaviour which is generated in a supervisor in response to the perception of imitation and is perceived by the learner as a reward. Such a system reduces the degree to which the brain depends upon pure portfolio similarity.

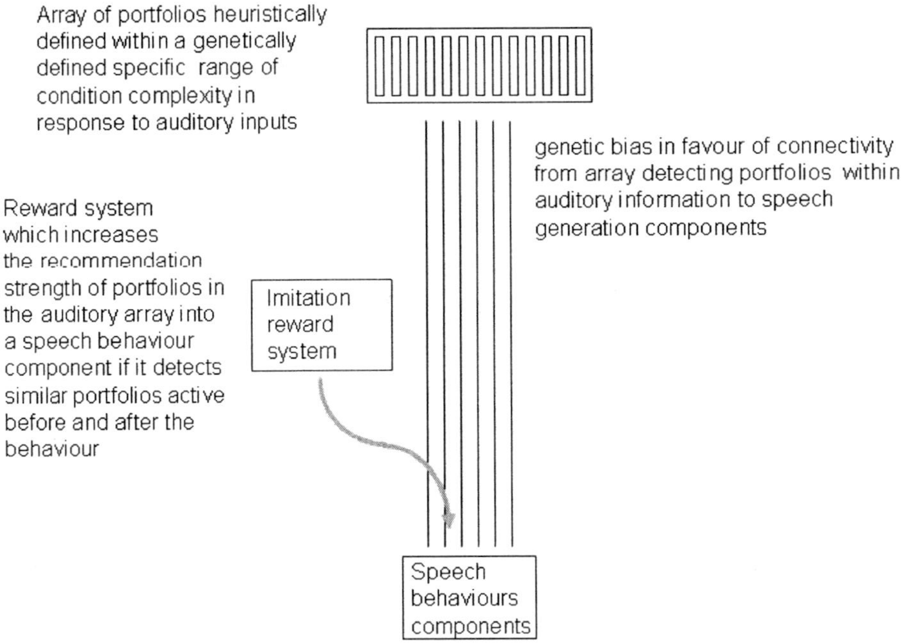

Figure 9.13. Speech imitation architecture. Genetic information specifies that an array of portfolios will be defined within auditory inputs with outputs within a specific range of condition complexity. This level of complexity provides good discrimination between different speech sounds. Portfolios are defined heuristically within this range. Genetic information also specifies a speech behaviours component which includes a set of speech generation behaviours corresponding with a range of phonemes. Genetic information also specifies connectivity between the array and the components. Genetic information specifies the existence of a reward system which increases the recommendation strengths of recent array portfolios into recently active speech components if portfolios activated in the array in response to hearing an external sound (i.e. one without accompanying speech generation behaviours) are similar to portfolios activated shortly afterwards in the array following a speech generation behaviour.

Such a supervisor reward system would be even more important for the imitation of facial expression. In this case there is little comparable sensory input because input from the supervisor is visual and input from the learner is only propioceptic. The degree of a priori information needed to support imitation is therefore greater. Portfolios able to detect different general expression types within visual inputs must be specified in advance, along with connectivity to behaviours generating the corresponding expressions. Supervisor feedback as illustrated in figure 9.15 could reduce the required degree of a priori information.

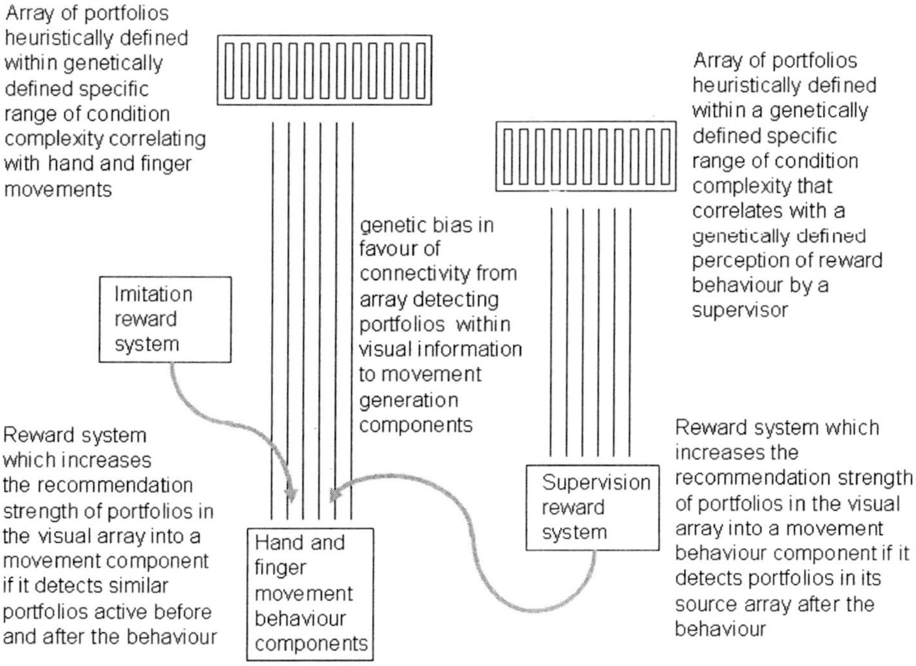

Figure 9.14. Physical manipulation imitation architecture. Genetic bias creates portfolio arrays at predetermined levels of complexity, although the portfolios in the arrays are defined heuristically. In addition to the reward system detecting similarities between externally observed and self behaviour, there is a reward system which detects conditions correlating with a type of behaviour by an external supervisor (e.g. a smiling face) and the resultant portfolios have recommendation strengths in favour of increasing the recommendation strengths of recently performed behaviours.

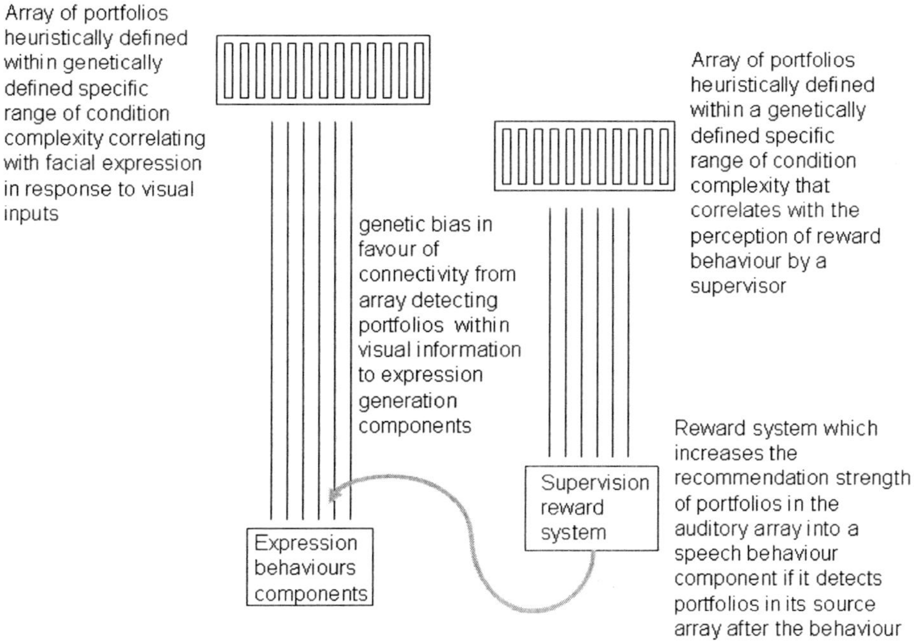

Figure 9.15. Facial expression imitation architecture. The required degree of a priori bias on structure is greater than for imitation of physical manipulations (figure 9.14) because there is no direct perception of self facial expression.

Bootstrapping of Memory and Behaviour

Behaviours can be defined heuristically with limited a priori guidance. Such definition will be illustrated by describing a possible process for acquisition of simple speech, with careful attention to the nature of the a priori (genetic) guidance needed. Learning goes through a series of steps generally consistent with observation of how humans learn to speak and understand words, but the purpose of this section is not to offer a formal model for speech acquisition but rather to demonstrate that speech can be acquired in the recommendation architecture model with only limited and plausible genetic guidance.

Genetic information specifies creation of a set of detailed competition components which drive muscle movements contributing to sound generation. Every possible such movement have a corresponding genetically specified competition component, and activation of the component results in the movement. Genetic information also specifies the existence of intermediate competition components which activate randomly selected sequences of detailed components and therefore generate sounds. Learning proceeds in a series of partially overlapping steps.

The first step is creation of an array of portfolios in clustering level A1 of figure 9.2, in response to hearing sounds. Because speech is somewhat different from other sounds, there will be a tendency for speech related portfolios to be somewhat separate from the portfolios created in response to other sounds. This tendency could be reinforced by genetically determined connectivity biases within clustering. However, these tendencies are not essential, although they will improve learning efficiency.

The next step is generation of sounds using intermediate components. Initially a component is randomly selected after any activation of a portfolio population in the array of sound related portfolios in level 1. A positive consequence feedback is genetically programmed to be generated if the portfolios activated in response to hearing an external sound (i.e. not self generated) are similar to the portfolios activated shortly afterwards in response to hearing a self generated sound. One effect of this consequence feedback is that the sequence of detailed components activated by the intermediate component is fixed long term. In other words, the presence of a sound in the environment results in production of that sound becoming instantiated in an intermediate component. If there is no feedback within some period of time, the intermediate component is reconfigured with a different randomly selected sequence of sound generating muscle movements, or deleted. A second effect of the consequence feedback is that the activated portfolios acquire recommendation strength in favour of activating the intermediate component. In other words, the behaviour of imitating sounds which are heard is acquired.

The next step in learning is that portfolios are created in level A2 of figure 9.2 in response to sequences of sounds which are heard. These portfolios will correlate partially (or ambiguously) with frequently heard sequences, and therefore with words which are heard. Higher level competition components are defined which activate randomly selected sequences of intermediate components. If a self generated sound sequence activates a similar portfolio population in level A2 to an immediately prior population activated by an external sound sequence, a genetically defined consequence feedback results in the higher level component being fixed and the active portfolios acquiring recommendation strength in favour of activating it. Thus the behaviour of imitating words which are heard is acquired.

The next step utilizes a genetically programmed tendency for the portfolios created in level A2 in response to sequences of sounds to have recommendation strength in favour of activation of portfolios created in levels V2, V3 and V4 in response to visual experiences, if the visual experience portfolios are often active at the same time as the sound sequence portfolios. The effect is that hearing the word will tend to activate a pseudovisual image of the type of object often seen when the word was present in the past. The portfolios making up a visual image will also recommend any other behaviours which have become associated with the object. In addition, the visual experience portfolios acquire recommendation strength in favour of activating the higher level component which tends to be activated by the sound sequence portfolios. The effect is that seeing the object will tend to result in speaking the corresponding word. Consequence feedback associated with the perceived behavior of adults in response to activating a higher level component (i.e. speaking a word) will affect the recommendation strengths of active portfolios in favour of the word just spoken.

Thus a set of genetically defined tendencies result in relatively efficient acquisition of simple speech behaviours. Learning does not require a priori internal definition of cognitive categories. Genetic information provided three types of information. Firstly, it indicates the available range of detailed muscle movements. Secondly, it biases initial connectivity in favour of the types of sensory inputs and portfolio condition complexity ranges which will most effectively drive those behaviours. Thirdly, it defines in general terms the circumstances in which consequence feedback will be generated and the effects of such feedback.

Heuristic Specialization of Clustering and Competition Resources

It is also possible for both clustering and competition resources to be optimized heuristically for particular types of behaviour. Such optimization means that a limited group of clustering portfolios on some levels of complexity provide more effective discrimination just between input states relevant to the behaviour type, a limited group of behaviour sequence components correspond with the behaviour type, and the portfolios preferentially target the competition components relevant to the behaviour type.

Such a specialization has a high resource cost, since clustering resources must be customized to the behaviour type. Portfolios in the customized resources may partially duplicate portfolios elsewhere used for other behaviours. To achieve such an optimization, a behaviour type which justifies the extra resource commitment must therefore be identified by the system. In other words, behaviours must be identified for which the importance of the behaviour justifies the increased resources.

The question then is, what is the appropriate criterion for "important" behaviours? If a behaviour is critical in relatively specific circumstances, its criticality can be handled by assigning a high recommendation weight to a limited set of general purpose portfolios and perhaps inhibiting future changes to the relevant portfolios. For example, learning to avoid the unpleasant consequences of a behaviour under specific circumstances simply involves assigning recommendation weights to appropriate portfolios, and the relevant portfolios may be active in response to many other circumstances.

A more relevant consideration is if a relatively limited range of behaviours frequently occurs in response to a relatively limited range of portfolios, and there is often contradictory consequence feedback in response to the behaviours. This circumstance indicates that special purpose resources would be both frequently utilized and that there is a need for such resources. The detection of the relatively frequent occurrence of a fairly small number of clustering columns activating a relatively small number of competition components is the criterion which triggers configuration of special resources.

Note that this criterion will not necessarily define a behaviour type which correlates exactly with a category of cognitive behaviour. For example, arithmetic processing is a category of behaviour which could potentially have a fairly small number of clustering columns activating a relatively small number of competition components, but even if resources were defined around such a behaviour, the definition process has no knowledge of arithmetic as "academically" defined, only of relative frequency measures. Some behaviours commonly regarded as arithmetic might be excluded, and other behaviours not regarded as arithmetic might be included in the behaviour type defined on the basis of relative frequencies.

The process of defining special purpose resources is illustrated in figure 9.16. The first step is detection of a limited set of clustering columns and competition components which are often active at the same time. The second step is assignment of clustering array resources at the levels of complexity at which the temporally correlated activity is occurring. These resources are given inputs which tend to be active at times when the behaviours are selected, and condition recording is forced at these times. The outputs of these arrays are targeted on the components corresponding with the behaviours, and given weak initial recommendation strengths. These recommendation strengths are rapidly increased by positive consequences resulting largely from behaviour selections on the basis of the recommendation strengths of the original array portfolios. As the special purpose recommendation strengths increase, connectivity from the original arrays is gradually disconnected. At this point, portfolios have been effectively cloned from the original group. The final step is to evolve these cloned portfolios on the basis of contradictory consequence feedback with respect to only the target behaviours. In other words, outputs to portfolios or new portfolios are added if the activation of the same portfolio followed by selection of the same behaviour within the type is sometimes followed by positive consequences, sometimes negative.

Use of contradictory consequence feedback in response to just the behaviour type will result in portfolios optimized for the behaviour type as less appropriate for other behaviours. Other behaviours will continue to be managed by the original resources.

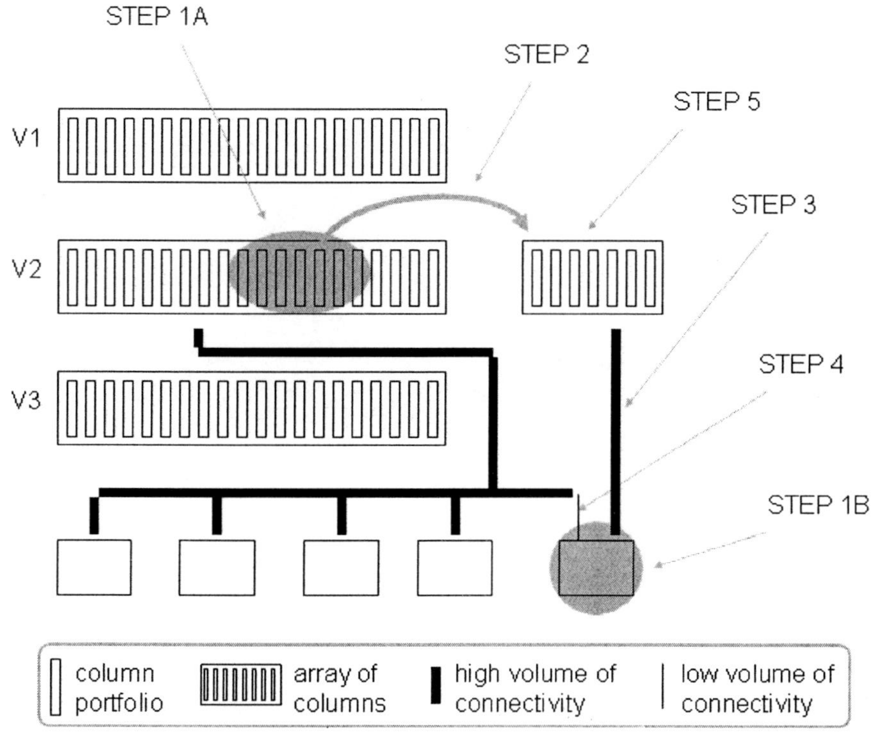

Figure 9.16. Process for configuring resources to support a frequently used behaviour type. A limited group of portfolios is selected on the basis that each one tends to be active frequently just before a limited group of similar behaviours are implemented. A limited group of similar behaviours is defined on the basis that they invoke a limited range of more detailed behaviours and tend to be recommended by a limited group of portfolios. The group of portfolios is cloned by creating new columns with a bias on inputs in favour of those active just before behaviours in the limited group are selected, and forcing condition recording when the original group is active. The outputs from the cloned group are given outputs targeted on the limited group of behaviours with initially weak recommendation strengths, increased rapidly by positive consequences. The recommendation strengths of inputs from the original array into the limited group are weakened and eventually many of the inputs are deleted. The portfolios in the cloned group are refined by contradictory consequence feedback from only the limited group of behaviours

Significance of the Cognitive Model

The essential requirement for a scientific theory, as discussed in chapter 2, is a hierarchy of causal descriptions on many different levels of detail which can be mapped into each other. The cognitive model described in this chapter is at a critical level of detail. It can be mapped into descriptions in terms of physiological concepts like cortex columns instantiating portfolios, and these physiological descriptions can be mapped into descriptions in terms of neurons. However, processes at the level of psychological observation can be described within the cognitive model. A key consideration to be addressed in later chapters is the degree to which the cognitive model is consistent with the wide range of experimental psychological observations of memory and cognitive processing in human beings.

References

Coward, L. A. (2001). The Recommendation Architecture: lessons from the design of large scale electronic systems for cognitive science.
2(2), 111-156.

Satori, G., Job, R. & Coltheart, M. (1993). The organization of object knowledge: evidence from neuropsychology. In D. E. Mayer & S. Kornblum (Eds.).
. Cambridge, MA: MIT Press.

Chapter X

Semantic and Episodic Memory

The term "memory" is applied to a very wide range of different phenomena in human beings. Experimental approaches to the scientific study of memory have included laboratory studies, collection of data on real life memory, study of the effects of disease and physical damage, and imaging the brain during memory processes. Theoretical approaches have included proposals of relatively independent memory subsystems, and creation of detailed computer models.

There have been many attempts to categorize experimentally observed phenomena into different types. One distinction made early on in the field is between declarative and procedural memory, where declarative memory is the memory for objects, facts and events and procedural memory is the memory for how to perform skilled behaviours. A closely related distinction is between explicit and implicit memory. This distinction had its origins in the types of tasks which amnesiacs cannot and can perform. For example, patients with Korsakoff's syndrome have short term memory for objects and events, but have lost the ability to create new long term memories [Butters 1984]. Such patients, if introduced to someone, can have an apparently normal conversation but will have no memory of the person after a separation of more than a few minutes. However, such patients can acquire and retain new motor skills.

Different memory systems have been proposed on the basis of different classifications of observations of memory phenomena, but such classifications have been criticized. For example, it has been argued that implicit memory as generally defined is just a collection of phenomena which are not linked by any common neural processing [Willingham and Preuss 1995].

An influential approach to organization of memory phenomena into independent subsystems is that of Tulving [1984, 1985], developed further by Schacter and Tulving [1994]. This approach suggests three criteria for identifying separate memory subsystems. One criterion is the existence of a group of memory tasks which have some common characteristics. The second criterion is the existence of an identifiable list of system features which is different from the list for any other system. The third is the existence of multiple dissociations between the systems, or multiple ways in which similar manipulations of tasks performed by different systems produce different effects upon performance. In the view of

Schacter and Tulving, unless all three criteria are met the system cannot be regarded as functionally separate. Given these criteria, they argued that there was evidence for the existence of five relatively independent memory systems: working memory; semantic memory; episodic memory; the perceptual representation system or priming memory; and procedural memory.

The spreading activation model for semantic memory proposed by Collins and Loftus [1975] and the working memory model proposed by Baddeley and his coworkers [Baddeley 1986] offer models for two of these memory systems at an intermediate level. However, in terms of the discussion in chapters 2 and 3, these models are user manual type and provide limited insight into system architecture. The consolidation model for memory proposed originally on the basis of the amnesic patient HM [Scoville and Milner 1957] attempts to model the flow of information within physiological structures. This model also attempts to address the issue of interference between early and later learning. Direct comparisons between all these models and the recommendation architecture model will be made later.

This chapter presents the case that the wide range of experimentally observed semantic and episodic memory phenomena can find consistent explanation in terms of the recommendation architecture cognitive model described in chapter 9. The following chapter will discuss procedural memory, working memory and priming. The account for a wide range of observed memory phenomena will be described. These accounts will be organized into commonly used categories like implicit and explicit memory, working memory, semantic memory, episodic memory etc. but without any assumption that these categories reflect separate underlying memory systems. It will be argued that differences between the information recording and access mechanisms used in the recommendation architecture for different types of memory account for the observation of separate memory systems as proposed by Schacter and Tulving [1994; see also Schacter, Wagner, and Buckner, 2000].

Memory within the Recommendation Architecture

The purpose of this chapter and chapter 11 is to support the thesis that the cognitive architecture described in chapter 9 can provide an account for a full range of experimentally observed human memory phenomena, and that deficits to individual recommendation architecture mechanisms can provide an account for the different specific combinations of memory deficits observed in human subjects with brain damage. To achieve this purpose, experimental work on a wide range of different types of memory phenomena will be summarized, including the evidence for dissociations between different types. The processes by which each memory type occurs in the recommendation architecture cognitive model will be described, and compared with the observed memory phenomena.

For the purposes of this comparison, memory phenomena will be discussed under a number of different major categories. However, because experimental evidence is often the existence of dissociations between different types of memory, the section devoted primarily to one memory type will contain some discussion of other types. As far as possible, the commonly used categories of memory phenomena have been used. However, this approach has some issues. The first is that many of the categories are associated with assumptions about memory systems supporting the category type.

For example, the use of the term implicit memory is controversial as discussed earlier. For the purposes of this paper the term explicit memory will be used to label a particular subset of experiments on episodic memory: word and object list recollection and recall. Similarly, implicit memory will be used to label a particular subset of experiments on procedural memory: speed of identification of letter strings as word or non-word. The recommendation architecture mechanisms which account for the differences between explicit and implicit memory will be described in relatively general terms, and more detailed mechanisms discussed under episodic and procedural memory.

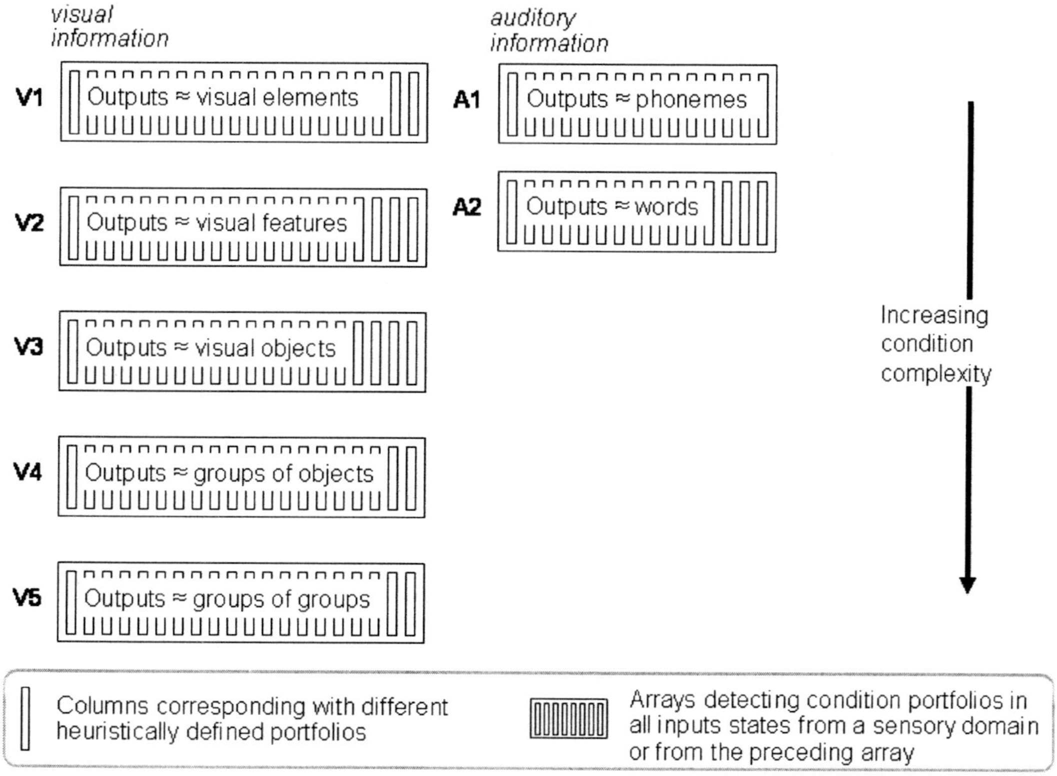

Figure 10.1. Clustering architecture. Parallel sequences of arrays of columns detect condition portfolios independently in visual and auditory input. Different arrays in a sequence detect portfolios within different behaviourally useful ranges of condition complexity. In the figure, these ranges of complexity correspond roughly with the complexity of cognitively relevant circumstances such as the presence of features. Portfolios at that level will have the best discrimination between such features. In practice sequences will have many more arrays of columns at different complexities and portfolios on several levels of complexity will be relevant to discriminating between cognitive features (see figure 9.4). However, to simplify explanations the number of arrays has been reduced.

Tulving and others have argued that a body of observational data labeled perceptual priming justifies the hypothesis of a separate perceptual representation system [Schacter and Tulving 1994]. Priming is the enhanced ability to respond to or remember an object when there has been prior unconscious exposure to the object or to a related object. Schacter and

Buchner [1998] have suggested that there are at least two types of perceptual priming: modality specific which is independent of semantic encoding; and modality independent which is enhanced by semantic encoding. Both of these types will be discussed in the perceptual priming section of chapter 11. However, in that chapter a further experimental distinction will be made. If consequence feedback was available at study for the same task as was performed at test, the phenomenon will be regarded as procedural memory rather than priming. Thus in the work of Cave [1997], the subjects at study were shown pictures and asked to name the pictures as fast as possible. At test up to 48 weeks later, the subjects demonstrated faster naming of the pictures encountered during study. These results are often cited as evidence for long term priming, but because subjects at study identified the pictures later seen at test, these results will be regarded as procedural memory. Priming will be limited to exposure to inputs relevant to a later task without successful completion of exactly the same task.

The important body of work on comparisons between subjects with cognitive deficits and normal subjects will be discussed in each section and more fully for the extensively studied subject HM in a separate section.

The cognitive architecture used in this chapter and chapter 11 is the model described in chapter 9. Figure 10.1 illustrates the somewhat simplified processing of visual and auditory sensory inputs in clustering. Figure 10.2 illustrates clustering arrays which detect information circumstances in arrays V3 and A2 of figure 10.1. Figure 10.3 and competition architectures. Table 10.1 is the list of important behaviour types supported by competition.

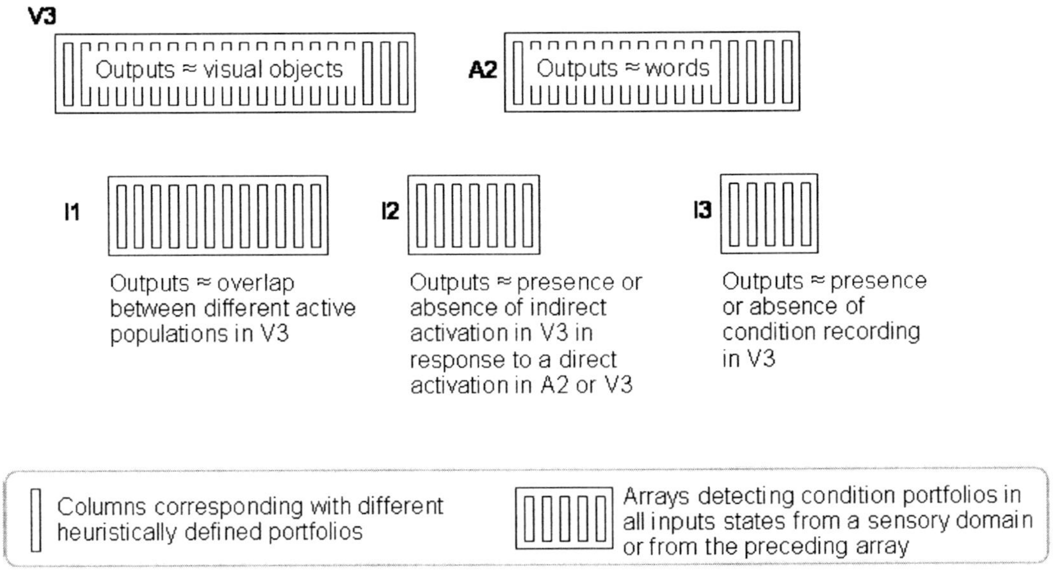

Figure 10.2. Some information state arrays relevant to understanding semantic and episodic memory. These arrays have inputs from sensory arrays and detect conditions correlating with different information states in those arrays. Different portfolios in array I1 are activated in response to different degrees of overlap between independent portfolio populations in array V3. Portfolios activated in array I2 discriminate between different degrees of indirect activation in array V3 in response to an existing activation in V3 or an activation in A2. Portfolios activated in I3 discriminate between different degrees of condition recording in V3. Similar information state arrays will exist for sensory arrays on other levels of complexity.

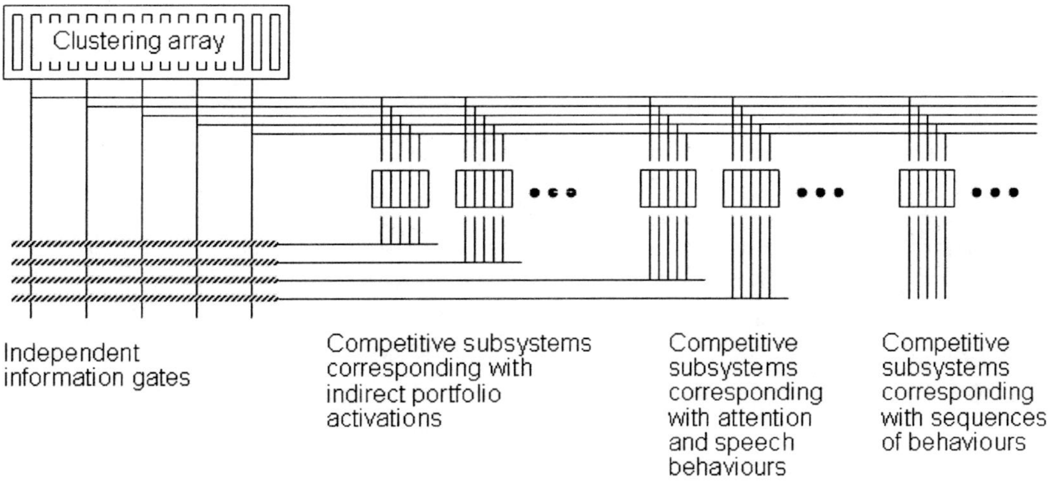

Figure 10.3. Structure of competition making use of clustering outputs from one array. Each output from clustering may have recommendation strength in favour of many different behaviours. Competition manages to selection of the most strongly recommended behaviours.

Table 10.1

A. Managing Information Availability

1. Prolong the activity of a currently active portfolio
2. Synchronize the activity of several different populations of currently active portfolios in the same array
3. Release outputs of one array to the next array

B. Activation on the Basis of Past Activity

4. Activate portfolios which have recently been active at similar times to the currently active portfolio
5. Activate portfolios which have often been active at the same time in the past as the currently active portfolio
6. Activate portfolios which have often been active just after past activity of the currently active portfolio
7. Activate portfolios which have often been active just before past activity of the currently active portfolio
8. Activate other portfolios not currently active but which have often been active in the past just before or just after the currently active portfolio
9. Reactivate recently active portfolios

C. Activation on the Basis of Past Condition Recording

10. Activate portfolios containing conditions recorded at the same time in the past as some conditions in the currently active portfolio
11. Activate portfolios containing conditions recorded just after some conditions in the currently active portfolio
12. Activate portfolios containing conditions recorded just before some conditions in the currently active portfolio
13. Reactivate portfolios which recently recorded conditions

D. Attention management

14. Perform a general sequence of attention and internal activation behaviours
15. Perform a specific sequences of attention and internal activation behaviours
16. Perform an individual attention behaviour

E. Speech Behaviour Management

17. Generate a sound
18. Speak a word
19. Say a phrase
20. Express meaning verbally by a sentence

F. Motor Behaviour Management

21. Perform a general sequence of motor behaviours
22. Perform a specific sequence of motor behaviours
23. Perform an individual motor behaviour

Recognition and Recall

Human beings have a capability to recognize the novelty or otherwise of an object which greatly exceeds their capability to recall details of the object when it is not present. For example, given a few seconds to examine each photograph in a set of 2500, subjects can later pick the familiar photograph from pairs in which only one came from the examined set at an accuracy level of 90% [Standing, Conexio and Haber, 1970]. The capability to recall details of previously seen objects in the absence of the object is much more limited.

In the recommendation architecture, a novel object will require recording of a significant number of new conditions in order to reach the minimum activated portfolio population and the degree of condition recording required is therefore a clear indicator of the degree of novelty. Discrimination between situations in which there is a high level of recording and situations in which recording is limited can be performed by clustering arrays which receive inputs indicating the presence or absence of recording in different locations. Such an array is illustrated in figure 10.2. Portfolios in such arrays can acquire recommendation strengths in favour of behaviours such as saying "I have never seen anything like that before".

However, as discussed in the section on episodic memory, recall of an object which is not present requires activation of some of the portfolio population which was active when the object was actually present. Such an activation can only be achieved by a much more complex process of indirect activation starting from whatever activation is currently present. For example, words describing where the object was seen may activate some of the portfolios which were active when at that place. These place related portfolios in turn may activate portfolios including information on the object. Verbal clues to the nature of the object may help to move the activated population towards the target. However, this indirect activation process is much less specific than the novelty indication, and recall of some objects may be impractical from the currently available activation.

Explicit and Implicit Memory

The experiments which represent the paradigm for the distinction between explicit and implicit memory measure the performance of human subjects in the laboratory in response to lists of words. The discussion in this section will be limited to these types of experiments, and a wider range of phenomena sharing some similar mechanisms will be discussed in separate sections on semantic, episodic and procedural memory.

In explicit memory experiments, subjects are presented with a list of words during a study phase. In a subsequent test phase, subjects are asked to recall the items on the earlier list (free recall testing) or are supplied with words and asked if they occurred on the earlier list (recognition memory). It is found that recall capability declines with time, but is somewhat improved by meaningful connections between words on the list. Recognition capability also declines, but is at a somewhat higher level than recall capability [Kirsner, Speelman, Maybery, O'Brien-Malone, Andersen and MacLeod, 1998].

In implicit memory experiments, subjects are presented with a series of letter strings, some of which are meaningful words, others are meaningless strings or non-words (e.g. cart, mude, blue, dace, ...). They are asked to classify each string as a word or as a non-word as rapidly as possible, and the time taken to make classifications is measured. In a study phase, subjects are asked to classify a series of strings. In a subsequent test phase, subjects are asked to classify another series which contains both strings from the study phase and new strings. It is found that reaction times to classify repeated strings are shorter than for new strings, for both words and non-words [Kirsner, Speelman, Maybery, O'Brien-Malone, Andersen and MacLeod, 1998].

Process for Explicit Memory

In the recommendation architecture cognitive model, during the study phase for explicit memory experiments, portfolios will be activated in arrays V1, V2 and V3 of figure 10.1 in response to objects in the laboratory environment including the room, the investigator, and the physical form of the lists.

If the words are presented verbally, portfolios will be activated in arrays A1 and A2 in response to hearing the individual words on the list. These portfolios will indirectly activate portfolios mainly in V3 which have often been active in the past when the auditory portfolios were also active. If the words are presented visually, the visual inputs will directly activate portfolios in V1, V2 and V3, and the portfolios activated in V3 will indirectly activate other portfolios in V3 which are again those which have often been active in the past when the directly activated portfolios were active. In both cases the indirectly activated portfolios in V3 will be those which would tend to be directly activated by visual inputs generated by the presence of the object or circumstances corresponding with the word. This correspondence exists as a result of an earlier learning period in which the object and the word were often present at the same time.

Because the individual words are familiar, there will be minimal condition recording within the portfolios activated directly or indirectly in response to the words. There may be novelty in the laboratory environment including the physical form of the lists, and therefore there may be condition recording in levels V1, V2, and V3 in response to objects in that environment. There may also be novelty in the objects in the laboratory environment, the form of the lists, and the juxtaposition of different such objects and words on the list. Such novelty would result in condition recording at the level of groups of objects or groups of groups in arrays V4 or V5 in figure 10.1. Hence conditions recorded on these levels could incorporate some information derived from individual words.

Because the portfolios activated in response to each word were active at similar times to portfolios activated in response to objects in the laboratory and in response to other words, all these portfolios will acquire recommendation strength in favour of activating each other on that basis at some later time. However, as discussed in chapter 9, such recommendation strengths decay fairly rapidly unless reinforced by usage and positive consequences following that usage.

In the test phase, the subject is asked to think about the earlier study phase. Portfolios activated in response to hearing words like "earlier" and "remember" have general recommendation strength in favour of activation of portfolios on the basis of past temporally correlated activity. In other words, these portfolios activate the competition components corresponding with these indirect activation behaviours, and such behaviours recommended by other portfolios will therefore tend to be accepted.

The subject sees some aspects of the same environment and/or hears words like "laboratory", "word list", studied", etc. Portfolios are directly activated in arrays V1, V2, V3, V4 and V5 in response to the visual perceptions and combinations of those perceptions, and indirectly activated by portfolios activated in array A2 on the basis of frequent past activity at the same time as the words were heard in the past. The resultant population will have significant overlap with the portfolio population active during the study phase, and the overlap will in general be greater than the overlap with the active population during any other past experience.

This resultant population will have recommendation strength in favour of activating any portfolios in V3, V4, and V5 which recently recorded conditions at similar times to any of the currently active portfolios and in favour of activating any portfolios which were recently active at similar times to any of the currently active. The active population will also have recommendation strength in favour of activating any portfolios in V3, V4, and V5 which recorded conditions at similar times in the past as any of the currently active portfolios; and recommendation strength in favour of activating any portfolios which were active at similar times in the past as any of the currently active portfolios. The presence of portfolios activated in response to hearing words like "earlier" and "remember" results in these types of recommendation being accepted.

Because there is a proportion of active portfolios in the primary active population which were active in the study phase, and because this proportion is higher than for any other previous experience, the indirect activation recommendation strengths of the primary population will tend to activate a secondary population which has a somewhat larger overlap with the population active during the study phase. This secondary population may then activate a tertiary population on the same basis which has even greater overlap with the population active during the study phase. This process could continue through further reactivation stages. The resultant derived population will be an approximation of the active population during the original study phase. In other words, the activation will be experienced as a partial re-experience of the study phase, but without the activation of portfolios close to sensory input in arrays V1 and V2.

However, activation on the basis of recent or frequent past temporally correlated activity is not as specific to the study phase as activation on the basis of recent or frequent past temporally correlated condition recording. The degree to which the reactivation matches the original will therefore depend upon the degree of condition recording in the individual portfolios during the study phase.

Some of the secondary and tertiary portfolios will have recommendation strength in favour of speaking the words on the studied list. The degree to which the words on the list will be recalled is determined by the degree to which portfolios with these speech recommendation strengths are activated. The relevant portfolios will be those in array V3 which have often been active in the past when either the word or the corresponding object have often been active. Such portfolios were active during the study phase, but as discussed earlier there will generally have been little condition recording in such portfolios.

The activation of these portfolios is therefore primarily dependent upon their activation on the basis of recent temporally correlated activity by the tertiary population approximating to the population during the study phase. Such recommendation strengths decay with time and memory on this basis will therefore decay. This decay accounts for the rapid decline over the first day or so.

The relevant portfolios could also be activated on the basis of temporally correlated condition recording, which would be in addition to activation on the basis of recent temporally correlated activity. However, such condition recording is limited in the study phase for two reasons. One reason is that recording is limited in array V3 because the words are familiar. The second reason is that recording in higher arrays like V4 and V5 is limited because the words are unrelated semantically. There are no phrases to activate portfolios in V4 or sentences to activate portfolios in V5 etc. A sequence of unrelated words will generate simultaneous activity in array V3 corresponding with the different words, but there will not be adequate recommendation strength to synchronize and release the outputs from V3 to V4, or from V4 to V5. There will therefore be no condition recording in these arrays. However, if there was some relationship between a group including some objects in the study environment and some words, condition recording in V4 and V5 might occur, making recollection of those specific words easier.

Recommendation strengths in favour of activation of the appropriate word related portfolios on the basis of frequent past temporally correlated activity will depend upon whether such activation has occurred and been followed by positive consequence feedback. Unless the word lists are recalled frequently following the study phase, this type of recommendation strength will not be present.

The recommendation architecture model therefore leads to the expectation that recall capability will generally decay with time, but decay will be reduced if there is some semantic relationship between the words, some meaningful relationship between words and novel elements in the study environment, or repeated recall within a short period of the study phase.

In the case of recognition memory, words are presented which may or may not have been on the study phase list. In this case the mechanism is a search for overlap between the portfolios activated in the primary, secondary and tertiary etc. populations and the portfolios activated in response to the words. Detection of such overlap is a less specific task than reactivation of the appropriate target portfolio population and recognition rates will be somewhat higher.

Process for Implicit Memory

The process for implicit memory is illustrated in figure 10.4. If a genuine word is presented, the activation of portfolios in array A2 containing conditions within the visual perception of the word will generate a strong secondary activation of portfolios in array V3. These V3 portfolios are those which were present in the past during a learning period when the word and the object corresponding with the word were both present, and will therefore tend to be those which are directly activated by the presence of the corresponding object.

As illustrated in figure 10.2, portfolios exist which discriminate between circumstances in which secondary activations in V3 are generated and those in which they are not generated in response to a direct activation in array A2. The activation of such portfolios are interpreted as recommendation to identify the current string as a word or as a non-word, for example by the behaviours of speaking the words "yes" or "no".

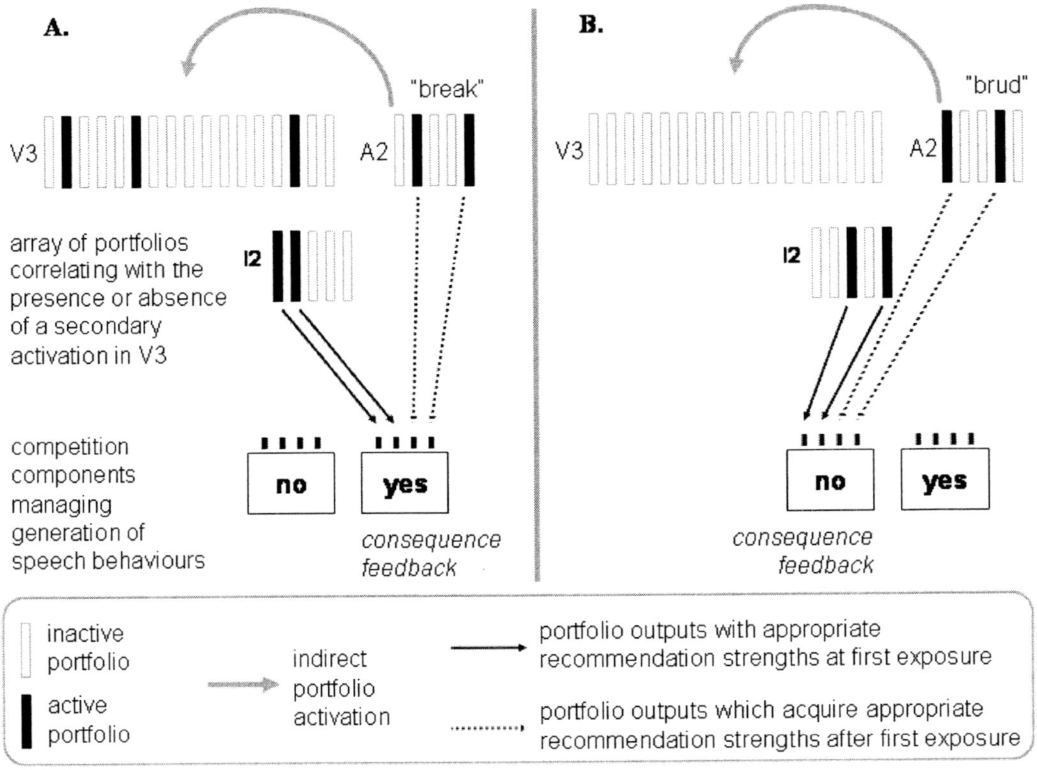

Figure 10.4. Process for implicit memory. In scenario A, a real word directly activates a portfolio population in array A2, and that population indirectly activates a population in V3. Portfolios activated in array I2 correlate with the presence of a strong indirect activation in V3 in response to the activation in A2, and these portfolios have recommendation strength in favour of saying that the word is a real word. In scenario B, the pseudoword does not generate an indirect activation in V3, and portfolios activated in I2 correlate with the absence of such an activation and have recommendation strength in favour of saying the word is not a real word. In both scenarios, positive consequence feedback followed the speech behaviour. The portfolios activated in A2 were active shortly before the speech behaviour and positive consequences, and therefore acquire recommendation strengths in favour of the speech behaviour. In future exposures to the words, some recommendation strength in favour of the correct behaviour will be available as soon as the direct activation in A2 is present, without the need to wait for the indirect activation in V3 and further activation in I2. Response time will therefore be shorter.

When a correct identification is made, the recommendation strengths for recently active portfolios into the accepted behaviour are permanently increased by consequence feedback. These portfolios include those activated directly by perception of the word in A2, and such portfolios therefore acquire some recommendation strength in favour of the appropriate response as illustrated in figure 10.4. Hence in a future exposure to the same string, there will be some recommendation strength in favour of the correct response as soon as conditions within the perception of the word are detected in A2, without waiting for detection or otherwise of the secondary population. The time taken to reach adequate recommendation strength to generate behaviour will therefore be reduced. A similar process will occur for visual presentation of the letter strings, but in this case the direct activation generated by the visual word will also be in V3.

Comparisons with Experiment

Consider now the more detailed experimental results for explicit and implicit memory. In explicit memory, the initial weights in favour of indirect activation are provisional and decline with time unless they contribute to a completed behaviour and therefore have consequence feedback applied. Explicit memory will therefore decline systematically as a function of time [see Scarborough, Cortese and Scarborough, 1977]. However, in implicit memory, consequence feedback is available for the completed behaviour, and decline will be much less [see Kirsner 1998]. Performance in explicit memory testing is strongly dependent on the level of condition recording incorporating information derived from list words in levels V4 and V5, which depends upon the degree of novelty in the experience in which the words participate. The degree of such recording in these levels will be strongly increased by semantic organization of the words. However, in implicit memory it is the specific portfolios activated in response to the letter string which determine the faster response. The identity of these portfolios is dependent on the details of the sensory experience and independent of any semantic relationships between words. Hence explicit memory will be affected by organization of the material to be remembered and implicit memory will not [Schacter, Bowers and Booker 1989]. Implicit memory will be more strongly reduced by changes to the lexical form of words than explicit memory [Forbach, Stanners and Hochhaus, 1974].

One of the origins for the distinction made between explicit and implicit memory was the types of tasks which could and could not be performed by amnesics [Dunn 1998]. The observed dissociations between normal and amnesic patients can be accounted for by the model that amnesics have lost the ability to record new conditions, specifically a failure of the mechanism which assigns and configures condition recording resources [see chapters 4, 9]. The effect of such a deficit would be that perceived objects would activate populations of portfolios based on portfolio definition prior to the onset of the deficit, and these portfolios would recommend all the behaviours recommended prior to amnesia, but portfolios could not add conditions. However, existing portfolios could acquire new or different recommendation weights. Hence the ability to activate portfolios active at the same time in the past and the ability to adjust the weights assigned to recently active portfolios on the basis of the success of current behaviour would be unaffected, but the ability to activate portfolios which recorded conditions at the same time in the past would not be present for circumstances after the onset of the deficit. Recognition and recall memory would therefore be lost, but the ability to classify words would be retained, and the ability to assign weights to the fairly simple sensory portfolios in favour of correct word classifications would also be retained, leading to the observed learning in implicit memory.

Dichotic Listening

A range of experiments have been performed to test subjects ability to remember when inconsistent information is provided to left and right ears. In the dichotic listening experiments of Triesman (1960), subjects were presented with two sections of very different text, one to each ear. They were asked to repeat aloud (to shadow) the text heard by one specific ear only. Part way through the presentation, the two texts were switched between ears. The subjects tended to respond by (erroneously) switching to repeating the text from the other ear (i.e. the meaningful continuation). At the end of the presentation, some content could be recalled from the shadowed text but not from the other text. Thus although the unshadowed text did not result in later recall, the ability to immediately switch implies that both ears were processing information in some fashion.

In the recommendation architecture model, there can be multiple populations of activated portfolios present at the same time in each of a number of different clustering areas receiving inputs from different sensory domains, such as different ears. For example, A1 and A2 arrays in figure 10.1 will be generated separately for each ear, and the separate sequences of arrays will detect portfolios within words presented to different ears. These portfolios will all have recommendation strengths in favour of activating portfolios in array V3 which occur within the visual objects corresponding with the words.

When different words are presented to the two ears, the indirect activation recommendations are in conflict, and the actual activation in V3 will depend upon the relative weight of the recommendation strengths.

If a semantically meaningful text is being heard, there will be indirect activations in V3, V4 and V5 resulting from words heard earlier. These indirectly activated portfolios will also have recommendation strength in favour of activating other portfolios often active at temporally correlated times in the past. If there are syntactical and semantic relationships between the words generating this indirect activation and the next word heard, these additional recommendations will reinforce the recommendations of the portfolios activated in A2 in response to that next word.

Hence in the dichotic listening experiment, the recommendation strengths of the existing indirect activation in V3, V4 and V5 will reinforce the recommendation strengths in A2 corresponding with the word which is a meaningful continuation of the text. If the meaningful continuation shifts to the other ear, the recommendation strengths of the indirect activation will reinforce those of the other ear and the words in that ear will drive subsequent indirect activations.

The attended text generates active portfolio populations in V3, V4 and V5, and condition recording in these active portfolios is the basis for later recall. The unattended text does not generate an active population in V3, V4 and V5 and therefore cannot be recalled later.

Semantic Memory

Semantic memory as defined by Tulving [1972] is the ability to recall a wide range of organized information about the world including facts, concepts and vocabulary, but excludes any recall of the specific contexts in which the information was learned. Ways of measuring semantic memory in the laboratory include sentence and category verification, and tip of the tongue experiments.

Sentence verification experiments measure the time for subjects to respond with the correctness or otherwise of sentences like “Is a pigeon a bird?”, “Is a penguin a bird?”, or “Is a cat a bird?”. Category verification experiments are essentially equivalent and measure such times for simple category-exemplar pairs like bird-finch or bird-tree. It is found that for different members of the same category paired with the correct category, responses are faster for more typical category exemplars. For example, the response to bird-robin is faster than to bird-chicken. However, responses to clearly incorrect category-exemplar pairs like “Is a turtle a bird” are also fast [Rips, Shoben and Smith 1973].

Tip of the tongue experiments investigate the common phenomenon of feeling that you almost know the answer to some question, but have not quite got it. Brown and McNeill [1966] read subjects the definitions of infrequently used words, and then asked if any of the subjects had a tip of the tongue feeling. Those who did were asked supplementary questions like “What is a similar word?”, “What does the word sound like?”, “How many syllables does it have?”, “What is first letter?” etc. They found that subjects could often supply such partial information on the word.

Mechanisms Supporting Category Verification

Category verification experiments can be understood by considering the portfolio populations activated in response to the words indicating category and exemplar. The processes by which such populations are learned and activated for the category can be understood by consideration of figures 10.5, 10.6 and 10.7.

In figure 10.5 the portfolios directly activated in V3 in response to different pigeon instances are conceptually illustrated, along with the portfolios activated directly in array A2 in response to hearing the word “pigeon”. The figure illustrates a learning period in which visual and auditory inputs occurred at the same time.

As illustrated in figure 10.5, different portfolios are activated in array V3 in response to each visual instance of a pigeon, but because of the fairly strong similarity between different pigeons there are some portfolios which are activated in V3 relatively frequently in response to different visual pigeon instances.

Also, there are portfolios in activated in A2 relatively frequently in response to different occasions on which the word “pigeon” is heard. To simplify the explanation, it is assumed that exactly the same set will be activated in response to each hearing of the word.

The learner will be looking at pigeons or pictures of pigeons when a teacher is saying the word “pigeon”. Portfolios activated frequently in V3 in response to different visual pigeon instances will therefore be the portfolios most often active at the same time as the portfolios activated frequently in A2 in response to hearing the word “pigeon”. Hence the portfolios frequently active in A2 will acquire recommendation strength in favour of indirectly activating the frequently occurring portfolios in V3. After the learning period, hearing the word “pigeon” will result in an indirectly activated portfolio population in array V3 which is made up of the portfolios most often activated in response to seeing a pigeon. Such an activation will be experienced as a pseudovisual perception of an "average" pigeon, but without the portfolios activated close to visual input in arrays V1 and V2.

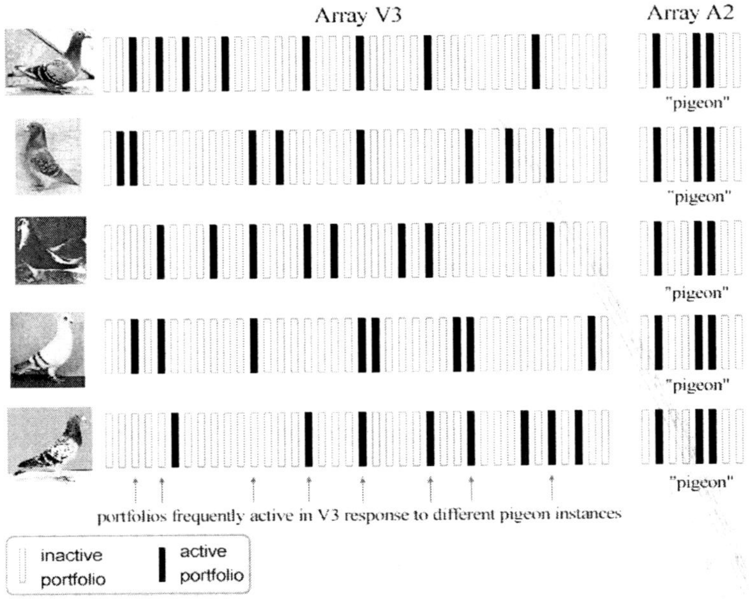

Figure 10.5. Acquisition of appropriate response to hearing the word "pigeon". Different portfolios are activated in array V3 in response to different visual pigeon instances. At the time of each visual instance the word "pigeon" is heard. Because there is some similarity between different pigeons, some portfolios in V3 tend to be activated relatively often in response to seeing a pigeon. The portfolios activated in A2 in response to hearing the word therefore acquire recommendation strength in favour of activating these frequently active portfolios in V3. After learning, hearing the word "pigeon" will therefore tend to activate a mental image of an averaged pigeon. The portfolios making up that mental image will have recommendation strengths in favour of any learned behaviours in response to seeing pigeons.

A similar association would be learned between the portfolios most frequently activated in V3 in response to visual instances of other types of bird, and the portfolios activated in A2 in response to hearing the word corresponding with the type of bird. The end point of such learning for a number of different types of birds is illustrated in figure 10.6.

However, because of the similarity between different types of bird, many portfolios activated in response to the perception of one type of bird will also be activated in response to the perception of other types of bird. There will be some portfolios which tend to be active relatively frequently for all types of bird as illustrated in figure 10.6. These portfolios will tend to be those most often active when the word "bird" is heard, because that word is heard at the same time as visual perceptions of many different types of bird. As illustrated in figure 10.7, these portfolios will therefore be the ones indirectly activated in V3 by portfolios directly detected in array A2 in response to hearing the word "bird".

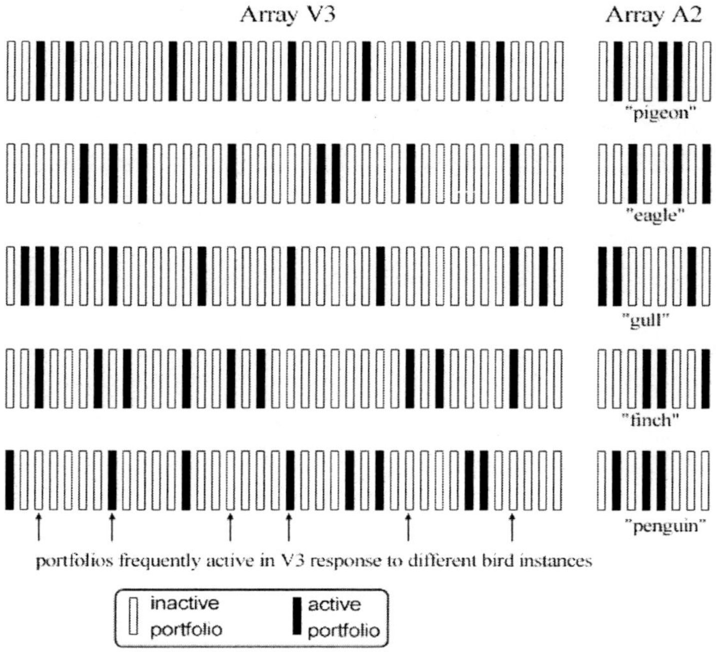

Figure 10.6. Results of learning appropriate responses to hearing the word for a number of different birds when a visual perception of the corresponding bird was present. The portfolios directly activated in At in response to hearing each word will tend to indirectly activate a group of portfolios in V3 which were often directly activated by visual perceptions of the corresponding bird. Because there is some similarity between birds, there will be some portfolios in V3 which will tend to be frequently activated in response to any bird instance.

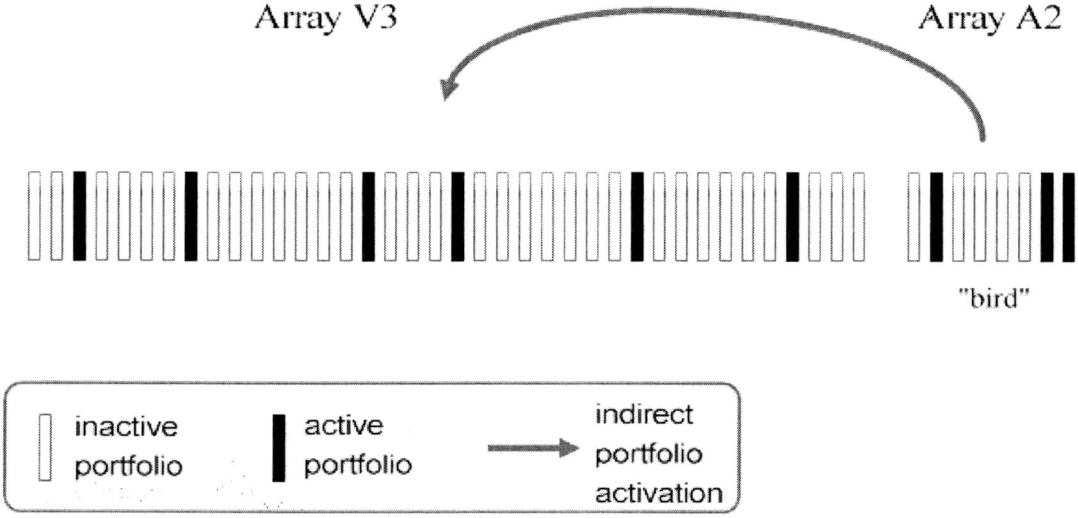

Figure 10.7. Result of learning the word "bird". Hearing the sound activates portfolios in A2, which indirectly activate the portfolios which tend to be directly activated in many different types of bird as illustrated in figure 10.6. The result will be a mental image of an averaged bird.

Such a "bird" activation will be experienced as a pseudovisual perception of an "average" bird, but again without the portfolios activated close to visual input in arrays V1 and V2. Some of these "bird" portfolios will also be activated in response to hearing the word corresponding with an exemplar of the category. However, as illustrated in figure 10.6, a word corresponding with a less typical exemplar of the category , such as "penguin", will tend to result in a lower proportion of the general "bird" portfolios being activated in V3 than a more typical exemplar like "pigeon".

The Category Verification Process

In category verification experiments, a subject is given the names of a category and an exemplar, and asked if the exemplar actually belongs to the category. Consider the example "bird/pigeon". The portfolios indirectly activated in V3 in response to hearing the word "bird" and the portfolios indirectly activated in V3 in response to hearing the word "pigeon" are illustrated in figure 10.8. There is significant overlap between the different portfolio populations, and this overlap is interpreted as a strong recommendations to say "yes, pigeon is a member of the bird category".

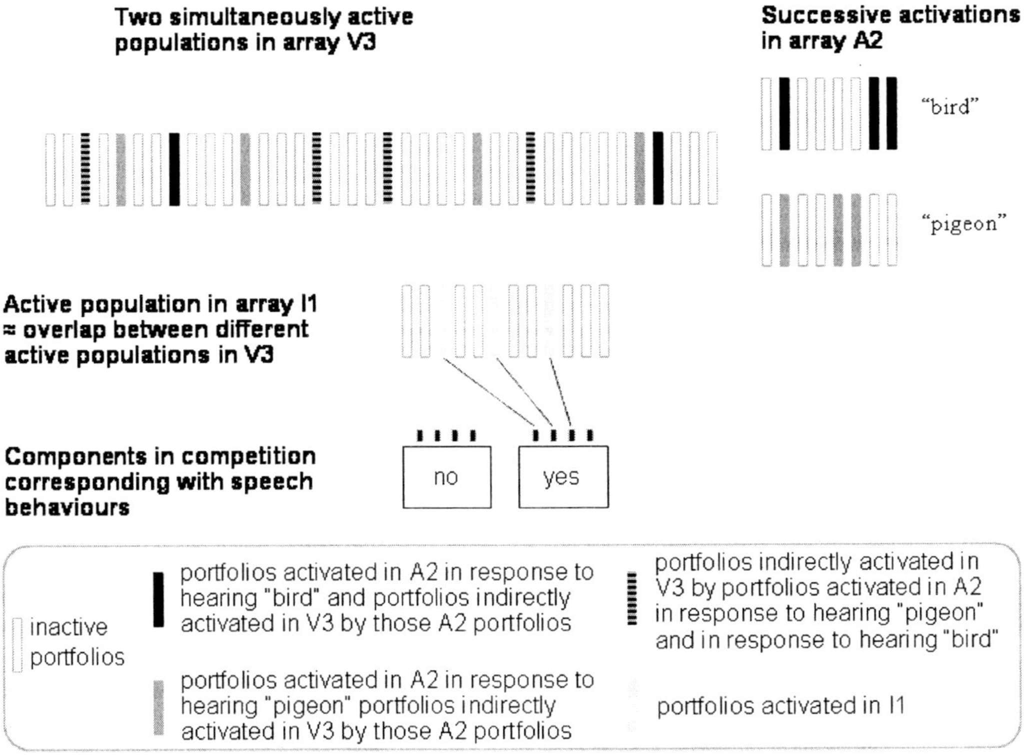

Figure 10.8. Generating a response to the bird/pigeon combination. The two words generate direct activations in array A2 which result in two simultaneously active indirect portfolio populations in array V3. There is significant overlap between these two populations. The portfolios activated in array I1 in response to the presence of this overlap have recommendation strengths in favour of saying "yes, a pigeon is a bird".

Consider next a less typical exemplar: a penguin. Because a penguin is a less typical bird, the overlap between the portfolio populations indirectly activated in V3 in response to the words "bird" and "penguin" is smaller but appreciable as illustrated in figure 10.9. The smaller overlap does not generate enough recommendation strength to immediately identify the penguin as a bird, but has enough recommendation strength to generate a further indirect activation. This indirect activation targets portfolios which have often been active in the past at the same time as the currently active "penguin" portfolios.

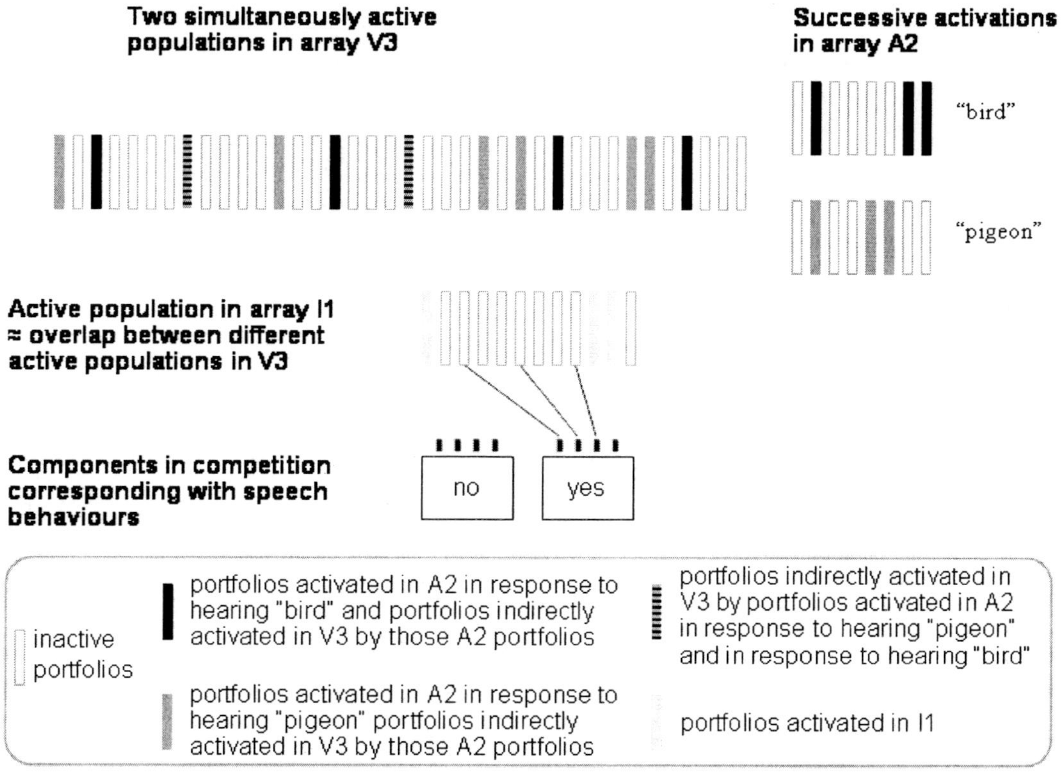

Figure 10.9. Generating a response to the bird/penguin combination. The two words generate direct activations in array A2 which result in two simultaneously active indirect portfolio populations in array V3. There is a small amount of overlap between these two populations. The portfolios activated in array I1 in response to the presence of this overlap do not have adequate recommendation strengths in favour of saying "yes, a penguin is a bird". However, there is recommendation strength in favour of activation other portfolios in V3 which have often been active in the past at the same time as the currently active "penguin" portfolios.

The effect of this recommendation can be understood by consideration of figure 10.10. This figure illustrates the portfolios activated in response to different types of birds. The portfolio population activated indirectly in V3 in response to "penguin" includes a few of the "bird" portfolios and some portfolios activated only in response to "penguin". The population also includes some portfolios activated in response to some other types of bird but not often enough to be a "bird" portfolio. These portfolios correspond with similarity conditions which are somewhat less typical of the category.

However, activation of additional portfolios on the basis of frequent past activation at the same time as the currently active "penguin" portfolios means that these less typical portfolios have recommendation strength in favour of activating other typical "bird" portfolios which were active in other types of bird instances when a less typical portfolio was also active. The existence of these portfolios therefore generates recommendation strength in favour of activating additional "bird" portfolios.

For example, figure 10.10 conceptually illustrates the frequency with which various other portfolios have been active in response to four other types of bird when a portfolio currently active in response to "penguin" are indicated. Adding these frequencies indicates that the "bird" portfolios will be the most strongly favoured for reactivation on this basis. Such reactivations increase the overlap between the "bird" and "penguin" populations to the point at which enough recommendation strength is present to identify as a member of category . The evolving population in array V3 is illustrated in figure 10.11.

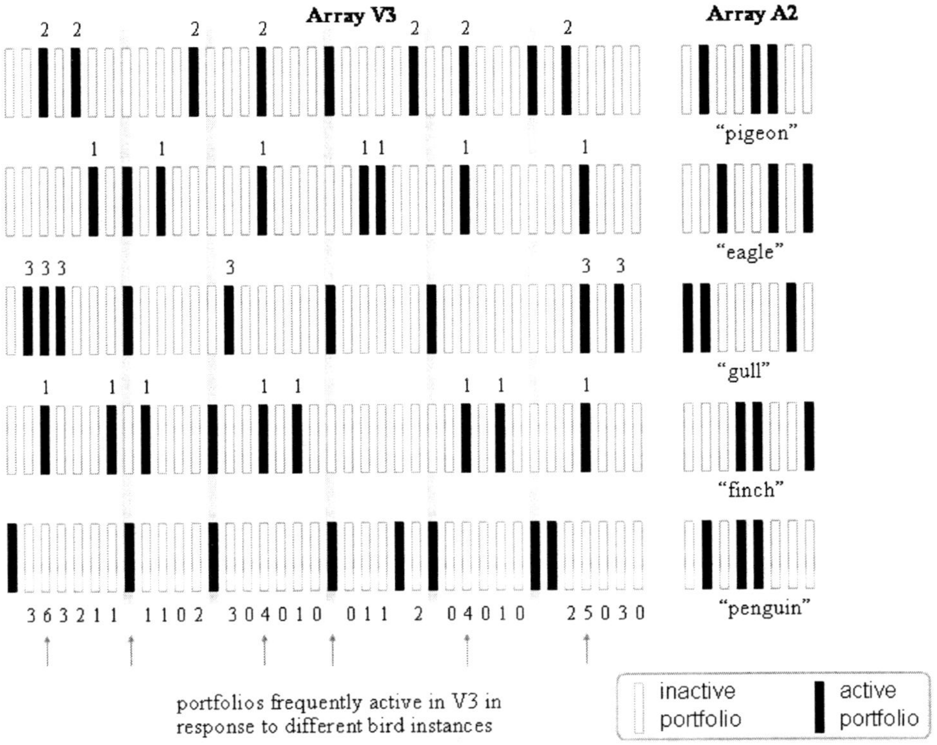

Figure 10.10. Activation history of portfolios indirectly activated in response to hearing the word "penguin". These portfolios have also been active in the past in response to other types of bird, as shown by the vertical grey bars. Other portfolios have also been active at those times. Some of these other portfolios have been active fairly often when the "penguin" portfolios have also been active. The numbers above the non-penguin active portfolios in each array response indicate the number of different penguin portfolios which were active at the same time within that other bird response. The total at the bottom therefore indicate how often a portfolio within the penguin activation was active in the past at the same time as the portfolio inactive in the penguin portfolio. The larger the number, the larger the recommendation strength possessed by the penguin portfolios in favour of activating the other portfolio.

If the category/exemplar "bird/cat" were offered, the overlap between the indirectly activated populations in V3 would be minimal and there would be immediate recommendation strength from activated portfolios in array I1 in favour of identifying that cat was not a member of category bird.

The additional indirect activation will require extra time. Identification of less typical exemplars will therefore take longer than typical and atypical exemplars, in agreement with the observations of Rips, Shoben and Smith [1973].

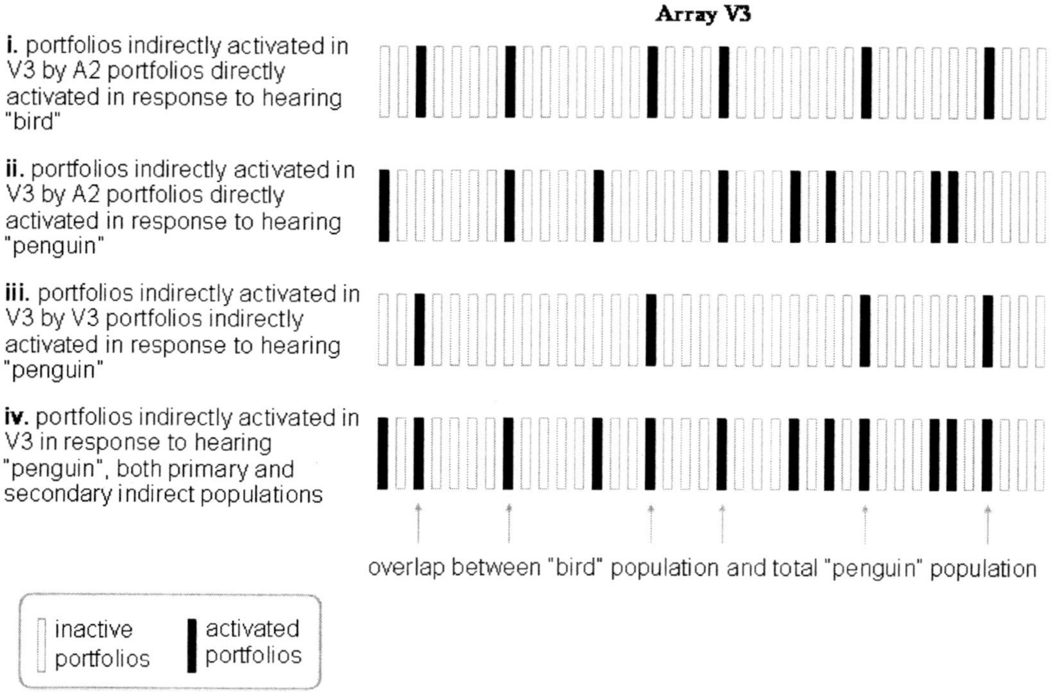

Figure 10.11. Steps to reach a response to an atypical category/exemplar combination bird/penguin. The population of portfolios indirectly activated in array V3 in response to hearing "bird" (i) has only a small overlap with the population of portfolios indirectly activated in array V3 in response to hearing "penguin" (ii). However, this small overlap results in indirect activation of additional portfolios (iii) which have often been active in the past at the same time as the "penguin" portfolios. When the additional portfolios are added to the "penguin" population (iv), the total population has significant overlap with the population of portfolios indirectly activated in array V3 in response to hearing "bird" (i). This overlap is the basis for identifying "penguin" as a member of category "bird".

In contrast with the spreading activation model of Collins and Loftus [1975], there are no units which correspond with concepts like categories or the features of categories. Portfolios are groups of conditions in which there has been a degree of randomness in the definition of individual conditions, but conditions within one portfolio have some similarity with each other and have tended to occur at similar times in the past. A portfolio may therefore have a probabilistic correlation with many different features and categories, with the probabilities expressed, for example, as recommendation weights into naming the features or categories. As suggested by Rosch [1973] therefore, natural categories have fuzzy membership.

Barsalou [1983] has emphasized that models of categorization must also account for the existence of ad hoc categories which are not based on similarity. Examples include "birthday presents for your mother" and "things to take on a camping trip". The recommendation architecture model provides a straightforward explanation for such categories. For example, in the case of "things to take on a camping trip", the name of each thing will activate a population of portfolios. Because the thing has been taken on a camping trip in the past, the population will include a subset which would also be activated while camping. The portfolios in this subset will overlap with portfolios activated in response to the words "camping trip". This overlap is the basis for categorization as described earlier.

In the tip of the tongue experiments [Brown and McNeill 1966], hearing the definition of the infrequent word activated a population of portfolios. Because the word was infrequent, activation of portfolios often active at the same time was not quite adequate to generate enough recommendation weight to speak the word. However, the active population contains enough recommendation strength in favour of different aspects of the word, and some of these aspects can therefore be named.

Semantic Memory Deficits

Damage to the brain can lead to loss of certain types of semantic memory but not others, demonstrating some degree of dissociation between the memory types. In the recommendation architecture model, semantic memory is dependent on two factors. One is the presence of appropriate information, for example in the form of portfolios with recommendation strengths to generate correct verbal responses. The other is the ability to activate the appropriate population of portfolios, generally indirectly through a series of intermediate active populations beginning with a population activated directly by conditions within auditory or visual inputs. Local damage could remove a specific group of portfolios or remove the capability to activate some groups in response to activity of other groups.

Because portfolios are partially ambiguous in cognitive terms, with no one portfolio or group of portfolios correlating exactly with a cognitive category, local damage would be very unlikely to remove all the portfolios relevant to any one such category. However, as discussed earlier, substantial condition recording and appropriate connectivity resources can be assigned a priori to support important types of behaviour. This a priori assignment will result in portfolios relevant to one such behaviour type being concentrated in one physical region. Portfolios and their associations with behaviours are defined heuristically, and therefore all the portfolios relevant to such a behaviour type will not necessarily be within that region, and some portfolios relevant to other behaviours will be within the region. However, physical damage to the region will result in a deficit predominantly in the behaviour type.

Consider for example the observations of deficits in which a subject can draw inanimate objects in response to the name of the object but when asked to draw animals can only generate a generic sketch [Satori, Job and Coltheart, 1993]. The model in this case is that because of the behavioural importance of responses to animals of all kinds, substantial resources are targeted a priori on conditions which are comparable to some degree with a generic sketch. Drawing behaviours develop by associating hand movements with conditions within these resources, and have the effect of reproducing the sensory inputs which generated the conditions. If extensive damage to the condition recording resources left only a small proportion, the residual conditions detected would tend to include a high proportion of the genetically programmed type corresponding with the generic sketch, and drawings would therefore be limited to that sketch.

Local damage could also limit the ability of portfolios on one specific level to activate portfolios on a specific other level. Consider for example the observations of mental deficits in which the ability to generate a verbal definition of a concrete word (e.g. needle, goose, acorn) is severely impaired, but the ability to generate the definition of an abstract word (e.g. supplication, satirical) is much less impaired [Warrington, 1975]. As discussed in section 3.1, portfolios activated directly by auditory input of a more abstract word will tend to indirectly activate portfolios in a higher level than portfolios indirectly activated in response to a more specific word. Thus in figure 10.1, indirect activation would occur in array V4 in response to an abstract word but array V3 in response to a specific word. Local damage interfering with indirect activation of portfolios in array V3 would therefore generate the observed deficit.

Similarly, local damage to one form of indirect activation but not others can account for deficits in which a subject can provide information in response to a picture of an object but not in response to the name of the object [McCarthy and Warrington, 1988], or subjects who can generate the appropriate word in response to a non-perceptual definition (e.g. "lion" in response to "king of the jungle") but not in response to a perceptual definition (e.g. "zebra" in response to "black and white striped horse") [Silveri and Gainotti, 1988].

Episodic Memory

Episodic memory is defined as the memory of the past with a context of what else happened at a similar time, in contrast with semantic memory in which, for example, memories of facts are detached from memories of where those facts were learned [Tulving 2002]. As a preliminary step towards describing the recommendation architecture processes which support episodic memory, the types of question which can be answered by such memory will be considered. There are perhaps three such types of question, all dealing with temporal relationships between experiences. Experiences in this context includes perceptions of the outside world, perceptions of internal feelings and emotions, and reminiscences of earlier perceptions (i.e. experiences of episodic memory or other types of memory).

The first type of question addresses the relative timing of different experiences, in other words "Did experience X happen before, at the same time as, or after experience Y?" or "What happened before, at the same time as, or after experience X?". The second type of question addresses the timing of experiences relative to some reference time sequence, in other words "What happened at time T ?" or "At what time T did event X happen?". This type of question requires a reference time sequence which can be accessed by memory. The third type of question addresses the amount of time which passed between two experiences, in other words "How long ago did experience X occur?" or "How long was the period between experience X and experience Y?". This type of question requires some reference time units. If the reference time units are defined by a reference time sequence, this question may reduce to the second type.

Examples of questions on the relative timing of experiences include "What did you do after breakfast this morning?" or "What were your feelings when you heard that you had got the new job?". The subjective memory experience is that the reminiscence of one experience generates the reminiscence of another.

Questions of timing relative to a reference sequence require such sequences. Examples of such sequences include the conceptual clock and calendar, we have frameworks of seconds, minutes, hours, days, weeks, months and years into which we fit memories. In addition to this "public" framework, we each have a number of frameworks derived from personal experience. One such framework is the sequence of places where we have lived. Another is the sequence of major relationships (marriage, divorce, births of children, deaths of family members, important friendships etc.). A third may be the sequence of jobs held. A fourth may be the semi-standard sequence of activities on a typical work day. Mapping exists to some degree indicating points of correspondence between different sequences. Examples of questions which can be answered are "did you get the red coat before or after you moved to the new house ?" or "What time did you place the phone call yesterday?". The subjective memory experience is that the reminiscence of one experience is compared with reminiscences of a number of reference experiences, and perhaps other reminiscences generated by reminiscences of reference experiences, to find a match.

Questions of duration between experiences require reference time units to provide a sense of duration. Examples of such questions might be "How long have you been waiting?" (in the absence of a timepiece), or "How long has it been since you were last here?" These units of duration anecdotally appear somewhat less reliable than the use of reference sequences in terms of match to objective scientific time. For example, we often have the impression that time passed more slowly earlier in our lives, and that the perceived duration of an experience depends on the nature of the experience (e.g. depending on the presence or absence of boredom). The experience is of assessing the quantity of something which exists between two experiences.

In practice, some accesses to episodic memory could incorporate several types of question. Thus a question like "Was the first landing on the moon before or after the assassination of President Kennedy?" might involve accessing memories of watching the landing (type 1) and fitting those memories on a reference time sequence (type 2), then doing the same for hearing news of the Kennedy assassination, and comparing the reference sequence times. Alternatively, generating a response might involve comparing the subjective length of time since the memories of the two events (type 3). Semantic memory could also be involved.

The point in discussing these three types of question is to suggest that they appear to indicate three types of capability. Type one allows access to memories relatively close in time to a memory which is already being accessed. Type two allows access to more distant memories, but requires use of a temporal mapping function. Type three allows subjective assessment of duration between experiences on a wide range of timescales, but with limited accuracy.

These three types of temporally related information access capabilities can be mapped into three recommendation architecture capabilities. The first capability is that there is recommendation strength associated with any active portfolio in favour of activating other portfolios which recorded conditions in the past at the same time as, or shortly before or after, an episode of condition recording in the active portfolio. The second capability is based upon the fact that a period of similar experience will tend to result in significant condition recording consistently across a limited set of portfolios throughout the period, and less recording in those portfolios outside the period. For example, living in one particular place for a period will result in an ongoing level of recording of conditions which are similar because they include information derived from perception of that one place. Similar conditions tend to be recorded in the same portfolios, which therefore correlate to a degree with the place. Any experience within the period which results in condition recording will therefore tend to have conditions recorded in the "place" related portfolios either shortly before, at the same time as, or shortly afterward experience related conditions. Hence activation of information related to the experience will tend to activate information related to the place and vice versa. The third capability is based on the fact that condition recording occurs continuously throughout life. Hence the volume of condition recording which has occurred between two experiences is a very rough indication of the elapse time between them.

The phenomena of episodic memory have been classified into a number of partially overlapping categories [Neisser and Libby 2000]. One category is recall, which can be targeted, cued, or involuntary. A second is autobiographical memory, with the particular case of major life narratives. A third is dating of events and establishing a temporal order. A fourth is memory for major events, with the particular cases of eyewitness and flashbulb memories. A fifth is creation of false memories. Techniques such as questionnaires and diary studies are used to investigate the general reliability and effectiveness of episodic memory.

Episodic Recall

In targeted recall subjects are asked to recall particular past events [Neisser and Libby 2000]. In cued recall, subjects are given a cue which may be a word [Robinson 1976] or a type such as "vivid memories" [Rubin and Kozin 1984]. In involuntary recall, some specific environmental stimulus such as a smell or a taste brings a memory to mind unsought [Berntsen 1996; 1998].

The processes resulting in episodic recall can be understood by consideration of figure 10.12. The scenario addressed in the figure is that of a subject being asked what they remember about the Bali bombing. The sounds of the three words "remember", "Bali" and "bombing" each activate portfolios in array A2 which in turn indirectly activate portfolios in array V3. The portfolios activated in V3 in response to hearing the word "remember" have recommendation strength in favour of a sequence of internal activation behaviours.

This sequence is to maintain the indirect activations in V3 in response to the next words simultaneously active at different modulation; then to synchronize the outputs of the populations and release to V4. Next to accept recommendations to activate portfolios in V3,

V4 and V5 which recorded conditions in the past at the same time as the active portfolios in V4, then to repeat this type of activation process several times.

If the target words have been chosen appropriately, the portfolio population initially activated in V4 in response to those words will have some overlap with the population which was active when the subject first experienced hearing about the target event, and this overlap will be greater than for any other prior experience. The process of successive activations on the basis of past simultaneous condition recording will therefore result in populations with successively higher overlap with the population during the target experience, provided that there was a significant level of condition recording during that experience.

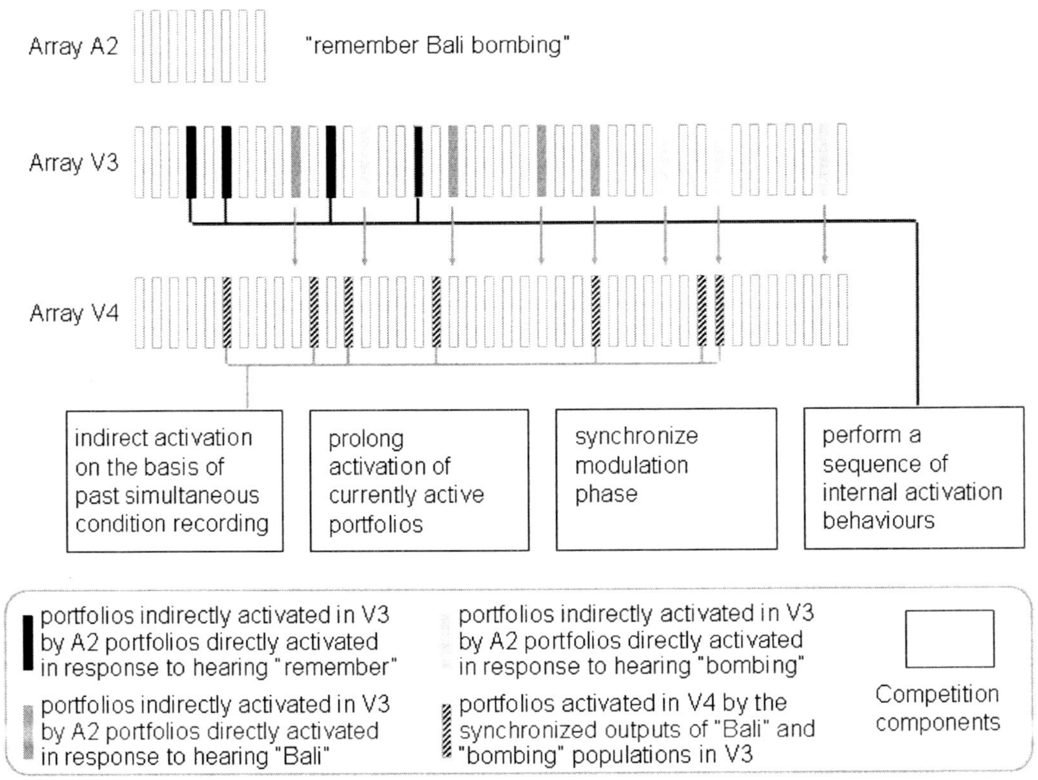

Figure 10.12. Process for episodic recall. The three words "remember Bali bombing" generate recall of what has been heard and seen about the Bali bombing. Portfolios activated in V3 in response to hearing "remember" have recommendation strength in favour of performing a sequence of activation behaviours. This sequence is to synchronize the outputs from multiple populations in V3 and release to V4, then to generate a series of indirectly activated populations in V4 from the resultant population activated in V4. These indirect populations are generated on the basis of simultaneous past condition recording. The effect is to move the activated population in V4 towards the population that was active when the subject first had news about the Bali bombing. This population has recommendation strengths in favour of speech behaviours describing that news.

The resultant population will be experienced as a general re-perception of the original experience, although the portfolios in V1 and V2 closest to input from the senses are not reactivated. The activated portfolios in this population have recommendation strength in favour of, for example, generating verbal descriptions of the event description. Use of recommendation strength in favour of activating portfolios which recorded conditions somewhat before or somewhat after condition recording in currently active portfolios allows the re-perception to be set at the beginning of the experience and to be moved through the experience.

Recommendation strengths will always also be present in favour of activating portfolios on the basis of simultaneous past activity and simultaneous past recording during other experiences. The activated population is therefore unlikely to be an exact match for the original, although in general the higher the level of condition recording during the event, the greater the probability of a close match.

Cued recall operates in a very similar fashion. However, the initial secondary population may contain portfolios which recorded conditions during a number of past events. In the absence of specific indication of one event in the verbal cue, the tertiary population will evolve towards the event which happened to be represented by the highest degree of condition recording in the initial population. Events which resulted in a high degree of condition recording across many portfolios will tend to be the end points of this process.

Involuntary recall is the result of strong condition recording in portfolios activated in response to a sensory stimulus (for example, a novel smell or taste) at the same time as strong condition recording in other portfolios in response to some event. A later repetition of the sensory stimulus activates the portfolios which originally responded to that stimulus. These portfolios in turn activate portfolios which recorded conditions at the same time in the past, resulting in a re-perception of the event.

Autobiographical Memory

Most people can provide some account of the course of their lives up to the present. This account is known as the life narrative [Neisser and Libby, 2000]. Such accounts contain information about places lived, jobs held, and major events like marriage and birth of children. Events we have been told about rather than remember personally (such as place of birth) may be included. Autobiographical memory can be defined to include all forms of self related memory [Brewer 1986], or restricted to particularly significant memories [Nelson 1993]. Some events are particularly important for the construction of the life narrative [Bruner, 1994; Pillemer, 1998]. The existence of a life narrative is important for locating particular events in time, because the particular event can be related to life narrative events with a known date [Friedman 1993].

A key element supporting autobiographical memory in the recommendation architecture model is the existence of consistent condition recording in particular groups of portfolios during time periods in which there is a major consistent factor in the perceived environment. For example, while living in a particular house in a particular city, there will tend to be condition recording in portfolios which resemble that environment, with a particularly strong period of recording when moving there for the first time. The existence of different such portfolio sets for different life periods constitutes a temporal map which can be used to access memories from different periods using mechanisms similar to those for episodic recall memory. Such maps may be based on any major, consistent element in the perceived environment, including employment and presence or absence of major family members or friends. Presence or absence of consistent simultaneous condition recording between portfolios corresponding with different maps provides correlations between the maps.

For example, an active population generated in response to some cue indicating a previous home can generate secondary populations corresponding with events which occurred while living there, and vice versa. Recommendation strengths in favour of activating portfolios which recorded conditions slightly before or slightly after recording in currently active portfolios makes it possible to activate populations corresponding with events along a timeline. Because of a high degree of condition recording during major life events, activations corresponding with such events can be most easily achieved, and in turn can be used to generate activations corresponding with slightly earlier and slightly later events. Recommendation strengths in favour of speaking dates can be associated with these populations, although as always, one portfolio alone may recommend a number of different dates corresponding with different events during which the portfolio was active, and a high integrity date is determined by adding recommendation strengths across all active portfolios.

The life narrative is therefore generated by activating an initial population corresponding with the earliest consistent period of condition recording on a map, then activating secondary populations corresponding with major episodes of condition recording which occurred at the same time. Activation on the basis of slightly later or slightly earlier condition recording than in the current population permits generation of a sequence of tertiary populations corresponding with major events in temporal sequence. The final major event is the move to the next period, after which populations corresponding with the major events during that period can be activated An activated population corresponding with an event has recommendation strengths in favour of various verbal descriptions of the event.

Flashbulb Memories

The concept of flashbulb memories derives from the observations of Brown and Kulik [1977] that memories for major news stories such as the assassination of President Kennedy were often associated with remarkably detailed recollections of the circumstances in which the news was first heard. However, as with episodic memory in general, the memories are not totally accurate: McCloskey, Wible and Cohen [1988] found that only about 60% of recollections were entirely consistent across tests performed 1 week and 9 months after a major news event.

In the recommendation architecture model, there are no special mechanisms for flashbulb memory. The phenomenon depends upon the novel and perceived importance of the news triggering a high level of condition recording across a number of portfolios. As a result, there will be a high level of initial indirect activation connectivity from these portfolios to portfolios recording conditions derived from the circumstances in which the news was heard, and in some cases conditions will combine information derived from the news and information derived from the circumstances.

Reminiscence of the news story will tend to be frequent, and such reminiscence is activation of the portfolios within which conditions were recorded when the news was first heard. Because of the very high level of condition recording, these portfolios will tend to activate portfolios which recorded conditions within the circumstances at the time. Because the news will tend to be recalled frequently in the days after the event, the indirect activation strengths wiil not decay rapidly. Hence at a much later date, reminders of the news item will retain a strong capability to activate reminiscence of the circumstances.

However, as for other forms of episodic memory, other portfolios often active at the time of reminiscences may record information at the time of the reminiscence and may later be activated as a result of detected simultaneous condition recording and become part of the active portfolio population which recommends verbalization of the memory of the original experience, reducing the accuracy of the memory.

Eye Witness Memories

In eye witness memory experiments [e.g. Thompson, Herrmann, Read, Bruce, Payne and Toglia, 1998], subjects are asked to recall as many details about an event as possible. Laboratory research typically uses videotapes or pictures and may not result in the same degree of emotional arousal as directly witnessing a crime. Such differences are believed to influence results [Yuille 1993]. Thus in laboratory studies, witnesses sometimes mistake an innocent bystander as the criminal in a photo lineup [Ross, Ceci, Dunning and Taglia, 1994], but field studies suggest that the presence of a bystander makes identification more rather than less accurate [Read, Tollestrup, Hammersley, McFadzen and Christensen 1990].

In the recommendation architecture model, eye witness memory depends upon similar mechanisms to episodic recall memory. However, the intense emotion of real life witness of a crime will be interpreted as a recommendation to record a high level of conditions. This high level of recording will make the reminiscence population constructed later a better approximation to the original population and real life eye witness memory will tend to be more reliable.

A major interest in the study of eye witness memory is the accuracy of such memories. In the recommendation architecture model, this accuracy in determined by the fact that recommendation strengths may be present for a wide range of secondary activation behaviours. An active population approximating the population during a witnessed event may therefore include some portfolios which were not active at the time of the event but have often been active at other times in the past at the same time as portfolios active during the event. Other irrelevant portfolios could be those which recorded conditions at the same time as portfolios active during the event, but during a different event. If the time between the events is short, it will be harder to keep total recommendation strength for these irrelevant portfolios below activation point. Thus if subjects are shown a video of a dramatic event, and half the subjects afterwards receive incorrect written information about the event (such as the colour of a vehicle), the false information is sometimes the information these subjects recall about the event [Loftus, 1979; Loftus, Levidow and Duensing 1992].

The cognitive interview approach was developed by Geiselman and Fisher and their collaborators [Geiselman, Fisher, MacKinnon and Holland, 1986] in an attempt to maximize recollection accuracy. This approach emphasizes witnesses recalling their personal and environmental context at the time of the event; recalling component events in different orders, and recalling the event from different perspectives. The benefit of this approach can be understood in recommendation architecture terms as approaching the activation corresponding with the eyewitness event from different directions, with the probable effect that consistent reports are more likely to correspond with the original event.

False Memories

The study of false memories had its origins in attempts to determine the validity of memories in which individuals report having discovered long forgotten memories of trauma. Such memories are referred to as "recovered", or in recognition that the accuracy of such memories may be in doubt, "discovered" [Schooler, Amador and Dendiksen, 1997].

Loftus, Coan and Pickrell [1996] demonstrated that is was possible to create the memory of an event in a subject which a close relative indicated had never taken place. In a more extensive study of this phenomenon, Loftus and Pickrell [1995] investigated 24 adult subjects. Descriptions of four childhood events were developed in consultation with a close relative of each subject, three which had actually occurred and one which had not. The false event was getting lost in a shopping mall, and the description included a shopping mall often visited in childhood, family members who often went on such visits, and information on the type of shops which interested the subject. Subjects were asked to provide additional written information (or write "I do not remember") in response to a written paragraph describing the event. They were then asked for supplementary memories of each event at two subsequent interviews. At the end of the second interview they were told that one memory was false and asked to identify it.

Five of the subjects chose a genuine event as the false memory at the end of the second interview. In some cases the subjects held strongly to the truth of the false memory, looking for additional confirmation that it was indeed false. However, in the original written response, the seven subjects who produced a full or partial false memory on average used many more words to describe their recollections for true memories than for false memories (138 words vs. 50 words). Only one produced more words for the false memory than average for the true memories (21 words vs. 20 words) and this was the subject using the least words overall.

In the recommendation architecture model, determination of whether a past event occurred depends upon the words describing it activating a strong secondary population on the basis of simultaneous past condition recording. The overall word count difference indicates that a stronger population was generated for the genuine memories. However, conditions will also be recorded during generation of the written response and during the subsequent interviews. As a result, the probability of a population of portfolios corresponding with a false event being activated on the basis of simultaneous past recording is increased. In most cases, a number of additional factors help to distinguish between such secondary recording and a real event. One factor is the presence or absence of activated portfolios containing relatively simple conditions close to sensory input, the absence of such conditions indicating that the event was only imagined. Another factor is the absence of portfolios containing conditions derived from primary environments at the time such as places lived in or often visited and family members. The use of family members and familiar places in the false memory would reduce the effectiveness of these factors and increase the chance of an imagining being confused with a real event.

Overall, recollection is not a precise reconstruction of a past activation, but the best approximation which can be achieved using available recommendation strengths. Even though there is a higher level recommendation in favour of activation on the basis of simultaneous past recording, there may be recommendation strength in favour of activation on the basis of simultaneous past activity which is sufficiently strong to activate some portfolios. Such portfolios will further dilute the accuracy of the memory. Thus a recollection is not an exact reproduction of the original activation, it is an active reconstruction of past events [Bartlett 1932] but can be overlaid with activation on the basis of later simultaneous condition recording and simultaneous activity.

Autonoetic Awareness

A critical element in Tulving's definition of episodic memory [Tulving 1983] is the presence of a type of awareness that includes idiosyncratic perspectives, emotions and thoughts at the time of the event being remembered. This type of conscious awareness accompanying retrieval is known as autonoetic awareness [Wheeler 2000]. The presence or absence of such personal recall is generally taken to be the primary distinction between episodic and semantic memory, although there are circumstances in which the distinction may not be completely clear and memory may shift from episodic to semantic [Conway, Gardiner, Perfect, Andresen and Cohen 1997].

In the recommendation architecture, the distinction between episodic and semantic memory is on the basis of whether the activation of the population corresponding with the memory is on the basis of past simultaneous recording of conditions or frequent past simultaneous activity. Activation on the basis of frequent simultaneous past activity will in most cases not favour portfolios corresponding with emotional states or thoughts at particular times, while activation on the basis of past simultaneous recording of conditions will tend to include such portfolios. The mechanism for access to episodic memories thus results in conscious awareness as discussed in chapter 12. A further factor is that conditions of a complexity most likely to be useful for activation on the basis of frequent simultaneous past activity may be different from conditions of a complexity most likely to be useful for activation on the basis of past simultaneous recording. Such a difference could result in regional specialization, with recording of conditions in response to many experiences occurring in both regions.

Perception of Elapse Time

The perception of elapse time can be based on the degree of condition recording over a period of time. However, because the degree of recording during an experience is dependent upon the novelty of the experience, this measure of elapse time gives an indication but is somewhat unreliable. During a period with a high degree of novelty (e.g. a period of intense learning) the degree of condition recording will be high. Later, such a period will therefore be perceived on the basis of degree of condition recording to have passed more slowly than a comparable time period during which novelty was lower. This perception accounts for the anecdotal observation that in retrospect for many people, time seems to have passed more slowly in childhood than later in life. During an experience with little novelty, fewer than average conditions are recorded per minute. Hence the time for recording the average number is perceived as significantly longer than a minute by the clock. This perception accounts for the anecdotal observation that subjective time passes very slowly during boring activities.

Management of Connectivity in Support of Episodic Memory

One issue is how connectivity to support activation on the basis of temporally correlated condition recording can be established. One way is to establish provisional connectivity from all portfolios to all other portfolios, but this approach is very resource intensive.

An alternative approach is to utilize the information implicit in the clustering resource management subsystem. This subsystem maintains a map of clustering device and column resources. Unused resources are assigned on the basis of recent condition recording in existing columns, using recent activity as the best available indicator of probable future activity. The resource management system also manages establishing appropriate connectivity to these newly assigned resources.

The resource management subsystem can therefore be used to efficiently target provisional connectivity to support indirect activation on the basis of temporally correlated condition recording. This efficient targeting is based on the assumption that the most probable locations for temporally correlated condition recording in the immediate future are the locations to which new resources have been assigned. Management of this provisional connectivity is therefore a plausible additional role for the resource management subsystem. The deficits resulting from damage to the Hippocampus system support the view that this system is the resource management subsystem performing the two roles. The evidence from deficits observed as a result of damage to the Hippocampus system in Korsakoff's Syndrome patients will be discussed in the next section. Evidence from the study of the patient HM will be discussed in more detail in a later section.

Episodic Memory Deficits

Because recording of new conditions is essential for creation of both new semantic and new episodic memories, loss of the ability to record conditions will result in loss of both types of memory. Thus in patients like HM [Milner, Corkin, and Teuber, 1968] and Korsakoff's Syndrome patients [Butters, 1984], both capabilities are lost.

However, damage affecting only the capability to activate on the basis of past simultaneous activity would be expected to affect only semantic memory, and damage affecting only the capability to activate on the basis of past simultaneous condition recording would be expected to affect only episodic memory. With brain damage primarily confined to front portion of temporal lobe, a patient no longer remembered the meanings of common words. All memory of historical events and famous people; of geography; of social customs; of attributes of animate and inanimate objects, their smells, colours, textures; was lost. However a vivid and accurate recollection of engagement and honeymoon; her father's death; her childrens' births and many incidents in the past remained [DeRenzi, Liotti and Nichelli, 1987].

The opposite case is source amnesia [Schacter, Harbluk and McLachlan, 1984] in which facts can be remembered, but not the circumstances in which the facts were learned. This phenomenon is in fact widespread in normal subjects, since most of us can remember, for example, that Paris is the capital of France but do not remember the circumstances in which this fact was learned [Wheeler, 2000]. This normal source amnesia is based on the initially rapid and subsequent more gradual decline of recommendation strength in favour of activation on the basis of simultaneous recording unless the recommendation strength is used on a number of subsequent occasions to activate portfolios. Recall of the fact based on simultaneous past activity results in preservation of recommendation strength in favour of such recall.

However, in some patients with lesions to the prefrontal cortex, source amnesia is much stronger and more pervasive [Janowsky, Shimamura and Squire, 1989]. Such patients when tested on recall of facts acquired in a laboratory learning episode one week before can remember the facts but are much less likely than normal subjects to remember that the fact was learned in the earlier laboratory episode, and more likely to attribute the learning of different facts to that episode.

The associations and dissociations between semantic and episodic memory deficits are thus consistent with the recommendation architecture model.

Memory and Sleep

A major requirement of a system with the recommendation architecture is for a process to heuristically minimize information exchange between modules at all levels. Such information exchange is reflected in the incorporation of conditions detected by one module in conditions detected by another module. The management of information exchange is therefore the management of the process by which provisional conditions are defined.

A compromise is needed between the operational advantages and the increased architectural complexity resulting from additional information exchanges between modules. As discussed earlier, the indicator of the probable functional advantage of an information exchange between two unconnected modules is frequent simultaneous activity of the two modules in the past. One way to determine the degree of such correlated activity which avoids the need to collect and store large amounts of information over a long period of experience is to perform an approximate input driven rerun of experience with a bias towards the more recent.

REM sleep closely matches this information exchange minimization process [Coward 1990]. It is an off-line process which includes an apparent rerun of experience, with a bias towards the more recent [Wilson and McNaughton, 1994]. Note that this process is radically different from the memory consolidation process proposed as the functional role of sleep with dreaming originally by Marr (1969 - 1971 [1991]). No memory consolidation is required in the information minimization process. Provided that provisional connectivity assignment continued on a random basis without the bias, the major effects of deprivation of REM sleep would be a long term increase in the difficulty of learning new behaviour without interference with existing behaviour. In other words, memory for objects and events would still operate. This is consistent with the absence of a major effect on memory following the reduction or elimination of REM sleep by antidepressant drugs [Vertes and Eastman, 2000].

The Amnesic Patient HM and the Memory Consolidation Model

The patient HM was treated for epilepsy in 1953 by resection of his medial temporal lobe. Although the epilepsy improved, the resection also resulted in severe memory deficits [Scoville and Milner 1957]. These memory deficits have been extensively studied over the past 50 years [see e.g. Milner, Corkin and Teuber 1968; Corkin 2002]. The resection was bilaterally symmetrical and included about half of the hippocampus proper and most of the entorhinal cortex and amygdaloid complex [Corkin, Amaral, Gonzalez, Johnson and Hyman 1997].

HM shows global semantic and episodic anterograde amnesia [Corkin 2002], but his comprehension and use of language is relatively undisturbed [Kensinger, Ullman and Corkin 2001] He is impaired in the acquisition of new words but retains the vocabulary acquired prior to 1953 [Gabrieli, Cohen and Corkin 1988]. He also exhibits a retrograde amnesia back to age 16, or 11 years prior to his operation [Sagar, Cohen, Corkin and Growden 1985]. He retains a good capability to learn new motor skills [Corkin 1968]. He exhibits almost normal perceptual priming, and his word stem completion priming is normal for pre 1953 words but impaired for post 1965 words [Postle and Corkin 1968].

Observation of HM's specific combination of deficits led Scoville and Milner [1957] to propose an influential model for memory. In this model, information from the senses is registered in the neocortex, then initially bound into a memory trace by the hippocampal system. This initial binding is completed in the short term (i.e. seconds to at most tens of minutes). A subsequent long term consolidation process reduces the role of the hippocampal system until the neocortex sustains the memory trace alone.

This model was proposed to account for the loss of the ability to create new semantic and episodic memories observed to occur after damage to the hippocampal system. However, the model has the issue that retrograde amnesia also associated with the same damage can affect memories as far back as 30 years or more [Nadel and Moscovich 1997]. Furthermore, there have been no proposals for why in the case of HM episodic memory is lost for 11 years prior to the resection but semantic memory (e.g. word meanings) for the same period is not lost, or for the mechanisms by which perceptual information is initially coded in the neocortex, transferred partially to the hippocampal system, and then transferred back again.

In the recommendation architecture model, the understanding of the deficits observed in HM is more straightforward, and the nature and location of the information recorded is clear. There are three related roles for the hippocampal system discussed in chapter 9. The first is to assign cortex condition recording resources as required to cortex modules, perhaps using a map of such resources. The second is a subset of the first role, and is to manage the creation of appropriate connectivity defining provisional conditions, using a partial rerun of recent experience to bias the connectivity. The third is to maintain the connectivity between cortex modules which recorded conditions at the same time. This long term connectivity maintenance is a natural role, because resources assigned at the same time (and generally used shortly after assignment) are more likely to record conditions at similar times.

On this model, damage to the hippocampal system could have two effects. One is to destroy the condition recording resource assignment process. The effect of such a destruction would be an inability to create new semantic or episodic memories. The second is to destroy the connectivity which allows indirect portfolio activation on the basis of time related condition recording. The effect of such destruction would be an inability to converge a secondary portfolio population to approximate a particular time period on the basis of simultaneous condition recording. Episodic memory for the time period corresponding with the damage would be lost. However, portfolios could still be activated on the basis of frequent simultaneous past activity, and semantic memory acquired in the same period would be accessible. Thus as in the case of HM, memory of word meanings would still be present for the period in which access to episodic memory had been lost. After damage, a population of portfolios created before damage would still be activated directly by sensory inputs and then other portfolios indirectly on the basis of frequent simultaneous past activity. Any behaviour which could be managed by existing portfolios, such as language comprehension and use to the level present before damage, would therefore be unaffected. Furthermore, Consequence feedback could associate whatever portfolios were activated with motor behaviours. However, portfolios could not be evolved to discriminate more effectively between perceptions corresponding with different skilled behaviours. Hence learning motor skills would be possible, but learning the meaning of new words would not. Similarly, procedural priming and word-stem completion priming for pre-damage words would be unaffected but word-stem completion priming for novel words would be impaired.

The recommendation architecture model can therefore give a simpler and more accurate account for the specific combination of deficits observed in HM without the need for memory related information to be moved back and forth between the cortex and hippocampal system.

References

Baddeley, A. D. (1986). Oxford: Oxford University Press.

Barsalou, L.W. (1983). Ad hoc categories. , , 211-227.

Bartlett, F. C. (1932). .
Cambridge: Cambridge University Press.

Berntsen, D. (1996). Involuntary autobiographical memories.
10, 435 - 454.

Berntsen, D. (1998). Voluntary and involuntary access to autobiographical memories.
6, 113 - 141.

Brewer, W. F. (1986). What is autobiographical memory? In D. C. Rubin (Ed.).
. New York: Cambridge University Press.

Brown, R. and Kulik, J. (1977). Flashbulb memories. 5, 73 – 99.

Brown, R. and McNeill, D. (1966). The "tip of the tongue" phenomenon.
5, 325 – 337.

Bruner, J. (1994). The "remembered self". In U. Neisser and R. Fivush (Eds.).
. New York:
Cambridge University Press.

Butters, N. (1984). Alcoholic Korsakov's syndrome: an update, 4, 2,
226-244.

Cave, C. B. (1997). Very long-lasting priming in picture naming. 8,
322 – 325.

Collins, A. and Loftus, E. (1975). A spreading-activation theory of semantic processing.
82(6), 407 – 428.

Conway M. A., Gardiner, J. M. Perfect, T. J., Anderson, S. J. and Cohen, G. (1997). Changes in memory awareness during learning: the acquisition of knowledge by psychology undergraduates. 126, 393 - 413.

Corkin, S. (1968). Acquisition of motor skill after bilateral medial temporal-lobe excision.
6, 225 - 264.

Corkin, S. (2002). What's new with the amnesic patient HM?
3, 153 - 159.

Corkin, S., Amoural, D. G., Gonzalez, R. G., Johnson, K. A. and Hyman, B. T. (1997). HM's medial temporal lobe lesions: findings from magnetic resonance imaging.
17(10), 3964 - 3979.

Coward, L. A. (1990). . New York: Praeger.

De Renzi, E., Liotti, M. and Nichelli, P. (1987). Semantic amnesia with preservation of autobiographical memory. 23, 575 - 597.

Dunn, J. (1998). Implicit memory and amnesia. In Kirsner, K., Speelman, C., Maybery, M., O'Brien-Malone, A., Andersen, M., and MacLeod, C. (Eds).
, 99-117. New Jersey: Erlbaum.

Forbach, G. B., Stanners, R. F., and Hochhaus, L. (1974). Repetition and practice effects in a lexical decision task. 2, 337-339.

Friedman, W. J. (1993). Memory for the time of past events. 113 (1),
44 - 66.

Gabrieli, J. D. E., Cohen, N. J. and Corkin, S. (1988). The impaired learning of semantic knowledge following bilateral medial temporal lobe resection. 7,
157 - 177.

Geiselman, R. E., Fisher, R. P., MacKinnon, D. P. and Holland, H. L. (1986). Enhancement of eyewitness memory with the cognitive interview. 99, 385 - 401.

Janowsky, J. S., Shimamura, A. P. and Squire, L. R. (1989). Source memory impairments in patients with frontal lobe lesions. 27, 1043 - 1056.

Kensinger, E. A., Ullman, M. T. and Corkin, S. (2001). Bilateral medial temporal lobe damage does not affect lexical or grammatical processing: evidence from amnesic patient HM 11, 337 - 346.

Kirsner, K. (1998). Implicit memory. In Kirsner, K., Speelman, C., Maybery, M., O'Brien-Malone, A., Andersen, M., and MacLeod, C. (Eds.). . New Jersey: Erlbaum.

Kirsner, K., Speelman, C., Maybery, M., O'Brien-Malone, A., Andersen, M., and MacLeod, C. (1998). . New Jersey: Erlbaum.

Loftus, E. F. (1979). . Cambridge, MA: Harvard University Press.

Loftus, E. F., Levidow, B and Duensing, S. (1992) Who remembers best? Individual differences in memory for events that occurred in a science museum. 6, 93-107.

Loftus, E. F., Coan, J. A. and Pickrell, J. E. (1996) Manufacturing false memories using bits of reality. In L. M. Reder (Ed.). . Mahwah, NJ: Erlbaum.

Loftus, E.F. and Pickrell, J. E. (1995) The formation of false memories. 25, 720-725.

Marr, D. (1991). . Edited by L. M. Vaina. Boston: Birkhauser.

McCarthy, R. A. and Warrington, E. K. (1988). Evidence for modality-specific meaning systems in the brain. 334, 428 – 430.

McCloskey, M. Wible, C. G. and Cohen, N. J. (1988). Is there a special flashbulb-memory mechanism ? 117, 171 – 181.

Milner, B., Corkin, S. and Teuber, H.-L. (1968). Further analysis of the hippocampal amnesic syndrome: 14-year follow-up study of HM 6, 215 - 234.

Nadel, L. and Moscovitch, M. (1997). Memory consolidation, retrograde amnesia and the hippocampal complex. 7, 217 - 227.

Neisser, U., and Libby, L. K. (2000). Remembering life experiences. In E. Tulving and F. I. M. Craik (Eds.). . Oxford: Oxford University Press.

Nelson, K. (1993). The psychological and social origins of autobiographical memory. 4(1) 7 - 14.

Pillemer, D. B. (1998). . Cambridge, MA: Harvard University Press.

Postle, B. R. and Corkin, S. (1998). Impaired word-stem completion priming but intact perceptual identification priming with new words: evidence from the amnesic patient HM 36, 421 - 440.

Read, J. D., Tollestrup, P. Hammersley, R., McFadzen, E. and Christensen, A. (1990). The unconscious transference effect: Are innocent bystanders ever misidentified? 4, 3 - 31.

Rips, L., Shoben, J. and Smith, E. Semantic Distance and Verification of Semantic Relations. 12, 1 – 20.

Robinson, J. A. (1976). Sampling autobiographical memory. 8, 578 - 595.

Rosch, E. (1973). On the internal structure of perceptual and semantic categories. In T. M. Moore (Ed.) . Academic Press.

Ross, D. F., Ceci, S. J., Dunning, D. and Toglia, M. P. (1994). Unconscious transference and lineup identification: towards a memory blending approach. In D. F. Ross, J. D. Read and M. P. Toglia (Eds.). . New York: Cambridge University Press.

Sagar, J. H., Cohen, N. J., Corkin, S, and Growdon, J. H. (1985). Dissociations among processes in remote memory. 444, 533 - 535.

Satori, G., Job, R. and Coltheart, M. (1993). The organization of object knowledge: evidence from neuropsychology. In D. E. Mayer and S. Kornblum (Eds.). . Cambridge, MA: MIT Press.

Scarborough, D. L., Cortese, C., and Scarborough, H. S. (1977). Frequency and repetition effects in lexical memory. , 3, 1-17.

Schacter, D. L., Harbluk, J. L., and McLachan, D. R. (1984). Retrieval without recollection: an experimental analysis of source amnesia. 23, 593 - 611.

Schacter, D. L., Bowers, J., and Booker, J. (1989). Intention, awareness and implicit memory: the retrieval intentionality criterion. In S. Lewandowsky, K. Kirsner, and J. Dunn (Eds.). . Hillsdale NJ: Erlbaum.

Schacter, D. L. and Buckner, R. L. (1998). On the relations among priming, conscious recollection, and intentional retrieval. 70, 284 - 303.

Schacter, J. C. and Tulving, E. (1994). What are the memory systems of 1994 ? In J. C. Schacter and E. Tulving (Eds.). . Cambridge, MA: MIT Press.

Schacter, J. C., Wagner, A. D., and Buckner, R. L. (2000). Memory systems of 1999. In E. Tulving and F. I. M. Craik (Eds.). . New York: Oxford University Press.

Schooler, J. W., Ambadar, Z. and Bendiksen, M. A. (1997). A cognitive collaborative case study approach for investigating discovered memories of sexual abuse. In J. D. Read and D. S. Lindsay (Eds.). . New York: Plenum.

Scoville, W. B. and Milner, B. (1957). Loss of recent memory after bilateral hippocampal lesions. 20, 11 - 21.

Silveri, M. C. and Gainotti, G. (1988). Interaction between vision and language in category specific semantic impairment. 5, 647 – 709.

Standing, L., Conexio, J., and Haber, R.N. (1970). Perception and memory for pictures: single-trial learning of 2500 visual stimuli. 19(2), 73 - 74.

Thompson, G. P., Herrmann, D. J., Read, D. J., Bruce, D., Payne, D. G. and Toglia, M. P. (1998). . Mahwah, NJ: Erlbaum.

Tulving, E. (1972). Episodic and semantic memory. In E. Tulving and W. Donaldson (Eds.). . New York: Academic Press.

Tulving, E. (1983). . New York: Oxford University Press.

Tulving, E. (1984). Multiple learning and memory systems. In K. M. J. Lagerspetz and P. Niemi (Eds.). . Holland: Elsevier.

Tulving, E. (1985). How many memory systems are there? 40, 385 - 398.

Tulving, E. (2002). Episodic memory: from mind to brain. 53, 1 – 25.

Vertes, R. P. and Eastman, E. E. (2000). The case against memory consolidation in REM sleep. 23 (6).

Warrington, E. K. (1975). The selective impairment of semantic memory. 27, 635 – 657.

Wheeler, M. A. (2000). Episodic memory and autonoetic awareness. In E. Tulving and F. I. M. Craik (Eds.). . Oxford: Oxford University Press.

Willingham, D. B. and Preuss, L. (1995). The death of implicit memory. 2(15). http://psyche.cs.monash.edu.au/v2/psyche-2-15-willingham.html

Wilson, M. A., and McNaughton, B. L. (1994). Reactivation of hippocampal ensemble memories during sleep. 265, 676-679.

Yuille, J. C. (1993). We must study forensic eyewitnesses to know about them. 48 (5), 572 - 573.

Chapter XI

Procedural Memory, Priming, and Working Memory

As discussed in the previous chapter, five memory systems were proposed by Schacter and Tulving [1994] based upon five types of memory phenomena. In the previous chapter the way in which the recommendation architecture provides an account for two of the types of phenomena (semantic and episodic memory) was described. In this chapter the account for the other three types of phenomena (procedural memory, priming, and working memory) will be described. The cognitive model is the same as that described in chapter 9 and used in chapter 10 for the discussion of semantic and episodic memory. The organization of clustering in this model is illustrated in figure 11.1. Each output from a clustering array may have recommendation strengths in favour of many different behaviours as illustrated in figure 11.2.

Procedural Memory

Procedural memory is defined as the ability to acquire skills. Observations of amnesics indicate that such memory is at least partially dissociated from semantic memory, since amnesics can acquire such skills at apparently normal rates. Thus amnesics can acquire motor skills such as mirror tracing tasks [Corkin 1968], perceptual skills such as reading mirror reversed words [Gabrieli, Merrill and Corkin, 1986], and cognitive skills such as improved speed of identification of letter strings as words or non words [Kirsner, 1998]. Korsakoff's syndrome patients with no apparent ability to acquire new semantic memories can learn to solve the Tower of Hanoi problem [Cohen, Eichenbaum, Deacedo, and Corkin, 1985].

Although the acquisition of semantic memory can speed up the learning process when constructed on-line during skill learning [Mathews, Buss, Stanley, Blanchard-Fields, Cho and Druhan, 1989], there can be inconsistencies between procedural and semantic knowledge of a skill, as revealed by studies of specialist expertise. "In some areas of expertise, there is a dissociation between what experts say they do and what they do ..... [An expert verbal] description typically bears only a superficial relationship to the expertise" [Speelman 1998].

Verbal descriptions by experts describing their expertise often correspond with beginner methods rather than actual methods, and requiring a verbal description can even result in the expert reverting to the less effective beginner method [Bainbridge 1977; Berry 1987].

In the recommendation architecture model for learning a skill, portfolios activated within clustering in environments where the skill is relevant must acquire weights in competition associated with skilled behaviours. These portfolios will generally include both new information elements resulting from novelty in the environments and information elements recorded in prior experiences. Although the new elements may be particularly useful for recommending the new behaviours, some skill learning would be possible using only previously recorded elements which happen to occur in the new environments. Thus skill acquisition could proceed in the absence of condition recording.

In the process of practicing a skill, an activated portfolio population recommends an action, and portfolio populations activated in response to sensory input from the experience of the action and its results interacts with existing populations to generates a new population recommending the next action and so on. At each stage the weights assigned to the elements in the population into the module corresponding with the accepted action in competition may be adjusted by consequence feedback. In verbally describing a skill, an activated population generates a verbal output plus an activated population corresponding with the next step, which in turn generates the next verbal output and the next activated population and so on. The populations activated in the course of generating verbal descriptions may not be exactly the same as those generated in the course of skilled behaviour, and the weights of the portfolios into verbal behaviour are not directly influenced by consequence feedback in response to skilled behaviour. Verbal descriptions can therefore be misaligned with the skill. However, when the skill is first learned, activated populations generated by verbal inputs may be a useful starting point from which to learn the most appropriate populations to generate skilled behaviour. Information on behavioural weights is within competition, and the information cannot be directly activated in clustering to generate verbal descriptions. Such verbal descriptions can only be updated by self observation during practice of the skill. No record exists of past values of weights, and past states of skill are therefore generally inaccessible.

Perceptual and Conceptual Priming

Priming has been defined as a nonconscious influence of past experience on current performance or behaviour [Schacter and Buckner 1998]. However, in this paper priming will be defined to exclude experiments in which the tasks completed at test were also completed in the same way at an earlier study phase, and consequence feedback (e.g. self recognition of response correctness or otherwise) was therefore available at study. If the influence of study was nonconscious in such experiments, they will be classified as implicit memory. The reason for this exclusion is that the presence of consequence feedback on the exact behaviour substantially changes the mechanisms available in the recommendation architecture. The effect of the exclusion is that experiments often cited as evidence for very long term priming [e.g. Cave 1997] are not regarded as priming but as implicit memory.

Following Schacter and Buckner [1998], a distinction will be made between perceptual and conceptual priming. Perceptual priming is modality specific (i.e. depends upon the degree of consistency between raw sensory presentations at study and at test) and does not benefit from semantic encoding. Conceptual priming is not modality specific and benefits from semantic encoding.

An issue with priming experiments is the degree to which there may be a semantic memory component in the phenomenon tested. The degree of semantic memory involved can be influenced by the instructions given to subjects, as will be discussed further in the later comparisons between normal and amnesic subjects.

An experiment which measures perceptual priming with minimal possibility of semantic memory involvement is tachistoscopic word recognition. In such experiments, subjects read a list of words at study, and at test are given brief glimpses of words and attempt to identify them. It is found that identification rates are higher for words read at study.

A second test of perceptual priming is word fragment completion. In these experiments a list of common words are read at study without instructions about any future test. Subsequently, without reference to earlier study, subjects are shown three letter sequences and asked to complete with the first word that comes to mind. Words on the studied list occur more often than equally possible alternative completions [Graf, Squire and Mandler 1984]. A third test of priming is presentation of masked pictures of objects, with such a brief presentation period that only ~13% can be named [Bar and Biederman 1998]. If a previous picture is shown again later, naming accuracy increases to ~35%.

An experiment type which can measure conceptual priming with minimal semantic memory component is a variant of the letter string classification technique used for implicit memory. In the experiments of Meyer and Schvaneveldt [1971], subjects were presented with a series of letter strings and asked to classify the strings as words or non-words as quickly as possible. They found that when a word was preceded by a semantically related word (e.g. nurse preceded by doctor), it was classified as a genuine word more rapidly than if it was preceded by an unrelated word (e.g. north) or a non-word (e.g. nuber).

In another type of conceptual memory experiment, subjects are given category cues (e.g. fruit) and asked to produce relevant exemplars (e.g. orange, pear). Previous exposure to the exemplars increases the probability of their production [e.g. Hamann 1990].

Within the recommendation architecture, the mechanism responsible for priming is activation of conditions which have recently been active in the past at the same time as currently active conditions. To understand the effects of this mechanism, it is important to remember that individual conditions are ambiguous and are only probabilistically related to sensory or cognitive features or objects.

In the case of tachistoscopic experiments on perceptual priming, initial brief exposure to a stimulus such as a word or picture activates a limited population of condition portfolios at a relatively simple level of complexity in array V2 of the architecture in figure 11.1. These portfolios in turn directly activate a very small population of portfolios in array V3. Some of these array V3 portfolios will have weights as recommendations to say the name of the object, but the small size of the population means that generally the weight into the appropriate naming will be too low to generate behaviour. In the second exposure, again a limited population of portfolios will be activated in array V2. However, a subset of these portfolios will in general have also been active in the recent earlier exposure, and therefore recommend activation of the other array V2 portfolios active at that time. The resultant larger population in array V2 directly activates a larger population in array V3, and the recommendation to say the name of the appropriate object is stronger.

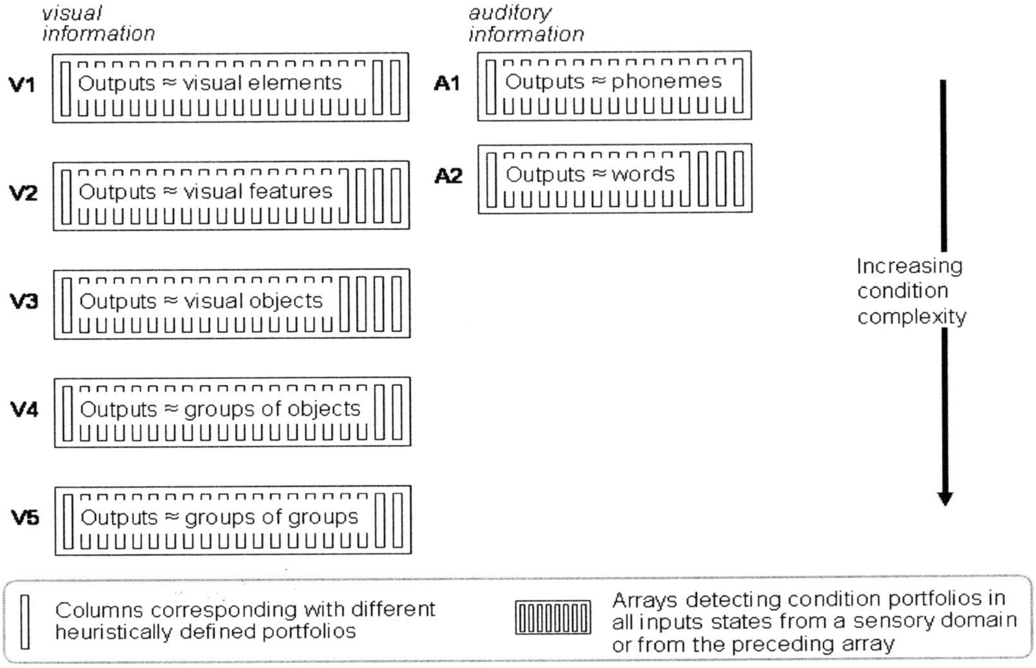

Figure 11.1. Clustering architecture. Parallel sequences of arrays of columns detect condition portfolios independently in visual and auditory input. Different arrays in a sequence detect portfolios within different behaviourally useful ranges of condition complexity. In the figure, these ranges of complexity correspond roughly with the complexity of cognitively relevant circumstances such as the presence of features. Portfolios at that level will have the best discrimination between such features. In practice sequences will have many more arrays of columns at different complexities and portfolios on several levels of complexity will be relevant to discriminating between cognitive features (see figure 9.4). However, to simplify explanations the number of arrays has been reduced.

In the case of word fragment completion experiments, the mechanism for completion is experimental activation of portfolios in array V2 which have been active in the past at the same time as the visual portfolios within the letter string in array V2, until the combined direct and secondary activation in array V2 generates an activation in array V3 which indicates that it corresponds with a word by generating a strong secondary activation in array V3. Among other recommendations, the portfolios within the strong secondary activation will recommend saying the word. For words not included in the study phase, the secondary portfolios in array V2 are activated on the basis of past simultaneous activity which has occurred multiple times and is independent of recency. For words included in the study, secondary portfolios in array V2 are also activated on the basis of recent simultaneous activity. Array V2 populations for previously studied words will therefore tend to be larger, increasing the probability of an array V3 activation having enough recommendation strength to say the word.

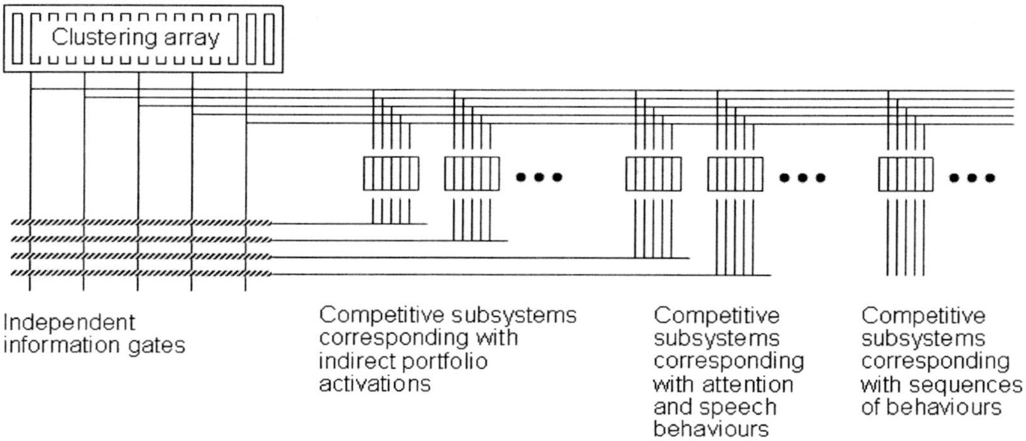

Figure 11.2. Structure of competition making use of clustering outputs from one array. Each output from clustering may have recommendation strength in favour of many different behaviours. Competition manages to selection of the most strongly recommended behaviours.

In the case of letter string identification experiments on conceptual priming, the main location for secondary activation of portfolios is array V3. For example, directly activated portfolios within the visual perception of the word in array V3 activate secondary pseudovisual portfolios present in the past when the topic of the word was present. The mechanism for distinguishing between a word and a non-word is detection of a strong secondary population. If the preceding word is semantically related, the populations of secondary portfolios generated in array V3 in response to the two words will have some overlap. The secondary portfolios in response to the second word which were also present in the response to the first word will recommend activation of other secondary portfolios recently present in response to the earlier word. The secondary population will therefore grow somewhat more rapidly, resulting in faster recognition that the letter string is a word.

In the case of priming of exemplar generation, the key point is that the population of portfolios in array V3 which is generated in response to the name of an instance of a category will have some overlap with the population activated in response to the category name. When given the name of a category and asked to generate exemplars, the mechanism is to generate portfolios in array V3 which were present in the past at the same time as the currently active (category) portfolios in array V3, until a recommendation to say a name is strong enough. For exemplars not in the study phase, this secondary activation is based only on the basis of past simultaneous activity which has occurred multiple times and is independent of recency. For exemplars in the study phase, there is in addition a secondary population based on recent simultaneous activity, and the probability of naming studied exemplars is increased.

Perceptual priming depends upon overlap between the populations of portfolios detected in array V2 in response to object perceptions at study and at test. Portfolios detected in this level are at a relatively simple level of complexity, and are specific to the exact sensory inputs. Changes to the exact sensory inputs between study and test will therefore reduce the overlap and therefore the priming. Thus priming is reduced by changes to the typography of words [Jacoby and Hayman 1987] or position of pictures [Bar and Biederman 1998] in tachistoscopic recognition experiments. Major changes to the exact sensory inputs means that priming can only occur as a result of array V3 (conceptual priming) overlap, which in the experimental conditions under which perceptual priming is tested will tend to be very weak. Thus there is no priming in same name, different picture tachistoscopic recognition experiments [Bar and Biederman 1998], However, when the modality is changed from verbal to visual between study and test in word tachistoscopic recognition experiments [e.g. Clarke and Morton 1983], there is more possibility of array V3 overlap and priming is considerably reduced but not eliminated. In word fragment completion experiments, auditory presentation of the words at study also reduces priming [Graf, Shimamura and Squire 1985].

The dissociation between explicit memory and perceptual priming is also of interest. As discussed in section 4.2, the major factor determining explicit memory at test is the degree of condition recording in levels 2, 3 and above at study, while changes to the exact sensory input will have little effect. However, for priming, the degree of condition recording has no effect and the determining factor is the degree of change to the exact sensory input. This conclusion is confirmed by the results of Jacoby [1983]. In his experiments, subjects were exposed to words in three different ways at study. One was to be shown the semantic opposite of the word and be asked to say the opposite (e.g. shown "light" and said "dark"). The second was to be shown the word and its opposite and be asked to say the word (e.g. shown "light-dark" and said "dark"). The third was to be shown the word and asked to say it (e.g. shown "dark" and said "dark"). A week later, subjects were tested in two ways. One was tachistoscopic recognition of studied and unstudied words. The other was to be given lists of mixed studied and unstudied words and be asked to circle studied words. It was found that priming was greatest when the word was seen, but explicit memory was greatest when the word was generated from its opposite.

As discussed in section 4.2, explicit memory is partially dependent on two types of mechanism: activation of portfolios containing conditions recorded at the same time in the past as conditions in already active portfolios; and activation of portfolios active at the same time in the past as already active portfolios. However, without activation on the basis of recording at the same time, recent simultaneous activity cannot be connected specifically with the earlier study episode. Activation on the basis of recent simultaneous activity can therefore enhance explicit memory but cannot alone support explicit memory. Thus the initial rapid decline in explicit memory over the first day or so can be attributed to the decline in the support provided by priming. This, however, is not the same as the proposal that explicit recognition memory could be supported in part by priming, with the fluency with which a subject processed the recognition cue being the indication of recognition [Mandler 1980; Jacoby 1983]. That proposal was discounted by the observation that recognition memory is not disproportionally spared in amnesia relative to recollection memory [Haist, Shimamura and Squire 1992]. In the recommendation architecture model, neither recollection nor recognition memory can operate in the absence of condition recording, which as discussed later is the mechanism affected in the recommendation architecture model for amnesia.

The observation that priming is typically present up to a few days [Tulving, Schacter, and Stark, 1982] is consistent with its being dependent upon the indirect activation conditions on the basis of recent simultaneous activity. Priming would not be expected to persist over long periods because the recommendation weights decline with time unless reinforced by participating in the generation of a useful behaviour. However, as discussed earlier, the presence of consequence feedback would invoke other, long term changes and the phenomenon is therefore not classified as priming in this paper.

Because priming depends upon activation on the basis of recent simultaneous activity and is independent of condition recording, the recommendation architecture model leads to the expectation that amnesics will show priming capabilities similar to normal subjects. Slightly poorer performance by amnesics would be expected based on normal subjects making some use of semantic memory capabilities. Word fragment completion experiments demonstrate this type of difference between normal and amnesic subjects [Warrington and Weiskrantz, 1968].

Working Memory

A range of experimental procedures are viewed as measurements of working memory. These procedures include verbal or visual list recall, short term forgetting, visual memory span, pattern matrix testing and Corsi block sequence testing.

In list recall experiments, subjects are presented with a sequence of objects and asked to immediately report the objects on the list, in any order. The longest list length for which all objects can consistently be reported correctly is measured. Typical objects are numerical digits, letters, or words, which are presented verbally or visually. Normal subjects can fully recall lists of about seven (plus or minus two) random digits, but only four or five random words or letters [McCarthy and Warrington 1990]. In a variation of this test, subjects are asked to report as many items as possible on longer lists. Normal subjects show enhanced recall for the first few items on the list (the primacy effect) and enhanced recall for the last few items (the recency effect) [Baddeley 2000]. A brief delay occupied with another task eliminates the recency effect but has much less influence on the primacy effect [Glanzer 1972].

List recall capability is considerably increased if sequential groups of objects within a list have some semantic connection. Such organization is called "chunking". For example, although list recall for random words is limited to four or five, recall of meaningful sentences can extend to sixteen words or more [Baddeley, Vallar, and Wilson 1987].

In the short term forgetting experiments of Petersen and Petersen [1959], subjects are presented with three consonants and required to retain them over a delay of up to 18 seconds while counting backwards in threes. Performance indicates a short term memory component which declines over about five seconds [see also Baddeley and Scott 1971].

Visual memory span is the number of different objects which can be reported after a brief (typically 100 - 200 milliseconds) exposure to a group of objects. Under these conditions, normal subjects can accurately identify between four and five objects [McCarthy and Warrington 1990]. A special case of visual memory span is when subjects are asked to report the number of objects after a brief exposure to a group of (often identical) objects. Normal subjects can report group sizes accurately up to about six [McCarthy and Warrington 1990].

Two tests evaluate visuo-spatial working memory, which is believed to be supported by a separate mechanism from working memory for objects with symbolic content, such as letters, words and numbers [Baddeley 2000].

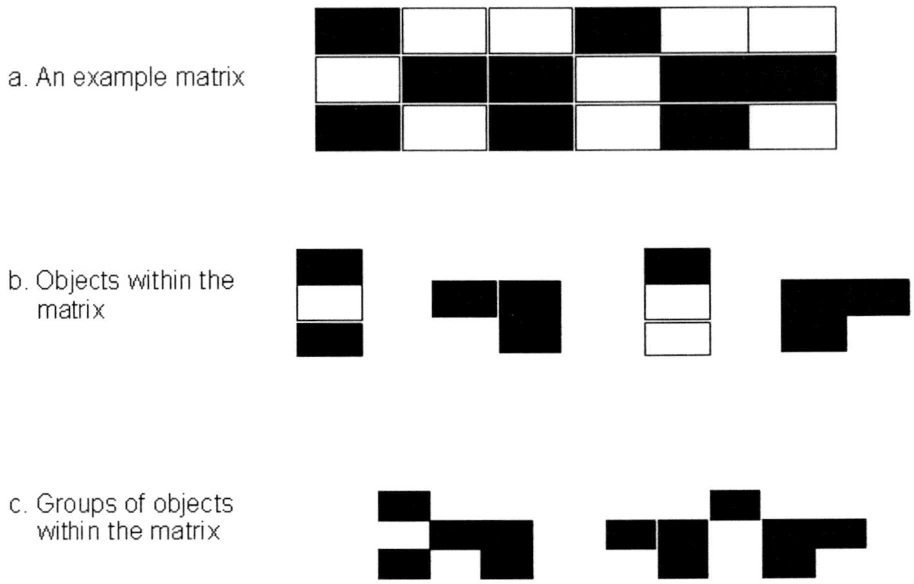

Figure 11.3. Pattern matrix test example. a. An example of a pattern matrix test with nine filled rectangles. b. Examples of visually objects correlating with portfolios activated in array V3 of the recommendation architecture model. c. Examples of groups of objects correlating with portfolios activated in array V4 of the recommendation architecture model.

Pattern matrix testing [Della Salla, Gray, Baddeley, Allamano, and Wilson 1999] investigates visual working memory. In this test, subjects are presented for a few seconds with a matrix of rectangles in which half of the rectangles are coloured black, and are immediately asked to indicate which rectangles were filled on an identical but unfilled matrix. Matrices contain from four to thirty rectangles, filled rectangles therefore range from two to fifteen. An example of a matrix with nine filled rectangles is illustrated in figure 11.3. The number of filled blocks on the most complex matrix accurately recalled is about nine for normal subjects.

In the Corsi blocks test [Corsi 1972; De Renzi and Nichelli 1975], a subject watches the performance of a sequence of taps on small blocks of wood. Nine such blocks are arranged randomly on a board in two dimensions. The subject is asked to immediately reproduce the sequence. Sequences are from two to eight taps, and normal subjects can accurately reproduce sequences of up to five or six taps.

A dissociation is observed between pattern matrix and Corsi blocks testing. A distracting task with a strong visual content degrades performance on the pattern matrix test more than on the Corsi blocks test, but a distracting task with a strong spatial content degrades performance on the Corsi blocks test more than on the pattern matrix test [Della Salla, Gray, Baddeley, Allamano and Wilson 1999].

Recommendation Architecture Mechanisms in Support of Working Memory

In the recommendation architecture model, the function of the mechanisms which give rise to many of the phenomena labeled working memory is to ensure that portfolio populations activated in response to different information sources can be kept separate, can be combined as required to directly activate higher level populations, and are active (by prolongation of activity or by reactivation) when their outputs are required to drive a selected behaviour.

For example, a number of different objects may be present in the visual environment, and the most appropriate behaviour may be determined by the presence of several (but not all) of these objects. As attention focuses on the different objects in turn, portfolio populations activated in response to one object must be kept active in array V3 in figure 11.1 until portfolios have been activated in response to other objects, and portfolios containing conditions made up of information derived from multiple objects can then be activated in V4. The reading process as described in chapter 9 is an example of this requirement. As discussed in that chapter, it may be necessary for several portfolio populations to be independently active within the same physical group of portfolios.

As time passes, the probability of an active portfolio being still relevant to behaviour decreases. Furthermore, there are limits to the number of independent active portfolio populations which can be distinguished within the same physical group. Hence active populations will need to decay with some time constant. However, it may sometimes be necessary for behavioural reasons to extend the activation of a population or to (perhaps partially) reactivate a recently active population. Such extensions and reactivations are behaviours which must have an adequate recommendation strength across currently active populations.

For example, an existing active population will have recommendation strength in favour of prolonging its own activity. The set of portfolios in one region which could be activated in response to an object is large. A subset of this larger set will be actually activated in response to a perception of the object. The identity of this subset could change throughout the period in which the activated population exists. Self activation recommendation strength is thus recommendation strength to activate other portfolios active in the past at the same time, prolonging the activity of the population although not necessarily the activity of specific individual portfolios. Such a recommendation strength will in general not be adequate to keep a population active indefinitely without the support of recommendation strength from other active populations. Self activation strength will therefore determine the time constant for decay of the population in the absence of supporting recommendation strength. Supplementary recommendation strength from other populations might for example be in favour of prolonging the activity of any population in the array in which the population to be prolonged is located, and could prolong such activity beyond normal decay times.

When several active populations on one level generate a combination population at a higher level, the higher level population is active at a similar time as its constituent populations, and therefore has recommendation strength in favour of reactivating these populations based on recent simultaneous activity. Again, supplementary recommendation strengths will in general be required for such reactivations to become accepted behaviours.

Limits to working memory are therefore imposed by the decay time of active populations, the number of populations which can be active simultaneously in the same physical region, and the availability of recommendation strength in favour of prolonging population activity or of reactivating population activity on the basis of recent simultaneous activity.

List Recall

When a random verbal list is presented, portfolio populations are directly activated in arrays A1 and A2 in response to the auditory inputs. If an item on the list corresponds with a meaningful concept, a portfolio population is indirectly activated in array V3 by the population in A2. The population directly activated in A2 in response to auditory input and the population indirectly activated in V3 by the A2 population will both have recommendation strength in favour of speaking the corresponding verbal item. The ability to report the list therefore depends primarily on the number of independent populations which can be sustained in array A2 and in array V3.

Population independence is based upon different phases of frequency modulation applied to different populations. The different phases limit the interaction between the populations as described in chapter 5. To preserve independence, the minimum phase separation between two populations must be of the order of the synaptic decay time. The number of possible independent populations will therefore be of the order of the modulation interval divided by the synaptic decay time, in other words about four.

When a list of letters is presented, the indirectly activated populations in array V3 correspond with past visual perceptions of the corresponding letter. Several independent populations will result in A2 and V3 in response to recent letters in the sequence. There is no meaningful (i.e. behaviourally relevant) relationship between the letters, and there will therefore be little recommendation strength in favour of synchronization of the outputs of the independent populations in V3 and release to V4. List recall is dependent on the number of independent populations which can be sustained at the same time in A2 and V3.

For lists of random words, the indirectly activated populations in array V3 are strong. However, the absence of semantic relationships between the words means that active populations in array V4 will be weak. List recall therefore again is dependent on the number of independent populations which can be sustained at the same time in A2 and V3.

If there is a significant semantic relationship between several sequential words in a list, the activated populations in array V3 corresponding with the words will have significant recommendation strength in favour of synchronizing the outputs of the separate populations and releasing them to array V4. The resultant population in V4 will correspond with the meaning of the phrase made up of the word sequence. Sequences of semantically related phrases (i.e. a sentence) would have recommendation strength in favour of synchronizing the outputs of the separate populations in array V4 and releasing them to array V5. The activation in V5 would correspond with the meaning of the sentence made up of the sequence of phrases.

The "word" populations in array V3 contributing to a "phrase" population in array V4 will have been recently active shortly before the population in V4, and that population will therefore have recommendation strength in favour of reactivating its constituent "word" populations on the basis of recent temporally correlated activity. Furthermore, any experience of similar phrases in the past will result in the population in V4 having recommendation strength in favour of reactivating its constituent "word" populations on the basis of frequent past temporally correlated activity. The "sentence" populations in V5 will similarly have recommendation strength in favour of reactivating their constituent populations in V4.

If four "phrase" populations could each regenerate four "word" populations, and in addition the "sentence" activation could reactivate the "phrase" populations, the observed reporting of sentence lengths of sixteen words or more [Baddeley, Vallar and Wilson 1987] could be supported.

Numerical digits represent an intermediate case. Because pairs of digits are themselves meaningful and generally familiar numbers, a list of digits will tend to activate populations in array V3 in response to individual digits and array V4 in response to sequences of digits. The V4 combination populations can in turn reactivate the individual populations in V3 which can recommend speaking the digits. Furthermore, some pairs of digits may activate populations in V3 because the pair forms a meaningful number in its own right.

As observed [McCarthy and Warrington 1990], lists of digits will therefore be recalled better that random letters or words but not as well as meaningful sentences.

For visually presented lists, input derived from individual items directly activates populations in array V3 in response to words as visual objects. The same limits will apply as for verbal lists.

For longer lists, a primacy effect and a recency effect are observed. The recency effect results from the list recall mechanism discussed, but the primacy effect is the result of an episodic memory mechanism. Because all the items on lists are familiar to the subject, condition recording is not triggered in response to individual items. However, the perceptual change corresponding with the start of a new task can trigger some condition recording, and this condition recording can be the basis for recall by activation of portfolios which recorded conditions in the recent past.

Short Term Forgetting

In short term forgetting experiments, a small group of objects is presented, such as three letters in the experiments of Petersen and Petersen [1959]. The recommendation architecture mechanism is that active portfolio populations corresponding the items are constantly refreshed by acceptance of recommendations to perform such refreshing. As discussed under list recall, recommendation strength comes from the populations themselves but requires activation of the appropriate competition component strength by other active populations. These other populations would be at higher levels can therefore be interpreted as recommending the type of task, in this case recall of the items. Self refreshing would be experienced as mental rehearsal. A distracting task of a verbal type means that the populations in the relevant higher level region recommend a different task, and item populations will therefore decay with a time constant determined by self refreshment recommendation strength of the item populations alone.

Visuo-Spatial Working Memory

Visual memory span depends upon activation of independent portfolio populations in array V3 in response to different objects in the visual field, and is again limited by the maximum number of such populations, or 3 - 4 as discussed earlier. The ability to count a group depends upon a different mechanism. A group of objects viewed as a single attention domain will result in direct activation of portfolios in array V3, some of which will have recommendation strength in favour of speaking the number of objects. As discussed in chapter 11, acquisition of such recommendation strength for a specific number depends upon development of a set of portfolios in V3 which are often activated when different groups of the corresponding number are present. As the number increases, the probability of such a consistent set forming decreases.

To understand the mechanisms supporting pattern matrix testing, consider the example matrix illustrated in figure 11.3a. Although the individual rectangles are conceptually objects, it is more plausible that the component objects perceived, or in other words the objects that generate active populations in array V3 are made up of several blocks as illustrated in figure 11.3b. Groups of such objects as illustrated in figure 11.3c might in principal result in active populations in array V4, and groups of groups in active populations in array V5. However, such V4 and V5 activations are likely to be weak because of the lack of meaning in the overall matrix.

The ability to recall block shading therefore depends upon the ability to support multiple activations in V3 corresponding with different objects made up of several blocks.

If three V3 populations can be supported, and a V3 population corresponds on average with three filled rectangles, the result would be the observed ability to accurately recall matrices with up to nine filled rectangles [Della Salla, Gray, Baddeley, Allamano and Wilson 1999].

The portfolios activated in V3 in response to pattern matrix perceptions will tend to be different from those activated in response to perceptions with semantic content, and some dissociation between the memory phenomena is therefore to be expected.

The mechanisms supporting performance in Corsi block testing are similar, but in this case the V3 populations correspond with one movement from one block to another and V4 populations correspond with groups of several such movements. Again, partial physical separation of the portfolio groups supporting pattern matrix and Corsi blocks testing will result in the observed dissociation between performance on the two tests with different types of distracting tasks [Della Salla, Gray, Baddeley, Allamano and Wilson 1999].

Working Memory Deficits

Observations of the cognitive deficits associated with brain damage indicates a wide range of dissociations between working memory and semantic and episodic memory and between different types of working memory. Unlike semantic or episodic memory, working memory does not depend upon recording of new conditions. Loss of the capability to record new conditions would therefore remove the ability to create new semantic and episodic memories but leave working memory unaffected. This dissociation is consistent with the observations that working memory generally operates normally in amnesic patients [Butters 1984; McCarthy and Warrington 1990; Corkin 2002]. A specific example of this dissociation is the disappearance of the primacy effect but preservation of the recency effect in recall of long lists [Glanzer 1972].

In the recommendation architecture model, there are two types of problem which could be caused by local damage and which would result in working memory deficits. One type of problem is loss of recommendation strength in favour of continuation of self activity for some populations. Such a problem would result from damage affecting the connectivity between the physical substrate for a group of portfolios and the competitive subsystem managing the behaviour type. The damage would affect only populations in arrays with affected connectivity. The result of damage would be faster decay of active populations containing affected portfolios.

A second type of problem is loss of recommendation strength of populations at higher levels in favour of reactivation of their source populations at lower levels. Again, such a problem would result from damage affecting the connectivity between the physical substrates for the portfolios on the different levels and the relevant competitive subsystem. For example, the ability of some types of activated populations in V4 of to activate source populations in V3 on the basis of recent simultaneous activity could be affected.

A range of cases in which list recall memory is severely affected have been reported [Caplan and Waters 1990]. In subject KF [Warrington and Shallice 1969; Shallice and Warrington 1970; Shallice and Warrington 1977], verbal recall for single numbers, letters or words presented verbally was good, but deteriorated considerably for two item lists. Recall of visually presented items was better, but visual recall of verbally presented items (by pointing to a visual array) was no better than verbal recall.

This combination of deficits would be generated in the recommendation architecture model by local damage affecting self prolongation recommendation strengths limited to A2 and V3. Such damage would result in the decay time of populations being faster than in normal subjects. These populations are the sources of recommendation strength in favour of speaking the corresponding word, and recall capability would therefore be less.

The observations that list recall deficits generally occur together with short term forgetting deficits [Warrington, Longue and Pratt 1971; Basso, Spinnler, Vallar and Zanobia 1982; McCarthy and Warrington 1987] are consistent with this interpretation of list recall deficits being generated by damage affecting population decay times.

In list recall experiments, lists are typically presented at one item per second. In normal speech, words are presented two or three times faster and higher level populations will encourage generation of strong secondary populations in array V3 corresponding with the meaning of the words. Hence the length of time required for word generated populations in V3 to remain active in order to generate "phrase" and "sentence" populations in arrays V4 and V5 is much shorter than the time such populations must remain active in list recall experiments. The ability to repeat sentences is dependent upon combination populations in levels 3 and 4 reactivating "word" populations in level 2, and will not be affected by damage to the capability of populations in array V3 to prolong their own activity. Thus patient IL [Saffran and Martin 1975] could repeat sentences much longer than her reduced digit span. Errors in sentence repetition were close approximations which retained the meaning of the original sentence. The model in this case is that loss of self prolongation strength in array V3 resulted in greater drift of such populations but the drift was constrained to similar meaning by reactivation recommendations generated by array V4 "phrase" populations into V3. Information on the sound of words is lost but meaning is preserved.

Deficits have also been observed in which repetition of lists is within normal span but sentence recall is impaired. An example is NHB [McCarthy and Warrington 1990]. This patient had lost the ability to provide a meaning for some words, but he was able to repeat lists of words whether or not the words retained their meaning for him. However, his ability to repeat sentences deteriorated when the sentence contained lost words. If none of the words had been lost, he could repeat sentences of six or seven words, but if one word was unknown he was only able to repeat the first words of the sentence and made sound based error on the remainder. For example, if in the sentence "The flag was coloured bright red" the word flag had been lost, his repetition attempt might be "The blag was fullered with a right breg". In this case the recommendation architecture model would be that this deficit is not one of the two types affecting working memory. Rather, in this case the mechanistic deficit is loss of the ability of some portfolios in V3 to be activated at all. The affected portfolios include many of those which recorded conditions in response to visual experiences which occurred at the same time as the lost words.

List recall for unrelated words would be supported by simultaneous populations in A2 and V3, or in A2 alone if the relevant portfolios in V3 had been damaged. However, sentence recall depends upon generation of appropriate populations in V4 and V5, and if the information from V3 is corrupted by damage, these populations will not be able to reactivate the appropriate populations in V3 and hence into level A2 to generate speech. The populations actually activated in A2 will reflect sounds and structure recently present, but not organized in a meaningful sentence.

The recommendation architecture approach is thus able to provide an account for a range of working memory deficits and dissociations.

Relationship with Alternative Models for Working Memory

An influential approach to modeling working memory phenomena and their role within cognition has been that of Baddeley and his coworkers [Baddeley 1986]. The approach was developed on the basis of an earlier model for the flow of information from perception of the environment to generation of behaviour proposed by Atkinson and Shiffrin [1968]. In the Atkinson and Shiffrin model, illustrated in figure 11.4a, the sensory systems provide information derived from the environment to a short term store (or working memory). Working memory uses this information plus information drawn from long term memory to generate behaviour. New information only reaches long term memory via working memory, by a process called consolidation.

The Baddeley approach models working memory in more detail within the Atkinson and Shiffrin framework. There are three major components to the Baddeley model as illustrated in figure 11.4b. These components are a phonological loop, a visuo-spatial sketch pad, and a central executive. Information is kept active within the phonological loop by a subvocal rehearsal process. The content of the phonological loop and of the visuo-spatial sketch pad is managed by the central executive.

There is a fundamental difference in modeling approach between these models and the recommendation architecture. This fundamental difference has its roots in the way in which modules can be defined in an operationally complex system. A module in an operationally complex system is a unit of information handling resources optimized for a particular type of operation. In order to limit use of resources, a high proportion of operations of the particular type will tend to be handled by the module independent of the features which are supported by the operations. Modules will therefore not in general correspond with features, but with groups of similar information operations. Such modules will not be an easy way to describe the features of the system. A description which focuses on simple explanation of how a system behaves will therefore not in general be a good explanation of how that behaviour is generated by the system. A relevant analogy here is the difference between the user manual and the system architecture for an electronic system. The assumption that components helpful for describing system behaviour are necessarily physical modules within the system architecture can only be valid if there are no limits to available resources. Otherwise such an assumption will lead to explanatory breakdown as the view taken becomes more detailed.

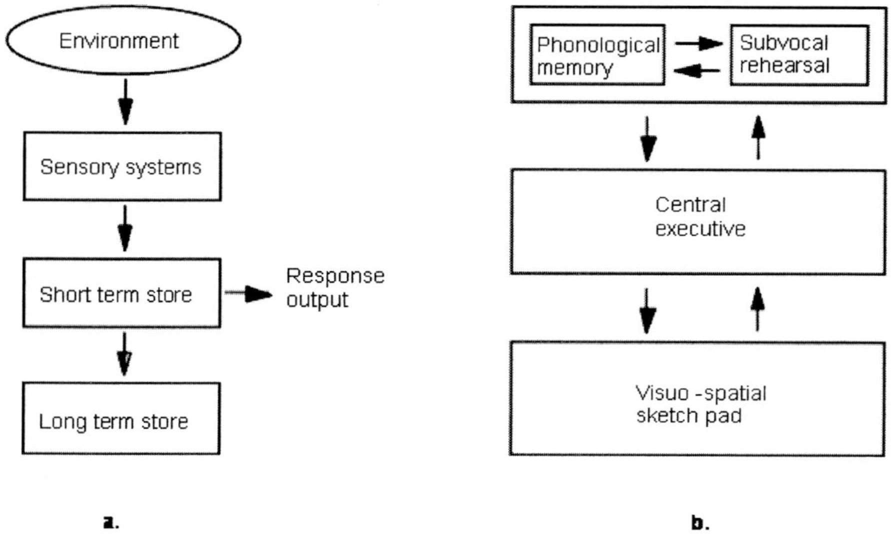

Figure 11.4. Prior models for the operation of working memory. a. Atkinson and Shiffrin [1968] proposed a model for the information flow from perception to response, and b. Baddeley and Hitch (1974) proposed a model for the internal operations of working memory within the Atkinson and Shiffrin model.

Although the Baddeley model provides a concise description for many of the phenomena of working memory, there is a major issue with the detailed definition of the central executive. As Baddeley [2000] has pointed out:

> "One danger with the concept of a central executive is that of postulating a homunculus that is simply assumed to have whatever capacities are necessary to account for the data. One response to this charge is to argue for the value of homunculi as a means of allowing the investigator to set aside some of the more intractable problems. The danger occurs when the theorist treats the homunculus as a solution rather than as a problem to be solved. The question of how to analyze the central executive remains a difficult one."

The more fundamental problem, of which the difficulty of analyzing the central executive is a symptom, is that the concept of "central executive" does not collect together a group of system operations which are supported by one module in the system, and that regardless of the descriptive value of the model, operations supporting the phenomena of working memory cannot be separated into modules corresponding with the components of the model.

In the recommendation architecture, there is no requirement for a homunculus like function. Prolongation or reactivation of the activity of a population corresponding with one set of information is determined by the current activity of all other populations. There is no physical separation between working memory storage resources and long term memory storage resources: the content of working memory is simply the currently activated information in long term memory. The different capabilities exhibited in working memory are the different capabilities by which one active population can prolong the activity of or reactivate another population. There is no consolidation process: any active information which is novel is made part of the long term store by its activation process. The limits to working memory are not defined by resource limitations to a separate working memory component but by limits to the number of different activations which can be active simultaneously in long term memory and by the decay time of such activations. Access to information newly recorded in long term memory after the decay of the portfolio population within which the information was recorded depends upon achieving a configuration of activity which can in turn reactivate the population.

Memory and the Recommendation Architecture

The memory systems proposal of Schacter and Tulving [1994] identified five independent memory systems on the basis that firstly they each corresponded with a group of memory tasks with some common characteristics; secondly they each had an identifiable list of system features different from the list for any other system; and thirdly multiple dissociations could be demonstrated between the systems. The five memory systems they identified were episodic memory, semantic memory, procedural memory priming memory and working memory.

In the recommendation architecture model, the observations supporting the existence of these separate memory systems can be accounted for by the different information recording and access mechanisms required. Creation of new episodic and semantic memories depend upon recording of new conditions. Creation of new procedural memories depend upon definition of appropriate recommendation weights in response to consequence feedback, while priming memory depends upon provisional assignment of recommendation weights in the absence of consequence feedback. Working memory does not require the recording of information. Access to episodic memory depends upon indirect activation of condition portfolios on the basis of the activity of other portfolios which recorded conditions at similar times in the past. Access to semantic memory depends upon indirect activation of condition portfolios on the basis of the activity of other portfolios which have often been active at the same time in the past. Access to procedural memory depends upon totaling the recommendation weights of the portfolios which are active. Access to priming memory depends upon indirect activation of condition portfolios on the basis of the activity of other portfolios which have recently been active at the same time in the past. Working memory is determined by limits on the number of independent portfolio populations which can be active at the same time in the same physical group of portfolios, and on the decay time for population activations.

The recommendation architecture model can therefore provide an account for the observations of different memory systems within an integrated cognitive architecture.

References

Atkinson, R. C. and Shiffrin, R. M. (1968). Human memory: A proposed system and its control processes. In K. W. Spence (Ed.). . New York: Academic Press.

Baddeley, A. D. (1986). Oxford: Oxford University Press.

Baddeley, A. D. (2000). Short-term and working memory. In E. Tulving and F. I. M. Craik (Eds.). . Oxford: Oxford University Press.

Baddeley, A. D. and Hitch, G. J. (1974). Working memory. In G. Bower (Ed.). . New York: Academic Press.

Baddeley, A. D. and Scott, D. (1971). Short-term forgetting in the absence of proactive interference. 23, 275 - 283.

Baddeley, A. D., Vallar, G. and Wilson, B. A. (1987). Sentence comprehension and phonological memory: some neuropsychological evidence. . London: Erlbaum.

Bainbridge, L. (1977). Verbal reports as evidence of the process operator's knowledge. International 11, 411 - 436.

Bar, M. and Biederman, I. (1998). Subliminal visual priming. 9(6), 464 - 469.

Basso, A., Spinnler, H., Vallar, G. and Zanobia, E. (1982). Left hemisphere damage and selective impairment of audio-verbal short term memory. 20, 263 - 274.

Berry, D.C. (1987). The problem of implicit knowledge. , 4, 144-151.

Butters, N. (1984). Alcoholic Korsakov's syndrome: an update, 4, 2, 226-244.

Caplan, D. and Waters, G. S. (1990). Short-term memory and language comprehension: a critical review of the neuropsychological literature. In G. Vallar and T. Shallice (Eds.). . Cambridge: Cambridge University Press.

Cave, C. B. (1997). Very long-lasting priming in picture naming. 8, 322 – 325.

Clarke, R. and Morton, J. (1983). Cross modality facilitation in tachistoscopic word recognition. 35A, 79 – 96.

Cohen, N. J., Eichenbaum, H., Deacedo, B. S., and Corkin, S. (1985). Different memory systems underlying acquisition of procedural and declarative knowledge. In D.S. Olton, E. Gamzu, and S. Corkin (Eds.), Memory Dysfunction: an Integration of Animal and Human Research from Preclinical and Clinical Perspectives. New York: New York Academy of Sciences.

Corkin, S. (1968). Acquisition of motor skill after bilateral medial temporal-lobe excision. 6, 225 - 264.

Corkin, S. (2002). What's new with the amnesic patient H.M.? 3, 153 - 159.

Corsi, P. M. (1972). Human memory and the medial temporal region of the brain. . McGill University.

Della Sala, S., Gray, C. Baddeley, A., Allamano, N. and Wilson, L. (1999). Pattern span: a tool for unwelding visuo-spatial memory. 37, 1189 - 1199.

De Renzi, E. and Nichelli, P. (1975). Verbal and nonverbal short term memory impairment following hemispheric damage. 11, 341 - 353.

Gabrieli, J. D. E., Merrill, F. D. and Corkin, S. (1986). Acquisition of perceptual skills in amnesia due to bilateral medial temporal-lobe (H.M.) or diencephalic lesions. 12, 20.

Graf, P., Squire, L. R. and Mandler, G. (1985). The information that amnesic patients do not forget. 10, 164 - 178.

Graf, P., Shimamura, A., and Squire, L. R. (1985). Priming across modalities and across category levels: Extending the domain of preserved function in amnesia. 11, 386-396.

Glanzer, M. (1972). Storage mechanisms in recall. In G. H. Bower (Ed.). . New York: Academic Press.

Haist, F., Shimamura, A. P., and Squire, L. R. (1992).On the relationship between recall and recognition memory. 18, 691-702.

Hamann, S. B. (1990). Level-of-processing effects in conceptually driven implicit tasks. 16, 970 – 977.

Jacoby, L. L. and Hayman, C. A. G. (1987). Specific visual transfer in word identification. 13, 456-463.

Jacoby, L. L. (1983). Remembering the data: analyzing interactive processes in reading. 22, 485 – 508.

Kirsner, K. (1998). Implicit memory. In Kirsner, K., Speelman, C., Maybery, M., O'Brien-Malone, A., Andersen, M., and MacLeod, C. (Eds.). . New Jersey: Erlbaum.

Mandler, G. (1980). Recognizing: The judgement of previous occurrence. 87, 252 - 271.

Mathews, R. Buss, R. Stanley, W. Blanchard-Fields, F. Cho, J. and Druhan, B. (1989). Role of implicit and explicit processes in learning from examples: a synergistic effect. 15, 1083-1100.

McCarthy, R. A. and Warrington, E. K. (1987). The double dissociation of short term memory for lists and sentences: evidence from aphasia. 110, 1545 - 1563.

McCarthy, R. A. and Warrington, E. K. (1990). . San Diego: Academic Press.

Meyer, D. E. and Schvaneveldt, R. W. (1971). Facilitation in recognizing pairs of words: evidence of a dependence between retrieval operations. 90, 227 – 234.

Peterson, L. R. and Peterson, M. J. (1959). Short-term retention of individual verbal items. 58, 193 – 198.

Saffran, E. M. and Marin, E. S. M. (1975). Immediate memory for word lists and sentences in a patient with a deficient auditory short term memory. 2, 420 - 433.

Schacter, D. L. and Buckner, R. L. (1998). On the relations among priming, conscious recollection, and intentional retrieval. 70, 284 - 303.

Schacter, J. C. and Tulving, E. (1994). What are the memory systems of 1994 ? In J. C. Schacter and E. Tulving (Eds.). . Cambridge, MA: MIT Press.

Shallice, T. and Warrington, E. K. (1970). Independent functioning of the verbal memory stores: a neuropsychological study. 22, 261 - 273.

Shallice, T. and Warrington, E. K. (1977). Auditory-verbal short term memory impairment and spontaneous speech. 4, 479 - 491.

Speelman, C. (1998). Implicit expertise: do we expect too much from our experts. In Kirsner, K., Speelman, C., Maybery, M., O'Brien-Malone, A., Andersen, M., and MacLeod, C. (Eds.), . New Jersey: Erlbaum.

Tulving, E., Schacter, D. L., and Stark, H. A. (1982). Priming effects in word-fragment completion are independent of recognition memory. 8, 4, 336-342.

Warrington, E. K. and Shallice, T. (1969). The selective impairment of auditory verbal short-term memory. 92, 885 - 896.

Warrington, E. K., Longue, V. and Pratt R. T. C. (1971). The anatomical localization of selective impairment of auditory verbal short-term memory. 9, 377 - 387.

Warrington, E. K. and Weiskrantz, L. (1968). New method of testing long-term retention with special reference to amnesic patients. 217, 972 – 974.

Chapter XII

Arithmetic Processing

The higher level processes by which human beings organize their understanding of the world and express that organization verbally are the capabilities regarded as most uniquely human. These capabilities are learned from experience with minimal a priori knowledge of the cognitive concepts which are the basis for the organization. In other words, understanding is bootstrapped with limited genetic guidance, and some teaching which can provide model behaviours to imitate and feedback on performance but cannot directly influence the internal information states or behaviours of the learning brain. The need for such bootstrapping without a priori knowledge means that the only basis on which information derived from the senses can be organized is temporally correlated activity as discussed in chapter 4.

The objective of this chapter is to describe in more detail how one type of cognitive process can be performed on this basis, and to demonstrate how the capabilities required for the process can be bootstrapped by a system with the recommendation architecture.

A higher level process of particular interest is arithmetical reasoning. It can be demonstrated that some concept of number exists in a range of animals from birds and rats to chimpanzees, and very limited arithmetical reasoning is also possible. However, even human beings have difficulty in acquiring a full capability in this area.

Arithmetical reasoning does not appear to be fully "natural" to human beings. By the time a child begins to learn formal arithmetic at around five years old, that child will have already learned sophisticated speech capabilities with much less apparent effort than will be required over the next five years to master the basic multiplication facts (0 x 0 = 0 to 9 x 9 = 81) [Dallaway 1994]. Children exhibit a large number of systemic mistakes when solving arithmetic problems, and adults continue to make plenty of mistakes when performing arithmetic. In this context Marr commented that "I have no doubt that when we do mental arithmetic we are doing something well, but it is not arithmetic" [Marr 1982].

Human beings appear to use multiple strategies for solving arithmetic problems. Approaches include counting, the use of relative magnitudes, memorization of a portfolio of facts, and standard algorithms and rules. The combination of strategies used to solve a particular problem depends on the problem and the individual. Thus finding the result of 6 + 5 may involve counting five steps up from six (common in children), recall of a memorized result, or conversion to a 5 + 5 + 1 problem in which 5 + 5 is easily recalled, and 10 + 1 can be solved by one counting step.

Arithmetic processing is of considerable interest because it illustrates a conflict between the architectural capabilities of the human brain and the capabilities needed for the performance of higher human thinking. As Rumelhart, Smolensky, McClelland and Hinton [1986] have commented, "we succeed in solving logical problems not so much through the use of logic, but by making the problems we wish to solve conform to problems we are good at solving".

In this chapter, the way in which arithmetical reasoning capabilities develop in a system with the recommendation architecture will be described, and the reasons for the observed limitations to those capabilities discussed. The approach will be to describe one example of arithmetic processing, discussing each step in the process in detail in terms of the architectural processes which support the step. Next a process by which the necessary capabilities could be bootstrapped using only genetically and physiologically plausible information will be described. In particular, learning must take place without a priori access to arithmetical category or process knowledge. The reasons for the experimentally observed errors and deficits will then be discussed.

Architecture for Arithmetic Processing

The critical requirement for handling behaviour in the recommendation architecture is definition of condition portfolios which can discriminate between behaviourally different circumstances. As discussed earlier, portfolios are simply groups of conditions which tend to be similar and/or to have often occurred at the same time in the past. Such portfolios do not correlate exactly with the relevant cognitive categories but can provide discrimination between such categories. To provide adequate discrimination, it may be necessary to define and use portfolios on many different levels of condition complexity. Furthermore, the most appropriate conditions may contain information derived from different senses such as visual and propioceptic. For example, the detection that two objects are very similar but at different positions requires detection of conditions which correlate with the presence of similar visual conditions (on some level of complexity) at two points in time when propioceptic conditions (on some level of complexity) derived from the position of the eye were different.

These considerations make the description of the sequence of clustering and competition operations supporting a behavioural process very complex. This complexity results because the system architecture is optimized for resource economy, and the available information sources from which arithmetic processing must be learned are limited to temporally correlated activity of sensory inputs and consequence feedback, plus some genetically defined mechanisms which provide positive consequences following imitation behaviours as discussed in chapter 9. As discussed in chapter 3, a simple to understand "user manual" description of a cognitive process will not in general reflect how system architecture operations support the process. The arithmetical processing architecture proposed by McCloskey, Caramazza and Basili [1985] and illustrated in figure 12.26 is an example of a user manual type of architecture which may be valuable for organizing the understanding of features but is of limited value for understanding how the underlying physiological operations support the features.

In order to make the description of arithmetic processing more comprehensible but still realistic, a number of simplifications to the architecture described in chapter 9 will be made. These simplifications can be understood with reference to the clustering architecture illustrated in figure 12.1. The simplifications are firstly that one array will be assumed to provide adequate discrimination between one level of cognitive circumstances. In other words, one array discriminates between objects, another between groups of objects, and a third between groups of groups. A second closely related simplification is that any one competition component will receive inputs only from one array. A third simplification is that portfolios with inputs from multiple sensory domains (i.e. combining visual, auditory and/or propioceptic information) will be omitted. A fourth simplification is that generally only visual inputs will be discussed. The description will go outside these simplifications only when it is essential for understanding the process.

In the clustering architecture illustrated in figure 12.1 there are three parallel sequences of arrays of columns. The first sequence defines and detects conditions within information derived from the eyes, the second sequence from auditory inputs, and the third from propioceptic inputs which indicate the position of the body, such as the orientation of the eyes relative to the head. Each array detects a set of condition portfolios within the outputs from the preceding array. It will be assumed for the purpose of description that conditions in the portfolios detected by visual information array V3 in figure 12.1 have a complexity comparable with objects in the visual field, and will therefore tend to provide the best discrimination between input states in which different object related behaviours are appropriate. An example of such an appropriate behaviour would be naming the object. Portfolios detected by visual information array V4 in figure 12.1 will tend to provide the best discrimination between different behaviours related to groups of objects, and those detected by array V5 discriminate between behaviours related to groups of groups of objects.

In the context of arithmetic, an object might be "3" and one appropriate behaviour might be to say "three". A group of objects might be "3 + 5" and an appropriate behaviour might be to say "that makes eight". A group of groups might be a complete equation such as "3 + 5 = 12 – 6", and an appropriate behaviour might be to say "that is incorrect".

In reality, a number of different arrays could all contribute to discrimination between two different arithmetically relevant circumstances. Genetic bias will tend to result in the competition subsystems initially receiving most of their inputs from portfolios on the levels of condition complexity most probably appropriate to their behaviour. However, this initial bias is not specific to arithmetic, and will require tuning by experience. The output of a portfolio on any level could in principle acquire provisional recommendation strength in favour of any behaviour. Such acquisition could occur randomly with a bias in favour of portfolios in arrays which had already acquired recommendation strength into the behaviour, or in favour of portfolios often active in the past when similar behaviours had been selected.

Acquisitions of this type can improve the ability of the system to discriminate between circumstances which are very similar at the simpler condition level. For example, the ability to distinguish between a handwritten S and a handwritten 5 could depend upon the higher level conditions correlating with word or number contexts (i.e. groups and groups of groups of characters). Such acquisitions can also make response times for learned behaviours faster. For example, if learning could first occur using relatively complex conditions but once learning had occurred some simpler conditions might provide adequate discrimination. An example discussed in detail later would be if conditions detected within "3 + 5" acquired recommendation weight in favour of saying "eight", and subsequently conditions within "3", "+" and "5" also acquired such recommendation weight, but such that the total recommendation weight would only exceed other alternatives if all three objects were present.

There is a level of condition complexity in the visual sequence of arrays at which the group of portfolios activated directly in response to visual input from an object is very similar to the group activated indirectly by the corresponding word. The arrays illustrated in figure 12.1 will be assumed to be at that level, and in most cases only the arrays processing visual information will therefore need to be discussed.

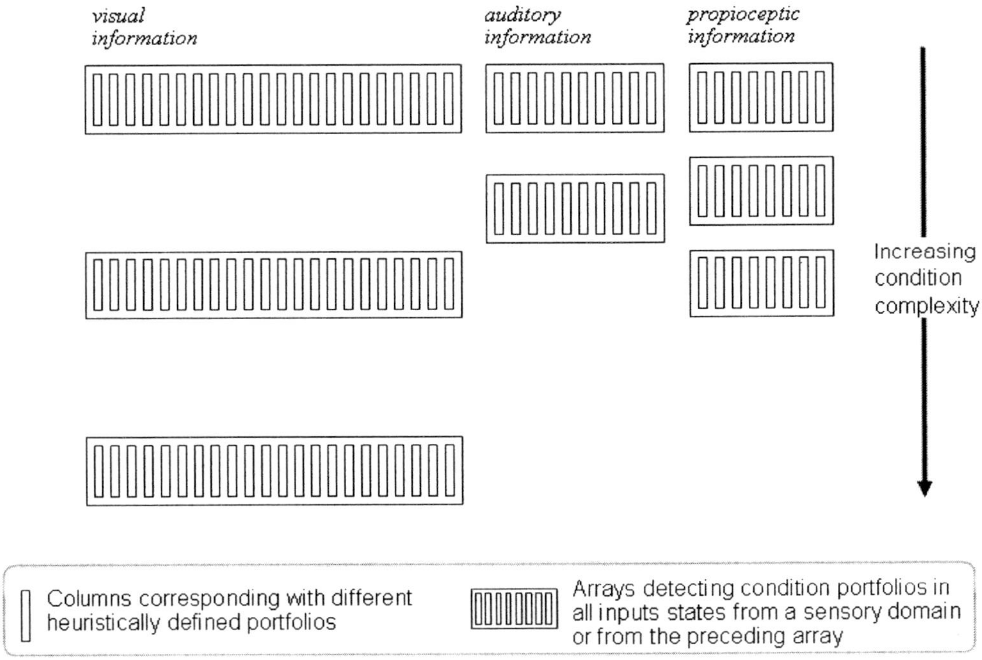

Figure 12.1. The clustering architecture to support arithmetic processing. This architecture is a simplified version of the architecture illustrated in figure 9.2.

The competition architecture for one array in figure 12.1 is illustrated in figure 12.2. The same active portfolio can recommend a range of different portfolio activation, attention and externally directed behaviours, including multiple different behaviours of the same type. Behaviours important for processing the equation 3 + 5 = 12 – 6 which could be recommended by an active portfolio in some array include synchronizing the activity of different portfolio populations within a specific array, releasing the outputs from one array to the next array, prolonging the activity of portfolios in the same array as the currently active portfolio or in a specific other array, activating portfolios in a specific array which have often been active at the same time in the past as the active portfolio, shifting the attention focus in some specific manner, or speaking a particular word.

The behaviours actually implemented will depend upon the current recommendation strengths of all currently active portfolios, and these recommendation strengths will be the result of past consequence feedback following the different behaviours. In addition, the behaviour selected will depend upon the relative activation state of different competition components. For example, if the component corresponding with a sequence of attention and internal activation behaviours has been selected at some point, the effect of the selection is that it lowers the thresholds of devices in the component corresponding with the first behaviour in the sequence, and lowers the thresholds for the second behaviour once the first behaviour has been performed and so on. Portfolios activated in response to, for example, sensory inputs at each point in time will have a range of recommendation weights including some into the components with lowered thresholds. The effect is that there is a bias in favour of the behaviours in the sequence, but a behaviour will only be selected once the portfolio populations in response to sensory inputs etc. have reached an adequate strength.

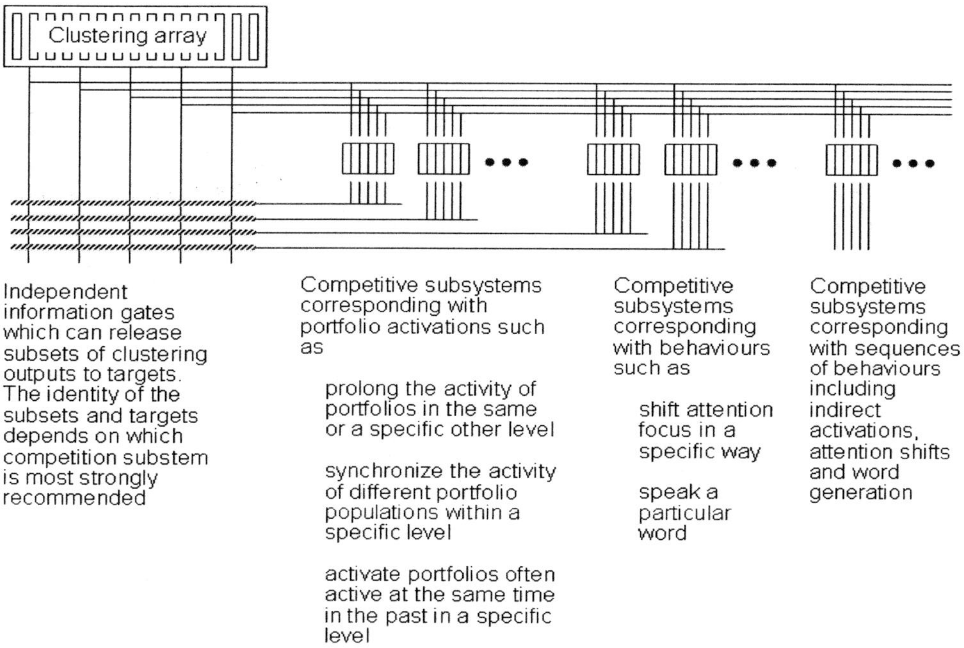

Figure 12.2. Structure of competition making use of clustering outputs from one array. Each output from clustering may have recommendation strength in favour of many different behaviours. Competition manages to selection of the most strongly recommended behaviours. In many cases a behaviour is the release of a specific subset of clustering outputs to another clustering array.

An Example of Arithmetic Processing

The performance of arithmetic processing in a system with the recommendation architecture depends upon the definition of an appropriate set of condition portfolios and the acquisition of recommendation strengths by these portfolios in favour of a range of attention and indirect activation behaviours. To provide an overview of the performance of arithmetic, the example of developing a response to the visual perception of the equation

$$3 + 5 = 12 - 6 \tag{1}$$

will be considered in detail. Appropriate responses include saying that the equation is incorrect, or suggesting that the 6 in the equation should be replaced by 4 to make the equation correct.

The way in which attention shifts within the visual field support the processing of the equation is illustrated in figure 12.3. As discussed in chapter 5, attention is the process of selecting a particular domain within the visual field and marking the information derived from that domain with a frequency modulation which ensures that portfolios containing conditions made up only of that information are preferentially detected. Different phases of the same modulation frequency can be used to maintain several independent populations of portfolios in the same array. Such independent populations could correspond with domains selected as the attention domain at different times. As illustrated in figure 12.3, the initial step is selection of the equation 3 + 5 = 12 – 6 as a single visual object. Condition portfolios detected by array V3 within the information derived from this object as a whole recommend that the subsequent attention behaviours will be the sequence of shifts illustrated in figure 12.3. In other words, the detection of condition portfolios within the equation as a whole will recommend a sequence of attention and indirect activation behaviours leading to an appropriate response. This sequence finds sub-objects on different levels of detail within the primary equation object and aligns the attention domain with these sub-objects as illustrated in figure 12.3.

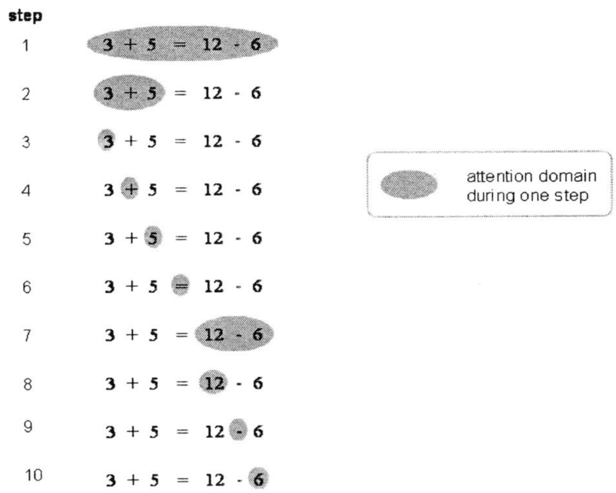

Figure 12.3. The sequence of attention domains used for generation of a response to the equation 3 + 5 = 12 – 6. At each step the domain is defined as a single object by detection of a boundary, then all sensory inputs within the boundary are tagged with a common phase of frequency modulation to ensure that condition detection is only within the information derived from within the boundary, as discussed in chapter 5.

Types of Behavioural Recommendations

The activity of any one activated portfolio will be interpreted by any competition subsystem which receives an indication of the activity as a recommendation with a weight in favour of the behaviour corresponding with the subsystem. One active portfolio in an array may simultaneously have recommendation strength in favour of different visual attention shifts, different sequences of attention and activation behaviours, release of outputs from a specific array into the next array, prolongation of the activity of currently active portfolios in the same or a different array, synchronization of different populations of active devices within the same or a different array, and a range of different speech behaviours. These recommendation strengths are instantiated in connection strengths of the outputs of devices in the column on to devices in the competition component corresponding with the behaviour.

Visual Attention Shifts

A visual attention shift might be to move and resize the attention domain to correspond with a visual object such as the equation as a whole. Portfolios correlating with a closed continuous boundary will have recommendation strength in favour of making the area within that boundary the attention domain. Such boundaries could be based on a relatively continuous line of contrast around an object or by band of empty space surrounding an object (i.e. a relatively continuous region in which contrast is absent). These two options are illustrated in figure 12.4. In the case of an equation, the relative isolation of the equation from other visual objects on the page is the initial definition of the object. Sub-objects are defined by (possibly narrower) bands of empty space.

Definitions on the basis of boundaries can be refined by the conditions detected within the information inside the boundaries. Thus once an attention domain roughly corresponds with a cognitively relevant object, some these conditions can further refine that correspondence. For example, if the correspondence is not exact, some of the activated portfolios will contain information mainly derived from the object, others will contain information from outside the object. However, portfolios containing information only from within the object will tend to have been active more often at similar times in the past. The presence of portfolios not often present together at the same time in the past is interpreted as a recommendation in favour of shifting the size, position and shape of the attention domain until the correspondence is more exact. This mechanism can even operate to separate objects which overlap in the visual field.

Activation Behaviours

As discussed in chapter 9, a portfolio may have recommendation strength in favour of activating other portfolios which have often been active or recorded conditions at similar times in the past. Such portfolios may be in its own or in other arrays. Prolongation or synchronization of the activity of already active portfolios in the same or another array are other examples of possible recommendation strengths. Release of outputs from one array to the next may also require adequate recommendation strength.

Speech Generation

A portfolio can also have recommendation strength in favour of different speech behaviours. Speech processing was discussed in outline in chapter 9, for the purposes of this chapter a portfolio will be regarded as recommending an utterance and the processes by which recommendations by different portfolios can result in comprehension and generation of complex speech will be omitted.

Sequences of Attention, Activation, and Speech Behaviours

The processing of an arithmetical equation is a sequence of attention, activation and speech behaviours for which the attention shifts illustrated in figure 12.3 provide the general framework. The sequence is invoked by activation of a competition component corresponding with the sequence, and such activation occurs when the portfolios detected within equation visual objects have adequate recommendation strength in favour of initiating the sequence.

The effect of activation of the sequence component is that the component first lowers the thresholds of devices in a more detailed component corresponding with the first behaviour in the sequence. When that first behaviour is complete, the thresholds of devices in the component corresponding with the second behaviour in the sequence are lowered and so on. However, the individual behaviour components will only generate their behaviour if there is also adequate recommendation strength in the portfolio populations activated in response to recent visual inputs. The lowering of thresholds biases the recommendation strengths of these populations in favour of the next behaviour in the sequence.

There are genetically defined components corresponding with simple behaviours, but components corresponding with sequences of behaviours are defined when a particular sequence frequently occurs, and is consistently rewarded by positive consequences. The two component sequences discussed below are examples of heuristically defined components.

The description of the arithmetic process is organized into sections corresponding with different areas of the equation being the focus of attention. The step numbers of the sections are the attention domain locations illustrated in figure 12.3.

Step 1: Attention Domain "3 + 5 = 12 - 6"

Perception of a broad visual domain (such as a room environment or a printed page) results in activation of condition portfolios containing conditions detected within that whole domain. Different portfolios activated under these circumstances in array V3 of figure 12.1 or in earlier arrays recommend focusing attention on different objects within that broad domain, where an object is defined as an area of the visual field which is within a boundary that can be continuously defined. Selection of an attention domain therefore depends upon the detection of condition portfolios in array V3 or earlier arrays which discriminate between the presence and absence of boundaries as defined by figure 12.4.

Portfolios activated by different object boundaries will compete, and to initiate arithmetic processing the recommendation strengths in favour of the domain containing the equation must be strongest. These strengths will initially recommend focusing attention on a domain roughly corresponding with the equation. As illustrated in figure 12.5, the portfolios detected within this roughly aligned domain will recommend shifting the size, position and shape of the domain until an exact correspondence is achieved. This process for achieving detailed correspondence will occur as part of every subsequent attention step but will not be described again.

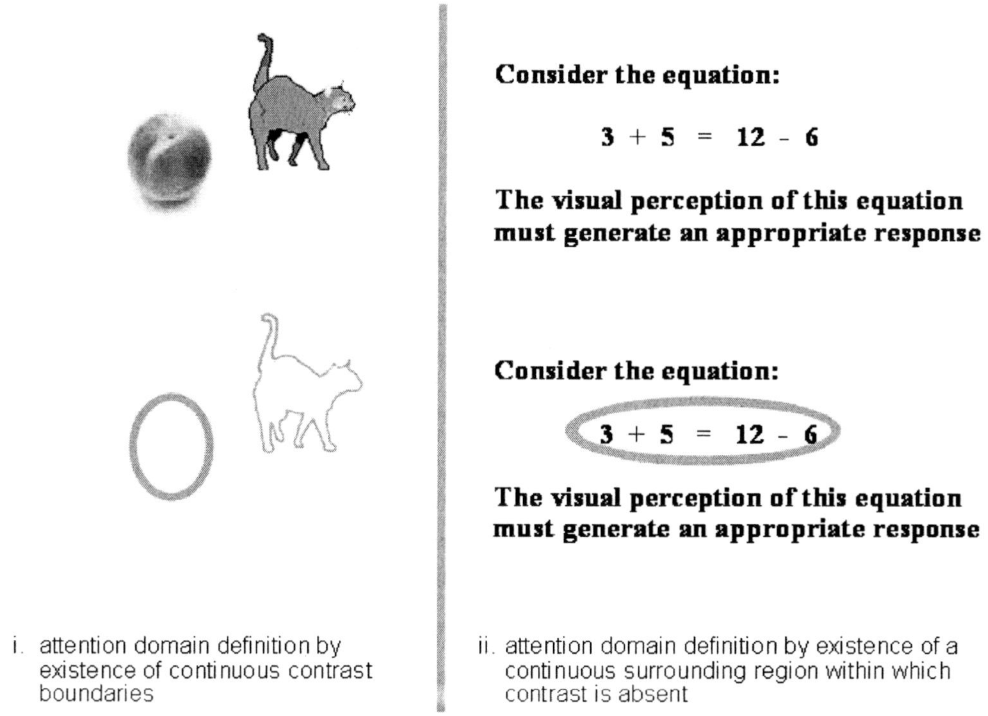

Figure 12.4. Examples of criteria by which attention domains can be defined within a larger visual field. Portfolios correlating with the presence of a closed, continuous contrast boundary (i) or with a closed, continuous boundary path within which contrast is absent (ii) have recommendation strengths in favour of imposing a common phase of frequency modulation on all sensory inputs within the boundary.

Once a close enough correspondence is achieved, as illustrated in figure 12.6, the portfolios detected in array V3 within the equation "3 + 5 = 12 – 6" viewed as a single object have recommendation strength in favour of a sequence of behaviours. The way in which they can recommend a sequence is by having adequate recommendation strength in favour of activating a general component driving every behaviour in the sequence in turn. The steps in the general sequence are:

I. Focus the attention domain on the left hand sub-group of the overall equation and perform the sequence of detailed behaviours recommended by the resultant portfolio population
II. Prolong the resultant activation in array V4
III. Focus the attention domain on the middle sub-group of the overall equation and perform the sequence of detailed behaviours recommended by the resultant portfolio population
IV. Prolong the resultant activation in array V4
V. Focus the attention domain on the right hand sub-group of the overall equation and perform the sequence of detailed behaviours recommended by the resultant portfolio population
VI. Prolong the resultant activation in array V4.
VII. Synchronize the outputs from array V4 and release to array V5
VIII. Perform the behaviour most strongly recommended by activated portfolios in array V5.

The first step in this sequence is to identify a sub-group within the left hand side of the equation "3 + 5 = 12 – 6". This sub-group is "3 + 5".

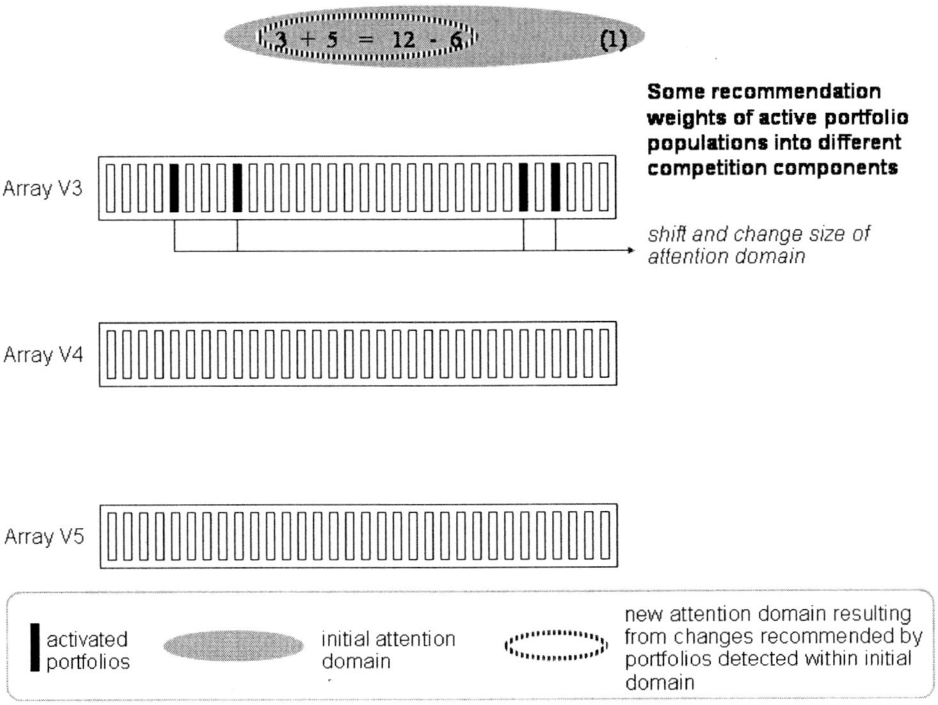

Figure 12.5. Portfolios detected within an attention domain with a poor match to an object defined by a boundary recommend changes to the attention domain.

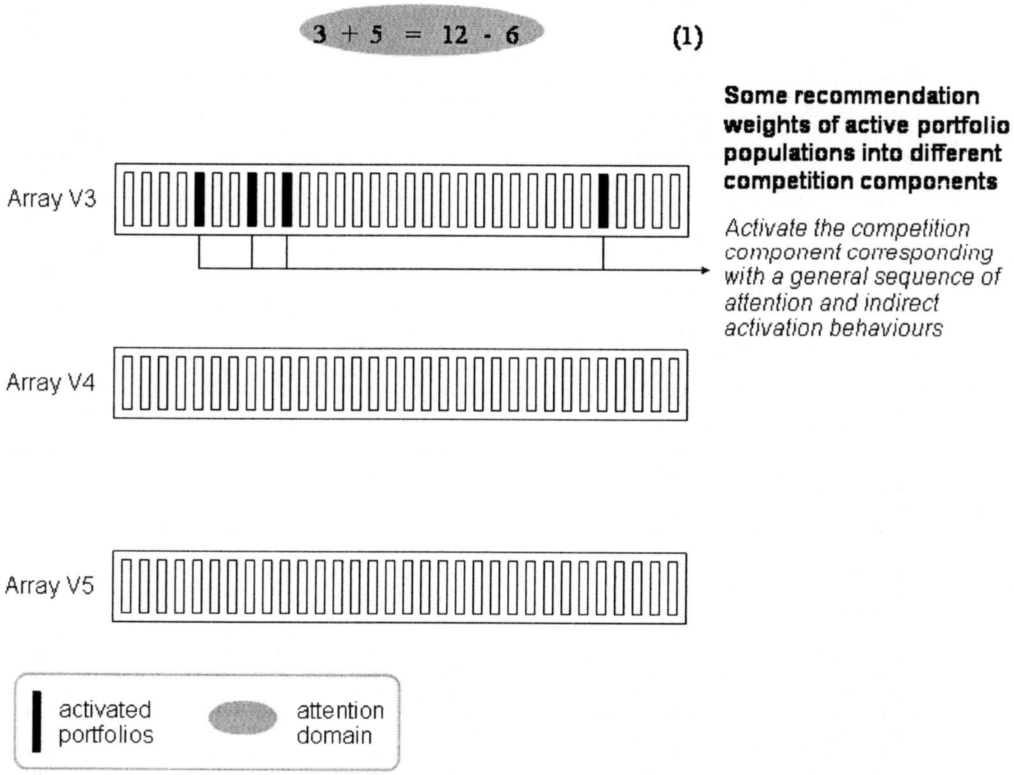

Figure 12.6. Portfolios detected within an equation viewed as a single visual object recommend activation of a competition component corresponding with a sequence of attention and indirect portfolio activation behaviours.

Step 2: Attention Domain "3 + 5"

As in the case of the full equation, the left hand sub-group is identified by activated portfolios correlating with a continuous surrounding boundary path within which contrast is absent. Again as for the full equation the attention domain may be adjusted to fit the sub-group by the recommendation strengths of the portfolios activated when the fit is initially less exact.

As illustrated in figure 12.7, the portfolios activated in array V3 in response to the first subgroup "3 + 5" of the equation have recommendation strengths in favour of activating a component corresponding with a more detailed sequence of behaviours. At this point the general component activated by the overall equation has reduced the threshold of devices in the more detailed component, increasing the effective recommendation strength of the population activated in response to "3 + 5" into the detailed component.

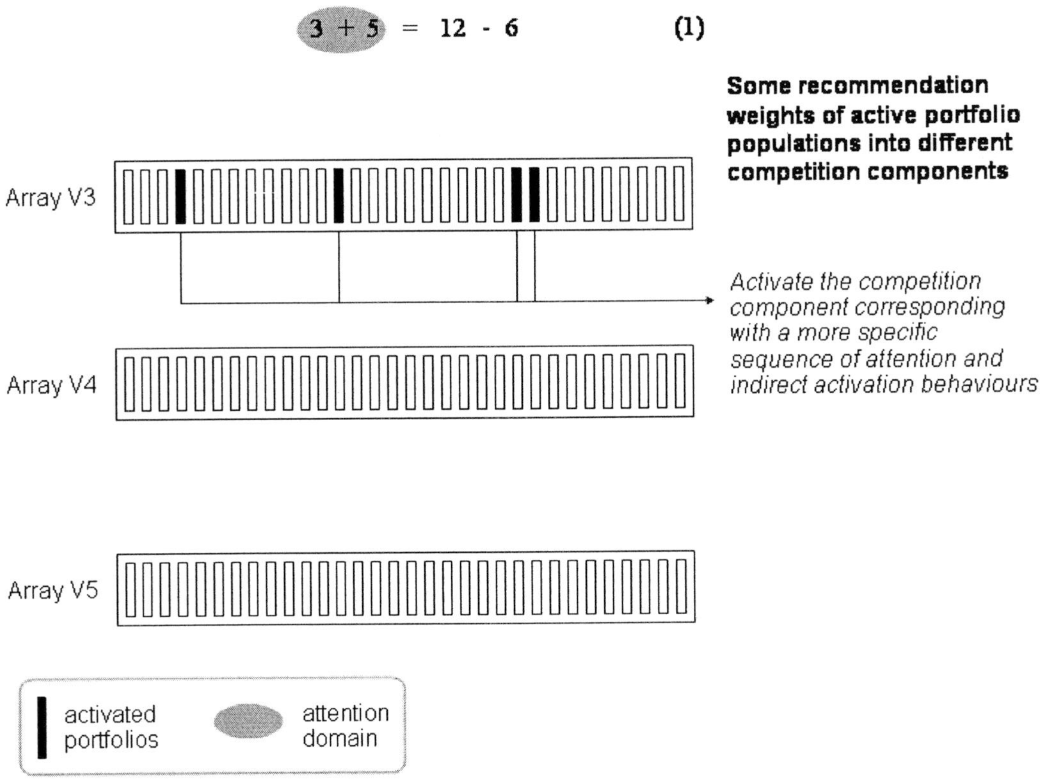

Figure 12.7. Portfolios activated in response to the attention domain covering the left hand side of the equation activate a competition component corresponding with a sequence of attention and internal activation behaviours. The first step in this sequence is aligning the attention domain with the first object in the left hand group as illustrated in figure 12.8

The more detailed sequence of behaviours is:

A. Focus attention on the left hand object in the subgroup
B. Prolong the resultant activity in array V3
C. Focus attention on the middle object in the subgroup and activate portfolios at a different frequency modulation phase
D. Prolong the resultant activity in array V3
E. Focus attention on the right hand object in the subgroup and activate portfolios at a different frequency modulation phase
F. Synchronize the outputs from array V3 and release to array V4
G. Extinguish activity in array V3

In the specific example, this sequence results in the attention domain being focused sequentially on the objects 3, +, and 5.

Steps 3 to 5: Attention Domains "3", " +" and " 5"

As a result of step A in the detailed sequence, the attention domain is focused on the visual object "3". The resultant portfolio activity is illustrated in figure 12.8. This activity has a range of recommendation strengths, including speaking the word "three", prolonging its own activity, shifting the attention domain to the next object on the right, and synchronizing and releasing outputs from array V3 to array V4. Because step A in the detailed sequence has been implemented, the component corresponding with prolonging activity (step B) is activated, and this recommendation is therefore implemented. Once implemented, the component corresponding with shifting attention to the next object on the right (step C) is activated, and this recommendation is therefore implemented and attention shifts to the object "+".

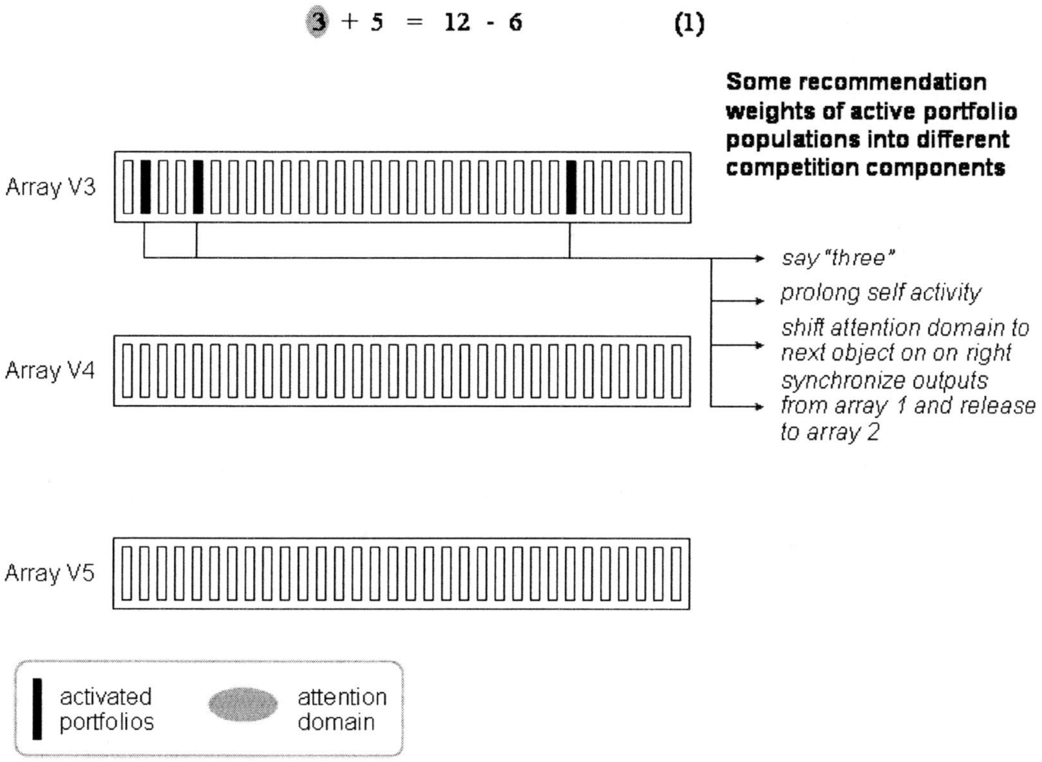

Figure 12.8. Portfolios activated in response to the attention domain covering the left character on the left side of the equation recommend a number of different behaviours. The behaviours with the strongest effective recommendation strengths will be those corresponding with the components currently activated by the sequence components which have themselves been activated by portfolios detected within the equation object (figure 12.6) and the left side of the equation object (figure 12.7).

When portfolios have been activated in array V3 in response to the third object, the pattern of activity is as illustrated in figure 12.9. Because step E in the detailed behaviour sequence has been implemented, the component corresponding with synchronizing and releasing outputs from array V3 into array V4 has been activated, and this behaviour therefore has the highest effective weight in the active population and is implemented.

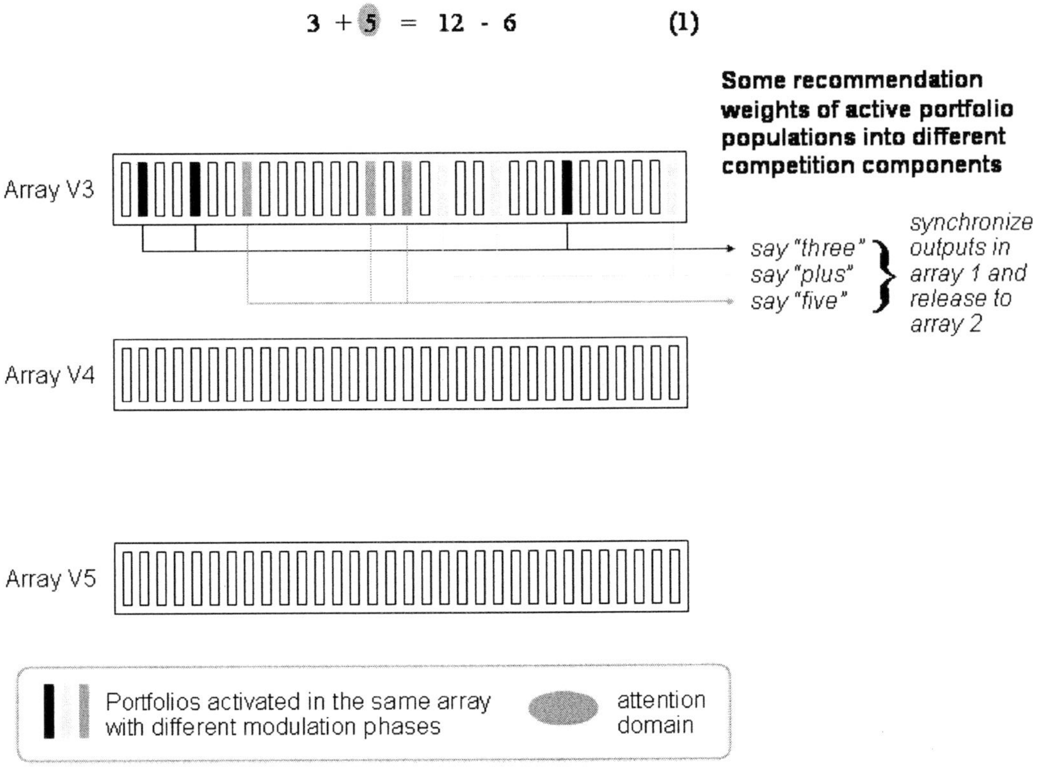

Figure 12.9. At this point in processing the equation, separate portfolio populations have been activated in response to the attention domain being focused sequentially on the three characters making up the left hand side of the equation. The populations have a range of character specific recommendation strengths such as speaking the name of the character. The populations also have recommendation strengths in favour of indirect activations of different types. Because sequence components activated earlier have at this point activated more detailed components corresponding with synchronizing outputs from array V3 and releasing to array V4, these type of recommendations are accepted

The resultant activity in array V4 is illustrated in figure 12.10. This active portfolio population has a range of recommendation strengths including saying "eight", extinguishing the activity in array V3, prolonging its own activity, shifting the attention domain to the next object on the right, and synchronizing and releasing outputs from array V4 to array V5.

Because step F in the detailed sequence has been implemented, the component corresponding with the behaviour of extinguishing activity in array V3 is activated, and it is this recommendation which is therefore implemented. The detailed sequence at this point has been fully implemented, which is step I in the general sequence. The component corresponding with step II in the general sequence is therefore activated. This is the component corresponding with the behaviour of prolonging activity in array V4, and this recommendation of the array V4 population is then implemented. Under the control of the general sequence component, the specific component corresponding with shifting the attention domain to the right is then activated, resulting in this recommendation of the active population in array V4 being implemented.

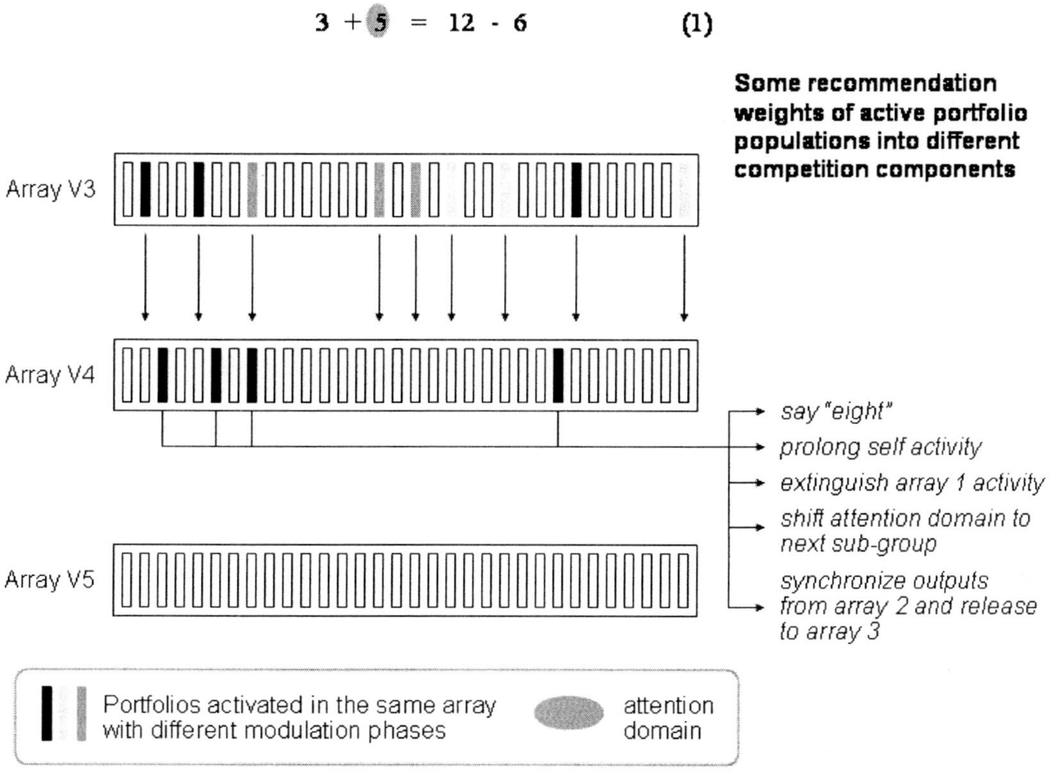

Figure 12.10. Acceptance of synchronization and release recommendations as illustrated in figure 12.9 results in activation of a portfolio population in array V4. This population again has recommendation strengths in favour of a range of behaviours. The behaviours with the greatest effective recommendation strengths will be those corresponding with components activated by sequence components. At this point these behaviours are prolonging activity in array V4, extinguishing activity in array V3, and shifting the attention domain within the equation to the middle character.

Step 6: Attention Domain "="

The detailed behaviour sequence component activated by the portfolio population in array V3 is a shorter version of the previous detailed sequence, and results in activation of a second portfolio population in array V4.

Step 7- 10: Attention Domain "12 - 6"

In these steps the attention domain is first centred on the visual object "12 – 6". The portfolios activated in array V3 in response are interpreted as a recommendation to activate the same detailed behaviour component as activated in response to "3 + 5".

Note that it was necessary in the course of this detailed behaviour sequence to focus attention on the visual object "12" rather than on the separate objects "1" and "2". This decision could be determined both by the relative proximity of the two individual numerals and by the presence of "=" and "-" objects on each side of it, both circumstances containing conditions with recommendation strength in favour of treating it as a single object within the arithmetic process.

This activation leads to the pattern of activity illustrated in figure 12.11. At this point behaviour VI in the general sequence has been implemented, and the specific component corresponding with synchronizing and releasing outputs from array V4 to array V5 is activated, As a result it is this recommendation of the active population in array V4 which is implemented, leading to the pattern of activation illustrated in figure 12.12.

In this step the attention domain is centred sequentially on the visual objects "12", "-" and "6". As described in detail for the domain "3 + 5", this process leads to activation of three independent populations of portfolios in array V3 corresponding with the three objects, which are then synchronized at the output level of array V3 and released as inputs to array V4. The activity in array V3 corresponding with the three objects and the higher level "12 – 6" object is then extinguished, leading to the situation illustrated in figure 12.11.

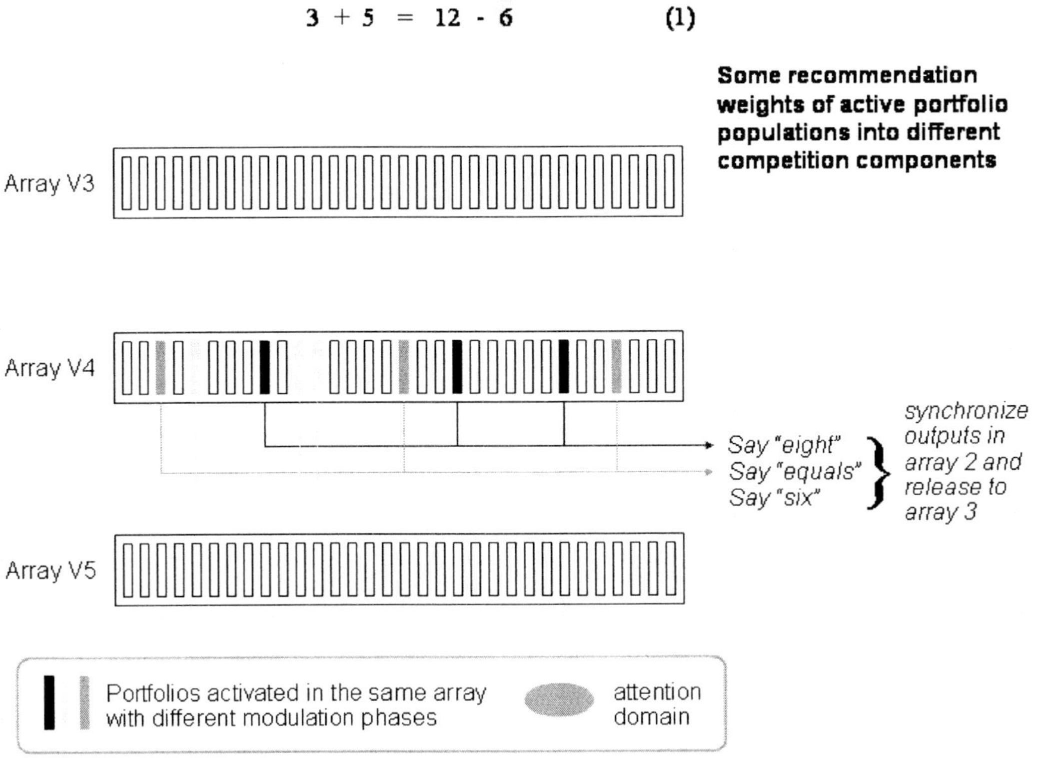

Figure 12.11. At this point in processing the arithmetic equation, separate populations of portfolios have been activated in array V4 in response to the three groups of characters making up the equation. Each of these groups is the result of separate activations in response to each character in array V3. The individual populations have recommendation strengths in favour of group specific behaviours and also indirect activation behaviours. Because components corresponding with some indirect activation behaviours have been activated by previous portfolio activations, those behaviours have the greatest effective recommendation strengths and are accepted.

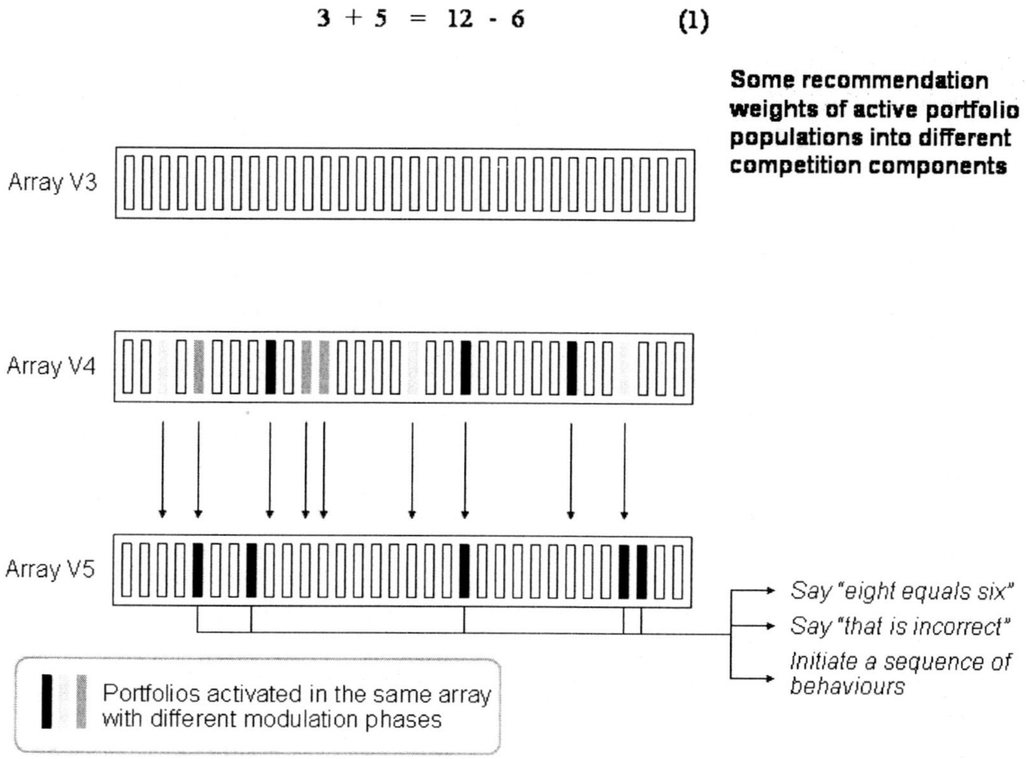

Figure 12.12. Following the synchronization and release of outputs from array V4 illustrated in figure 12.11, a portfolio population is activated in array V5. This population represents the end point of the sequence of behaviours processing the equation, and has a range of recommendation strengths. One of these behaviours is to initiate a modified version of the equation processing sequence.

Each of the three independent populations in array V4 has a significant total recommendation strength in favour of a different speech behaviour as illustrated. Thus the population in array V4 resulting from the combination of information derived from the objects "12", "-" and "6" in array V3 has significant total recommendation strength in favour of saying "six". However, any one of these recommendation strengths will in general be less than the strength generated in favour of a synchronization behaviour by the combination of the still active portfolio population generated in response to the whole equation in array V3 and the three populations in array V4. This behaviour is to synchronize the three populations in array V4 at the output level and release the outputs to array V5. The result of the synchronization and release behaviour is to generate an active population of portfolios in array V5 as illustrated in figure 12.12. This population has recommendation strengths in favour of a number of different behaviours including saying "eight equals six", saying "that equation is incorrect" and initiating a further sequence of attention and activation behaviours by activating a further general sequence component.

Further Sequence of Attention and Activation Behaviours

This further sequence component repeats the processing performed on the original equation until the attention domain is centred on the final numeral "6". At this point the pattern of activation is as illustrated in figure 12.13. The portfolios activated in response to the visual object "6" in array V3 have recommendation strength in favour of saying "six". They also have recommendation strength in favour of continuing the regular equation handling sequence, and additional recommendation strength of this type comes from the still active portfolio population in array V3 generated in response to the whole equation. The portfolios activated in response to the visual object "6" in array V3 also have recommendation strength in favour of activating other portfolios in array V3 which are not currently active but which have been often active in the past at similar times to past activity of the "6" portfolios. Because the specific component corresponding with this indirect activation behaviour has been activated at this point by the general sequence component, this recommendation has the greatest effective strength.

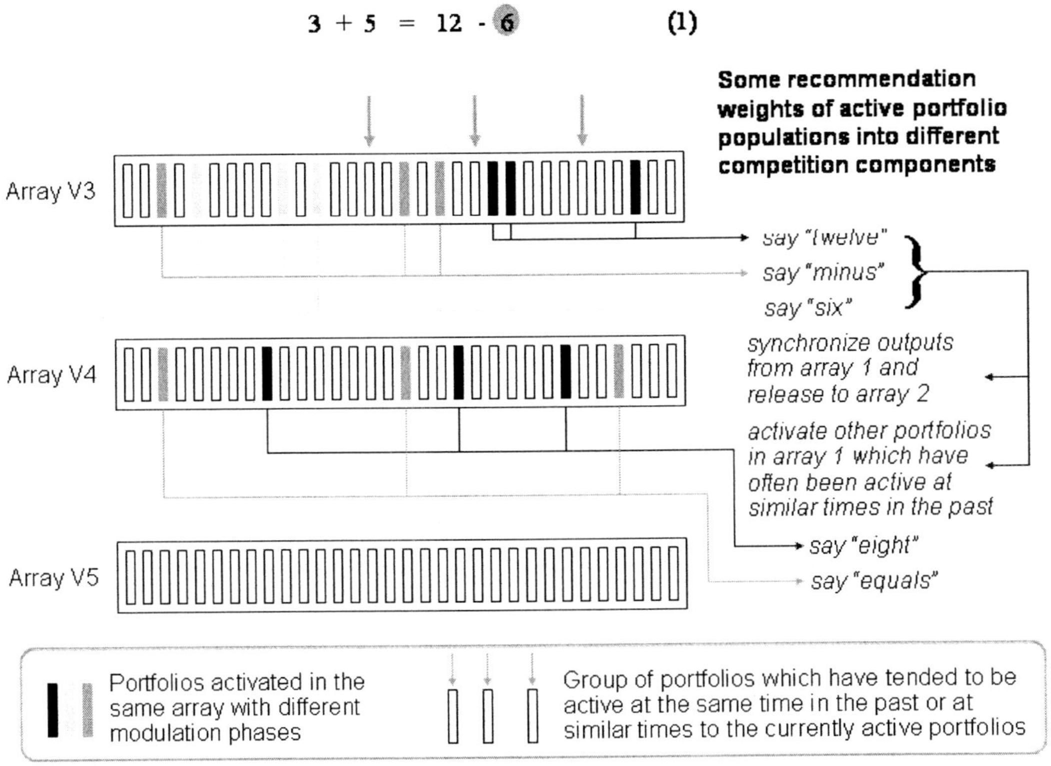

Figure 12.13. The illustration shows the different portfolio populations activated after the first two groups in the equation and the first two characters in the third group have been processed by the modified equation processing behaviour sequence(which to that point is the same as the regular sequence) and the third character in the right hand group is being processed by the modified sequence, which for that character differs from the regular sequence. The portfolios activated in response to the third character "6" as before have a range of recommendation strengths in favour of different behaviours. Because of activation of a specific component by the modified sequence component, the strongest effective recommendation strength is in favour of activating other portfolios in array V3 which are not currently active but which have often been active in the past somewhat before or somewhat after the currently active group. The effect is to activate portfolios which would have been activated in response to a different number from "6".

Because other numerals have often been present in the past at similar times to the numeral "6", this recommendation will tend to result in the activation of portfolios activated by some of these other numerals. Because these secondary activation portfolios have recommendation strength in favour of activating portfolios often active in the past at the same time, the tendency will be for a tertiary population to converge on a population typically activated in response to one numeral. Such a numeral might be "3".

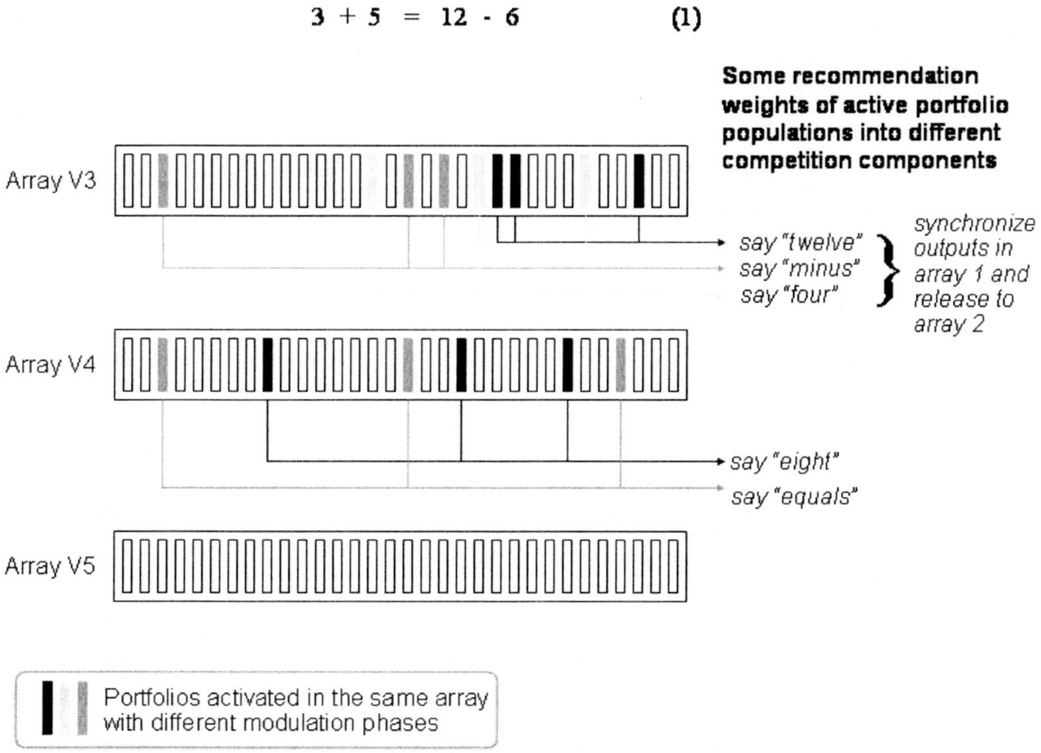

Figure 12.14. The situation soon after that shown in figure 12.13. The recommendation by portfolios directly activated in response to "6" to activate portfolios in array V3 which are not active but have often been active in the past just before or just after the portfolios activated in response to "6" has been accepted. The resultant population then activates a further population on the basis of frequent past simultaneous activity. The population therefore evolves towards a consistent set corresponding with perception of some number different from six. The diagram illustrates the situation, perhaps after a number of attempts, in which a population corresponding with the number four is activated.

Under the control of the sequence components the outputs from array V3 are synchronized and released to array V4, then the independent populations in array V4 are synchronized and released to array V5. At this point the active portfolio population in array V5 has recommendation strengths in favour of saying "eight equals nine", saying "that is incorrect" or of initiating another cycle of indirect activation by extinguishing the current indirect activation in array V3 and the most recent activation in array V4 and generating a new indirect activation in array V3 which contributes to activations in arrays V4 and V5 as before. Eventually the numeral selected in array V3 by this process is "4" as illustrated in figure 12.14. The three independent activations in array V4 then have recommendation strengths in favour of saying "eight", "equals" and "eight" as illustrated in figure 12.15. Note that the populations corresponding with the two "eight"s will generally overlap to some degree, and the overlapping portfolios could be indicated by different output devices modulated at two different phases or the same devices generating outputs modulated at two different phases and therefore at a higher average rate overall.

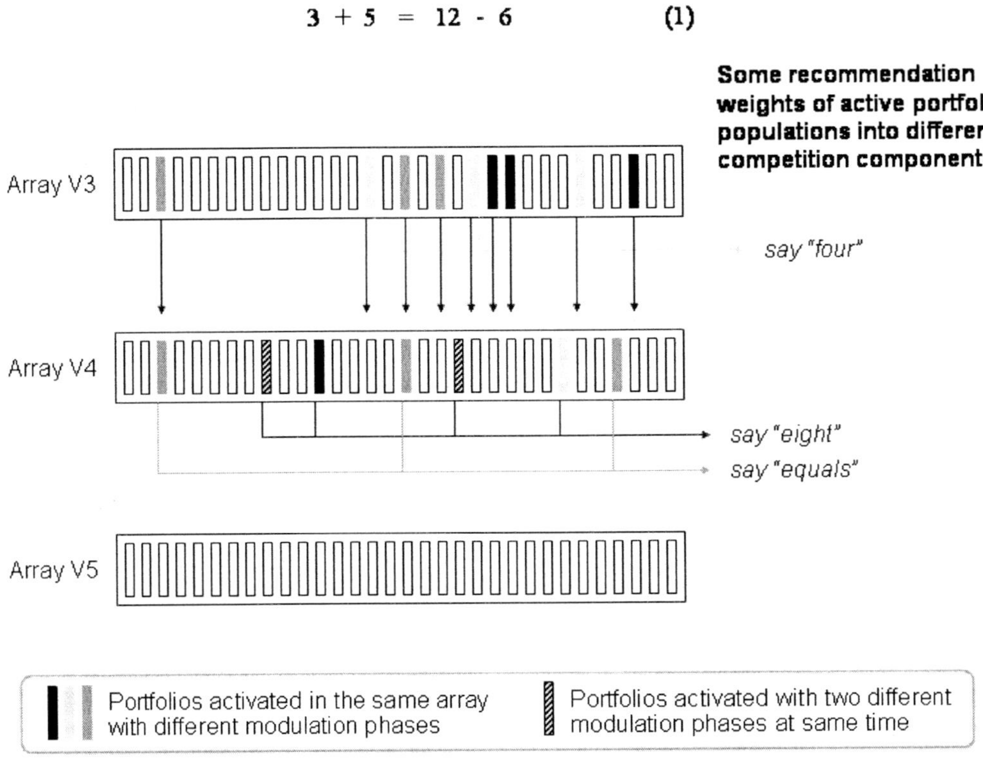

Figure 12.15. The situation soon after that shown in figure 12.14. Outputs from populations in array V3 corresponding with 12, - and an indirectly activated "4" have been synchronized and released to array V4 where they are about to generate a portfolio population.

An important difference between the original general behaviour component and the component invoked by the error condition is that at this point, as illustrated in figure 12.16, is that instead of all the activity being extinguished in array V3, the activity generated indirectly is retained active when the rest of the activity ends. The three populations in array V4 are then synchronized. The resultant activation in array V5 has recommendation strengths in favour of saying "eight equals eight" or "that is correct". When combined with recommendation strengths of the indirectly activated portfolio population in array V3 corresponding with "4", this activation in array V5 also has recommendation strength in favour of saying "the last number should be four".

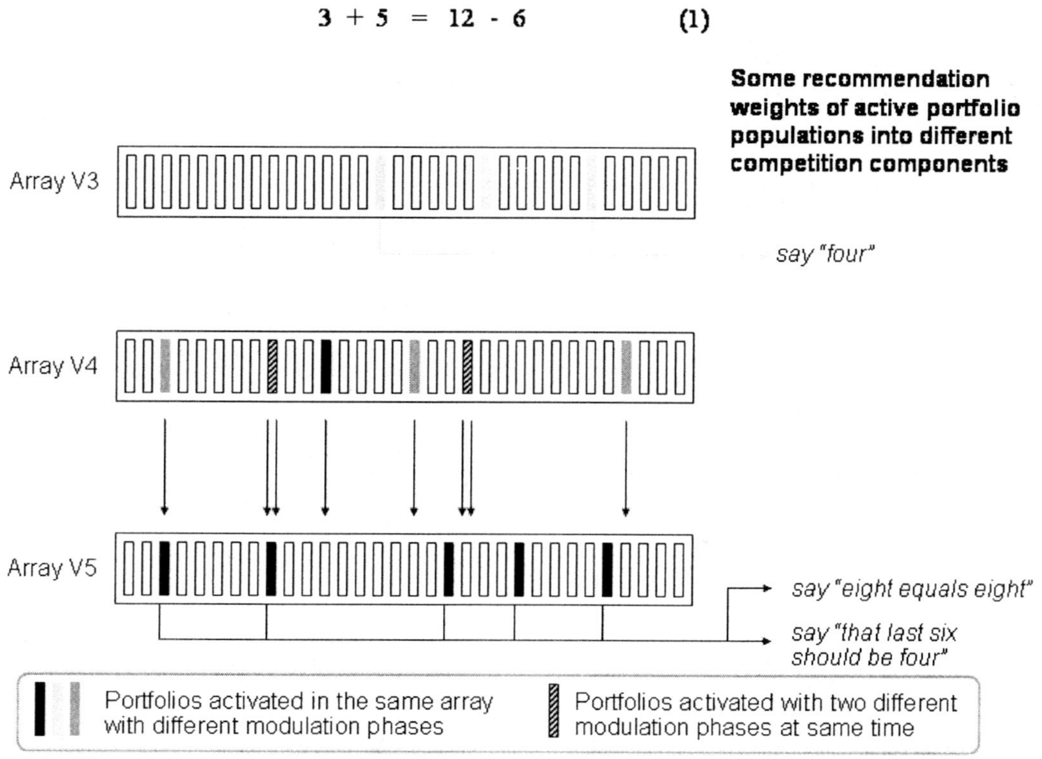

Figure 12.16. The situation soon after that shown in figure 12.15. Outputs from populations in array V4 corresponding with 3 + 5, + and 12 – 4 have been synchronized and released to array V5 where they generate a portfolio population. This population in array V5 combined with the residual population in array V3 has recommendation weights in favour of a number of different behaviours, including saying that the final six in the original equation should be four. If the population in array V3 had corresponded with an incorrect number, there would be recommendation strength in favour of repeating the indirect activation process in array V3 to generate a different number.

The Complexity of the Process

The process described for responding to an arithmetical equation can readily be generalized to include multiplication and division. However, the process appears much more complex than the algorithmic process followed by electronic calculators. As will be clear from earlier discussion, the reason for this complexity is that the need for learning without interference with prior learning but using resources also employed for other cognitive purposes largely defines how the brain must process equations or any other cognitively relevant objects. Consider now how the brain learns to process equations.

Learning of Arithmetic

This section will describe how arithmetic capabilities can be bootstrapped with limited and plausible a priori (genetic) guidance and plausible external supervision. Plausible genetic information can create a tendency for certain types of portfolio to recommend certain types of behaviour, but only if such a tendency is generally useful (i.e. not specifically for arithmetic). Plausible external supervision can provide behavioural models for imitation and reward certain behaviours, but cannot guide the internal states of the brain. The objective is to demonstrate that a process exists by which an arithmetic capability can be bootstrapped by a system with the recommendation architecture. It is not argued that the learning process in human beings is exactly the one described, although the process is generally modeled on the way in which children are taught arithmetic, and similarities between the model and the actual human physiological process can be expected.

A Priori Genetic Information Defining Competition

The need to generate appropriate behaviours in response to objects, groups of objects and groups of groups is a general need for any animal. A behaviour may be required in response to an individual other animal which could be different from the appropriate behaviour in response to a group of such animals. The behaviour in response to the group may be affected by the presence of other groups of animals and of objects or groups of objects in the environment. Thus the most appropriate behaviour of a deer in response to one wolf alone may be different from that in response to a pack, and will also depend upon the presence of other deer, including young, and upon the presence of trees, rivers etc. in the environment.

The existence of behaviours of finding closed boundaries of various types in the visual field, and using these boundaries to define domains within which conditions will be preferentially detected could therefore be expected to be specified genetically. Although finding such boundaries is a difficult algorithmic problem it is in principle solvable, particularly if refined by adjustments to the size and shape of current visual domain to eliminate sensory information activating portfolios not often active in the past at the same time as most other portfolios. Such an adjustment would have the effect of moving the definition of the boundary to exclude areas not consistent with the bulk of the area within the boundary. The existence of behaviours of shifting the visual domain to the next area within a continuous boundary on the left or right (or above or below) the current visual domain, or to an area within a continuous boundary but within the current visual domain can also be expected to be genetically programmed.

Equally, the existence of various synchronization and release behaviours can be expected to be genetically programmed, along with behaviours to utter the individual sounds out of which speech is constructed.

Existence of a behaviour means that there is a competition component corresponding with the behaviour which has appropriate output connectivity to drive the behaviour if the component is activated by greater recommendation strength than any other behaviour. Some of the competition components which can be expected to exist as a result of genetic specification are illustrated in figure 12.17. These behaviours will be relevant to a wide range of different behavioural features.

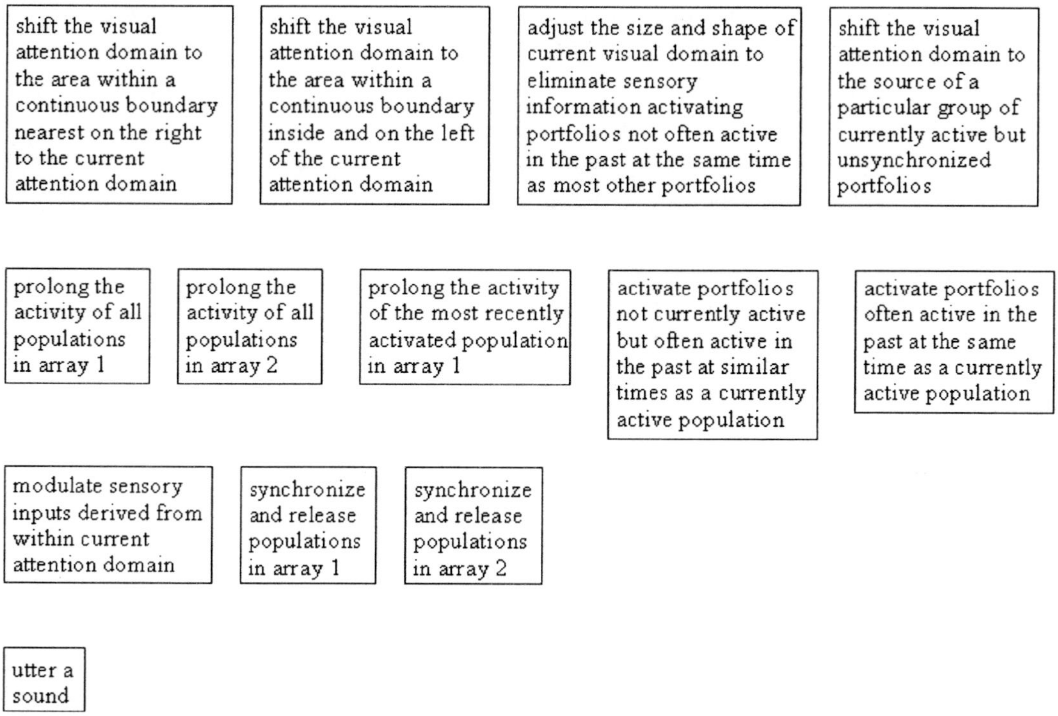

Figure 12.17. Available genetically specified indirect activation components which are utilized to construct arithmetic behaviours.

Some sequences of behaviours could also be defined by genetic specification. Such a sequence would be specified as a higher level component which if selected would excite the component corresponding with the first behaviour in the sequence and if that component was activated (i.e. the corresponding behaviour selected) would excite the component corresponding with the second behaviour in the sequence and so on. Components corresponding with sequences could also be defined heuristically by random selection modified by consequence feedback or by selection of sequences which frequently occurred. Higher level components managing sequences of sequences could be defined in the same way.

In general, components for arithmetic processing components or for any other higher cognitive process will need to be defined heuristically.

A priori Genetic Information Defining Clustering

All these genetically specified competition components would of course require recommendation strength from portfolios in clustering before they could drive their behaviour. Genetic specification could determine that arrays tended to be defined to detect conditions within certain ranges of complexity corresponding with portfolios discriminating most effectively between objects, groups of objects and groups of groups. Adequate discrimination might require portfolios on many different levels of complexity even for just objects, and genetic information could specify the existence of these levels. Such specification could be achieved by biasing of the number of device layers and the average number of inputs required to activate a neuron, the connectivity for internal column and array organization, and the initial sources of condition defining inputs to arrays.

Genetic bias could also ensure that portfolios in specific arrays had provisional connectivity and recommendation strength into specific behavioural components.

A priori Genetic Information Encouraging Imitation Behaviours

Another element of genetic information is an internal reward system which increases the recommendation strengths of behaviours which are followed by detection of conditions within sensory inputs that are similar on certain levels of complexity to conditions detected just before the behaviour was performed. One example, discussed in chapter 9, is that such a bias in favour of uttering sounds similar to sounds just heard is important in the process for speech acquisition. Genetic bias could specify the existence of a reward system that compared conditions detected before and after a behaviour and the portfolio array complexities which were compared.

A Priori Genetic Information Encouraging Behavioural Experimentation

There is survival value in a tendency to investigate a range of alternative behaviours for responding to the same, frequently occurring circumstances. This tendency is essential if no behaviour has been given a specific a priori bias. If a bias has been placed on a type of behaviour, investigation of alternatives may be limited to within the type. There is also value in a tendency to investigate a range of alternative behaviours even after one behaviour has been followed by positive consequences. Such a tendency reduces the risk of settling into a "local minimum" corresponding with a behaviour which is rewarded but is less efficient than other alternatives.

However, if a specific behaviour following a specific set of circumstances has been rewarded consistently over a long period, it is less likely that a better alternative exists, and such investigation will be less useful.

Such tendencies could be instantiated physiologically in a bias in favour of portfolios being given recommendation strength into a wide range of behavioural components initially, and this bias persisting even after portfolios have acquired substantial recommendation strength in favour of one behaviour following positive consequences. This tendency should decline with time, or with the number of times existing recommendation strengths have been rewarded.

There is a general resemblance between this semi-random selection of behaviours in early learning and the play phenomenon observed in higher mammals (Coward 1990).

A priori Genetic Information Defining First and Second Order Rewards

Genetic information could specify that certain types of conditions which exist within sensory perceptions of certain types of behaviour by other human beings result in release of signals which modulate the input strengths into recently performed behaviours of all the portfolios with such inputs that were recently active. In such a first order reward system, certain behaviours by a supervisor would be interpreted as rewards, others as punishments. Reward behaviour could be refined heuristically by allowing other conditions that occurred at the same time as the preprogrammed reward conditions to acquire recommendation strength on to releasing the reward.

A second order reward system makes it possible to reward a behaviour which produces a similar result to an already rewarded behaviour without supervisor intervention. A behaviour itself results in sensory inputs within which conditions could be detected. Higher order conditions corresponding with the sequential occurrence of the original sensory inputs and inputs corresponding with the correct behaviour could be defined and detected. The behaviour of generating a reward (i.e. release of signal to increase recommendation strengths) occurs shortly after the sequential occurrence. Hence conditions detected within the sequential occurrence could gain recommendation strength on to the behaviour of generating the reward signal.

A Process by which the Necessary Capabilities can be Acquired

The first requirement for a learning process is the definition of a set of portfolios which permit discrimination between circumstances in which different arithmetic type behaviours are appropriate. The process for improving the discrimination provided by an array of portfolios by adding portfolio outputs or portfolios when contradictory consequences are detected was described in chapter 4. Portfolios on some levels of complexity would be much more effective than those on other levels as a basis for discrimination. To avoid excessive demand for resources, some level or limited number of levels will need to be selected as the starting point for optimization. The simplest approach would be to optimize the portfolios on the level which is providing the best discrimination early in the learning process. However, there is some risk that selection of the array complexity on the basis of the best initial discrimination will result in an array complexity which is suboptimal for longer term learning of arithmetic processing. For behaviour types which have had an impact on natural selection over a long period, the choice might be influenced by genetic bias, but this is unlikely in the case of arithmetic. The selection may therefore be determined by random factors in early experience, resulting in considerable differences in the learned capability between individuals. To avoid excessive description complexity, the description of the learning process will only include one array, and the optimization process will be omitted.

The second requirement is that each portfolio must acquire an appropriate combination of recommendation weights into different behaviours so that each portfolio activation results in an appropriate behaviour. This requirement includes both the acquisition of the appropriate physical connectivity and the tuning of the weights of that connectivity. The selection of the array complexity optimized for arithmetic behaviour discussed earlier will to some degree depend upon which arrays have the greatest initial connectivity into the relevant behavioral components. Once optimization is under way, new outputs from the array will tend to be given provisional connectivity on to components frequently invoked in the past by existing array outputs. Again, to simplify the description, the process by which connectivity is assigned will be omitted.

Clustering array definition and competition connectivity and weight definition will of course occur in parallel, with contradictory consequence information influencing array definition.

The arithmetic learning process is made up of a series of steps which will be described in sequence. However, in practice several steps could occur at the same time or some steps could occur in a somewhat different order. The steps are listed in table 11.1

1. learning the concept of number
2. learning the number sequence
3. learning "number plus 1" including a behavioural interpretation of "plus" and "equals"
4. learning addition
5. learning long addition
6. learning the concept of multiplication
7. acquisition of recommendation strengths by input numbers

Step 1: Learning the Concept of Number

In this step, the brain of the learner is exposed to a large number of sensory experiences in which under the guidance of a teacher the attention domain is focused on a group of objects, and the teacher speaks the number of objects in the group. Conceptually, for one array, this process is illustrated in figure 12.18 for the example of three objects. There are some portfolios in the array which tend to occur relatively frequently when there are three objects in the visual domain, and there are some portfolios which tend to occur relatively frequently when the word "three" is heard. Some specific auditory portfolios are therefore frequently active at the same time as some specific visual portfolios as illustrated in figure 12.19, and the auditory portfolios therefore acquire recommendation strength in favour of activating the visual portfolios. These visual portfolios instantiate the concept of three. In addition, when the learner imitates speaking "three", the behaviour is rewarded. These visual portfolios were active just before the reward, and therefore acquire recommendation strength in favour of speaking "three".

Note that the visual portfolios activated in response to hearing the word "three" might include portfolios in a number of different arrays at different levels of complexity, but will not include portfolios at complexity levels close to raw sensory inputs. In other words there is no visual hallucination. However, it is of interest to ask the hypothetical question what actual visual input could result in activation just the "three" portfolios in some array. One answer might be three amorphous grey objects grouped close together in an otherwise blank visual field. In reality there is no such visual input, because any real input would almost certainly activate additional portfolio in the array. It is also important to note that any one of the "three" portfolios will also in general have recommendation strengths in favour of a wide range of behaviours unrelated to the number three.

In practice, the levels of complexity on which portfolios discriminate effectively between different numbers of objects in a group may not be the same as the levels of complexity on which portfolios discriminate between different types of object, and multiple levels of complexity may be required to achieve adequate discrimination. Genetic bias could favour the existence of arrays at levels of complexity which make number discrimination possible, and there could also be optimization of an array or arrays for number discrimination as discussed below. However, the observation that the average human cannot number with a single glance any group with more that about six members [McCarthy & Warrington 1990] indicates that there are limits to this discrimination at the level of condition complexity generally utilized. However, for numbers above the limit for precise verbal response, estimate generation behaviour could be adequately managed.

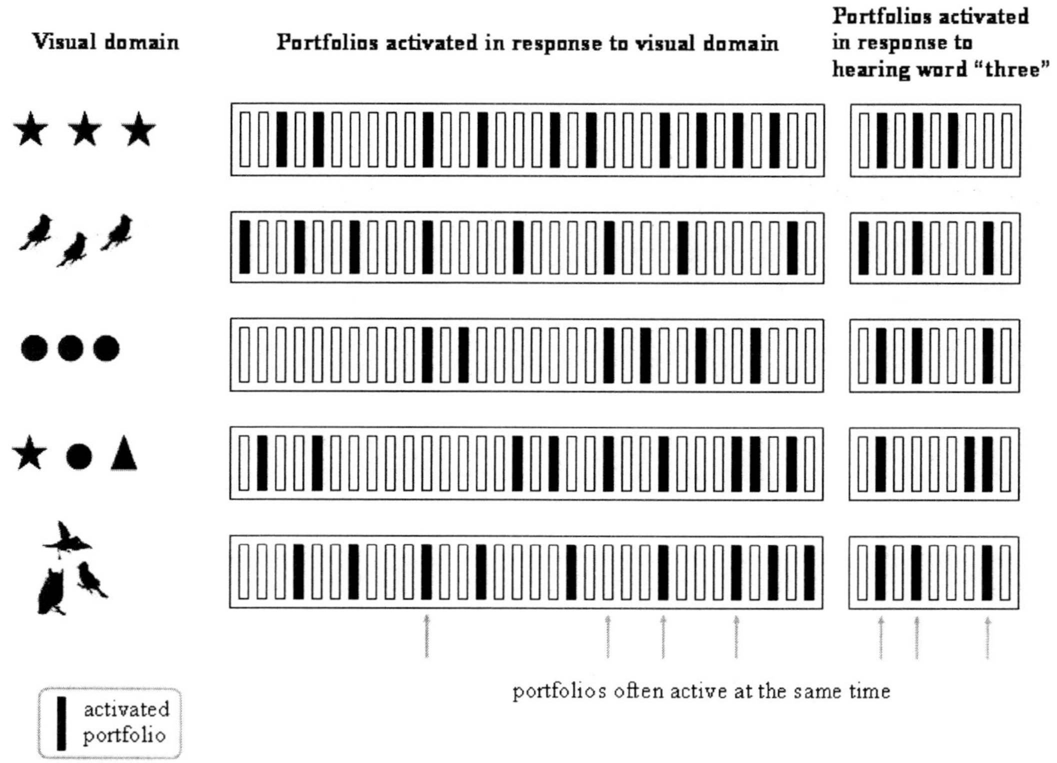

Figure 12.18. Learning the concept of number. On some level of complexity, similar portfolios tend to be activated when a group of three objects is visually perceived. In an auditory array, similar portfolios are activated whenever the word "three" is heard.

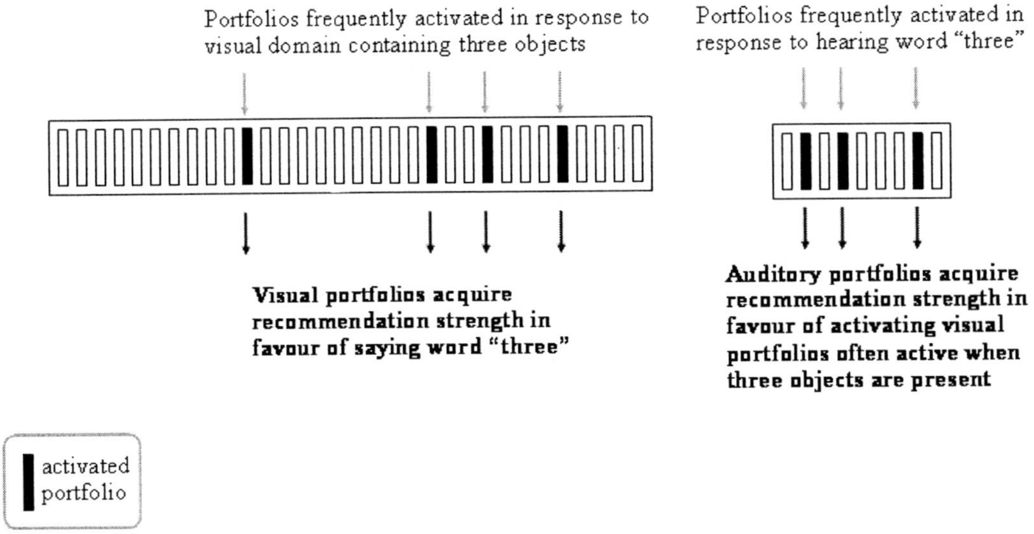

Figure 12.19. Given the experience illustrated in figure 12.18, auditory portfolios often active following the hearing of the number three acquire recommendation strength to activate visual portfolios often active following perception of groups of three objects. The portfolios often activated following perception of groups of three objects acquire recommendation strengths in favour of saying "three".

Step 2: Learning the Number Sequence and Counting Behaviour

In this step, the brain of the learner is repeatedly exposed to a teacher verbally repeating the list of numbers one, two, three, four, etc. Imitation of the sequence is rewarded. Each time a number is heard, it activates both visual and auditory portfolios, these activations precede the behaviour of speaking the next number in the sequence, and this behaviour is rewarded. As a result, the portfolios activated in response to one number acquire recommendation strength in favour of speaking the next number. They also acquire recommendation strength in favour of speaking the number after that, but not enough strength to implement the behaviour unless supplemented by recommendation strength derived from the middle number.

When the learner imitates the teacher pointing to the objects in a group in sequence, speaking the next number in the sequence as the next object is indicated, and speaking words like "there are items in the group", the learner behaviour is rewarded. Such counting behaviour in response to a group of objects is therefore acquired.

Note that there are now two alternative behaviour sequences by which the same external behaviour can be reached for small groups of objects. One is direct recognition of the number, the other is counting.

Step 3: Learning "Number Plus 1"

The brain of the learner is provided with numerous examples of visual scenarios such as the one illustrated in figure 12.20. In these scenarios, the learner is encouraged by a teacher to imitate the behaviour of aligning his attention domain to include both the group of objects and the isolated object, and generate the word recommended by the portfolios activated as a result. There may be some random behavioural experimentation until the learner succeeds to selecting this two step behaviour, but once it is selected it will be rewarded by the teacher. The group of portfolios often activated in response to hearing or seeing the word "plus" will often be active just before the rewarded two step behaviour, and will therefore acquire recommendation strength is favour of that behaviour.

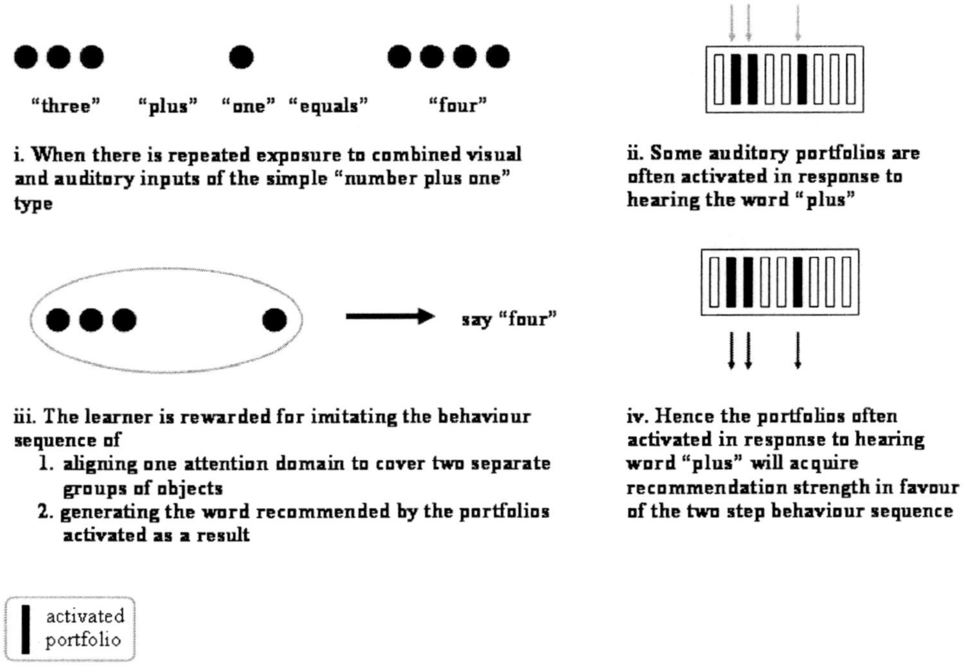

Figure 12.20. Process for learning simple addition by imitation of attention domain manipulations.

In parallel with learning this two step attention behaviour, an alternative behaviour illustrated in figure 12.21 may also be learned. In an example of this behaviour, attention is focussed sequentially on "four", "plus" and "one", and portfolios activated at different phases in array V3 in response to the three objects. Outputs from the three groups of portfolios are then synchronized as inputs to array V4. The portfolios activated in array V4 as a result are active just before the behaviour of speaking five, and because that behaviour is rewarded acquire recommendation strength in favour of it. Furthermore, portfolios activated in response to hearing the word "equals" are active just before the behaviour of synchronizing the separate outputs from array V3, and therefore acquire recommendation strength in favour of such synchronization behaviour.

The behaviour of counting all the objects is a third alternative sequence of behaviours generating the same final behaviour. Again, if the generation of the appropriate verbal response has been rewarded in the past, even for some other intermediate behaviours, the counting process will also be rewarded and vice versa.

Step 4: Learning Simple Addition

The three alternative behaviour sequences described in step 3 could all be applied to the addition of any pair of integers. Initiation of any of these behaviours could be recommended by detection of conditions within the objects "plus" and "equals", although the first sequence described in step 3 will only work for sums in which the total is less than six. A set of known sums of the type 2 + 3 = 5; 2 + 4 = 6; 2 + 5 = 7 etc. will accumulate for which the appropriate response has been acquired. For additions involving numbers larger than the limit of about seven discussed earlier, for number concept processes will only support estimate generation. However, counting can generate an exact response.

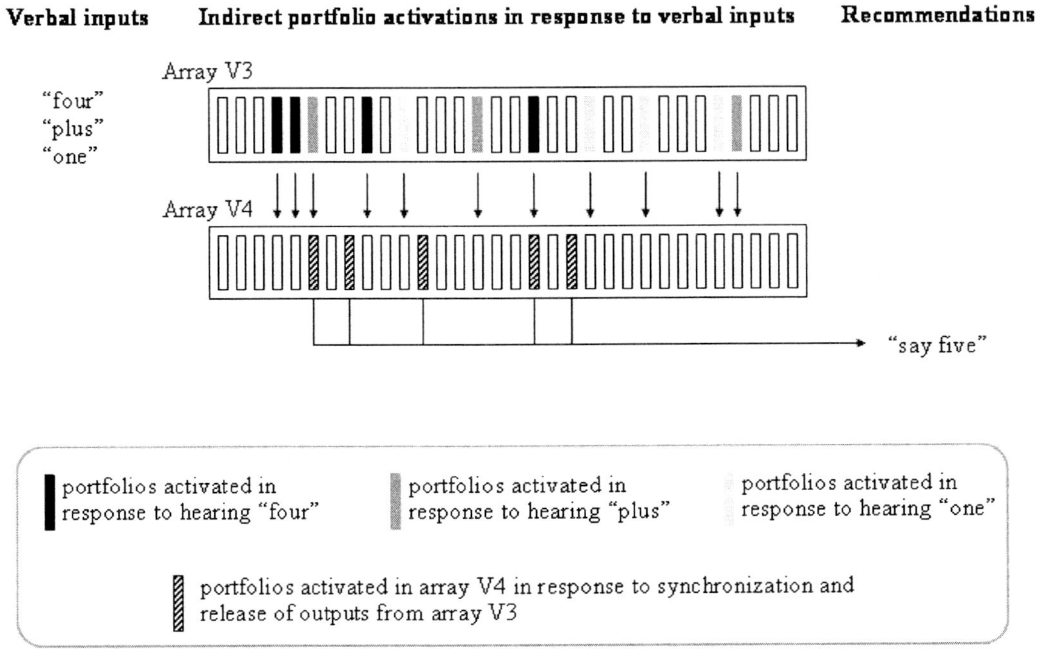

Figure 12.21 Process for learning simple addition by imitation of speech. Portfolios detected within auditory inputs corresponding with the three words "four", "plus" and "one" indirectly activate pseudovisual portfolios in array V3. The outputs of the three populations are synchronized and released to array V4 where they activate a combined population. Imitation of hearing the word "five" plus a reward following when the word is said results in the active portfolios in array V4 acquiring recommendation strength in favour of saying "five".

Encouraged imitation of a sequence behaviour of the type used in the equation example described earlier will result in portfolios containing information derived from both of the two numbers and the "+" symbol acquiring recommendation strength on to speaking the exact answer. Where counting has previously been used to generate an answer and the answer has been rewarded, learning can occur in the absence of a supervisor through generation of the same answer by the different process followed by a second order internal reward.

Step 5: Learning Long Addition

A preliminary to learning long addition is acquisition of recommendation strengths by the portfolios activated in response to seeing a number in favour of writing the number. The acquisition of these recommendation strengths will again involve imitation in order to gain efficiency in the learning process. Given this capability, the performance of long addition can be learned as the sequence of steps illustrated in figure 12.22.

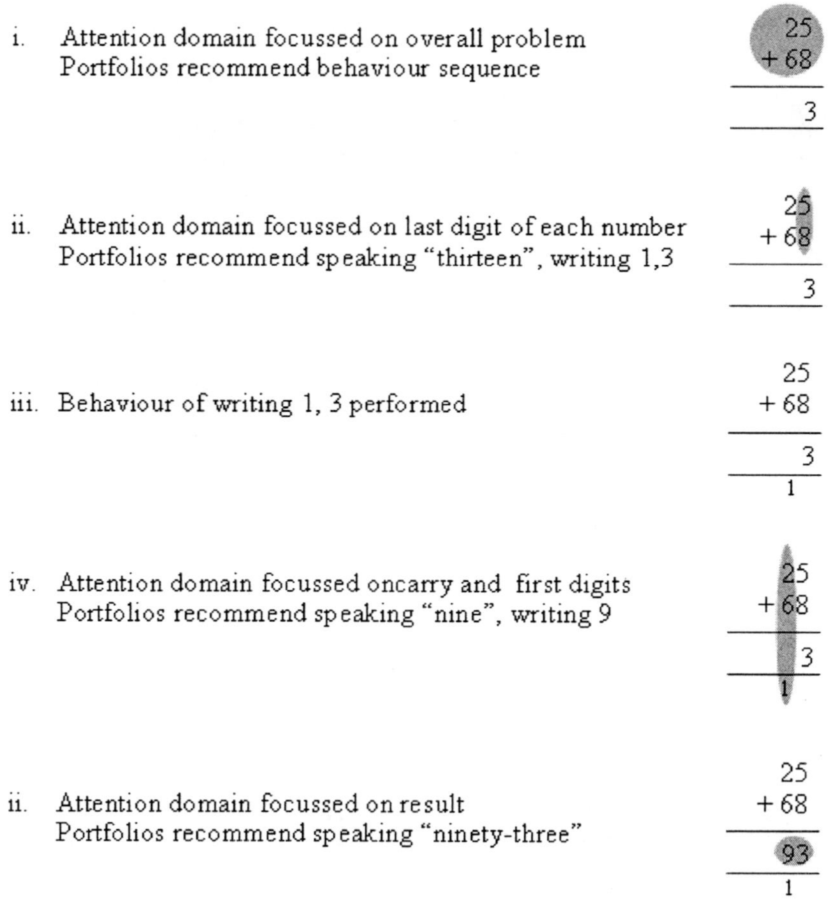

Figure 12.22. Process for learning long addition. The learning brain is encouraged to imitate a sequence of attention and indirect activation behaviours and rewarded when the sequence is followed and results in the appropriate behaviour.

Step 6: Learning the Concept of Multiplication

The description of how the concept of multiplication is learned requires reference to arrays of portfolios detecting conditions within propioceptic sensory inputs as illustrated in figure 12.1. These portfolios discriminate between different eye positions relative to an unmoving head. For descriptive simplicity, only one array of portfolios within visual information and one within propioceptic information will be included, although in reality portfolios detecting conditions on a number of different levels of complexity within visual, propioceptic and combined information may contribute to effective behaviour.

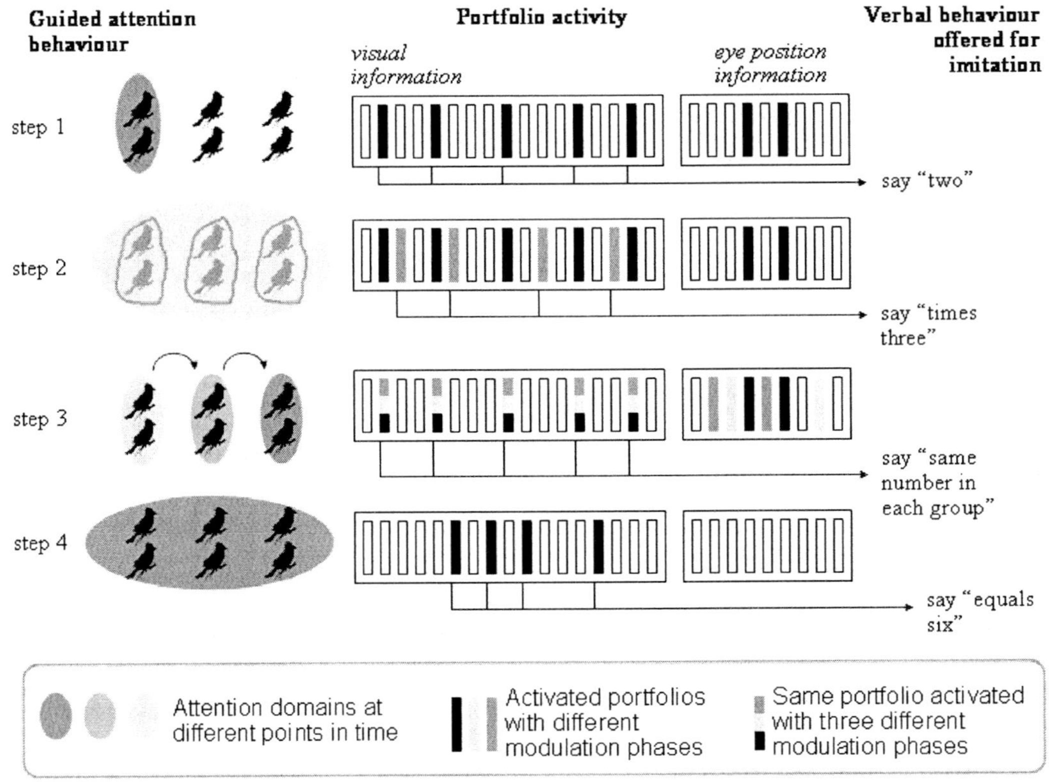

Figure 12.23. Process for learning the sequence of attention behaviours required for multiplication. In step 1 the learner is encouraged to focus attention on one of the smaller groups and say the number of objects in the smaller group, in the example to say "two". In step 2 the learner is encourage to say "times" and the number of smaller groups, ignoring their internal structure. In step 3 the learner is encouraged to focus the attention domain on each of the three smaller groups in succession and say "there is the same number of objects in each group". In step 4 the learner is encouraged to look at the larger group as a whole and say "equals" and its number of objects. These behaviours are rewarded by the teacher.

Learning a sequence of attention behaviours which instantiate the concept of multiplication is illustrated in figure 12.23. In the example in the figure, the learner is exposed to a picture of six objects arranged in three groups of two, and is encouraged by a teacher to follow a sequence of attention behaviours and to imitate a verbal behaviour after each attention behaviour. The first behaviour in the sequence is to focus attention on the first two objects and generate the behaviour of saying "two". The second behaviour is to look at the six objects as three groups of two and say "three". The third behaviour is to focus attention in quick succession on the three groups and observe that they are identical, for example generating the behaviour of saying "there are the same number of objects in each group". The fourth behaviour is to look at the group of six as a whole and say "six". The words "two times three equals six" are used repeatedly by the teacher at appropriate points. The teacher provides appropriate consequence feedback at different points.

The effect of imitating the sequence is that certain portfolios are active shortly before other specific portfolios, and recommendation strengths in favour of activation on the basis of subsequent activity in the past will be acquired.

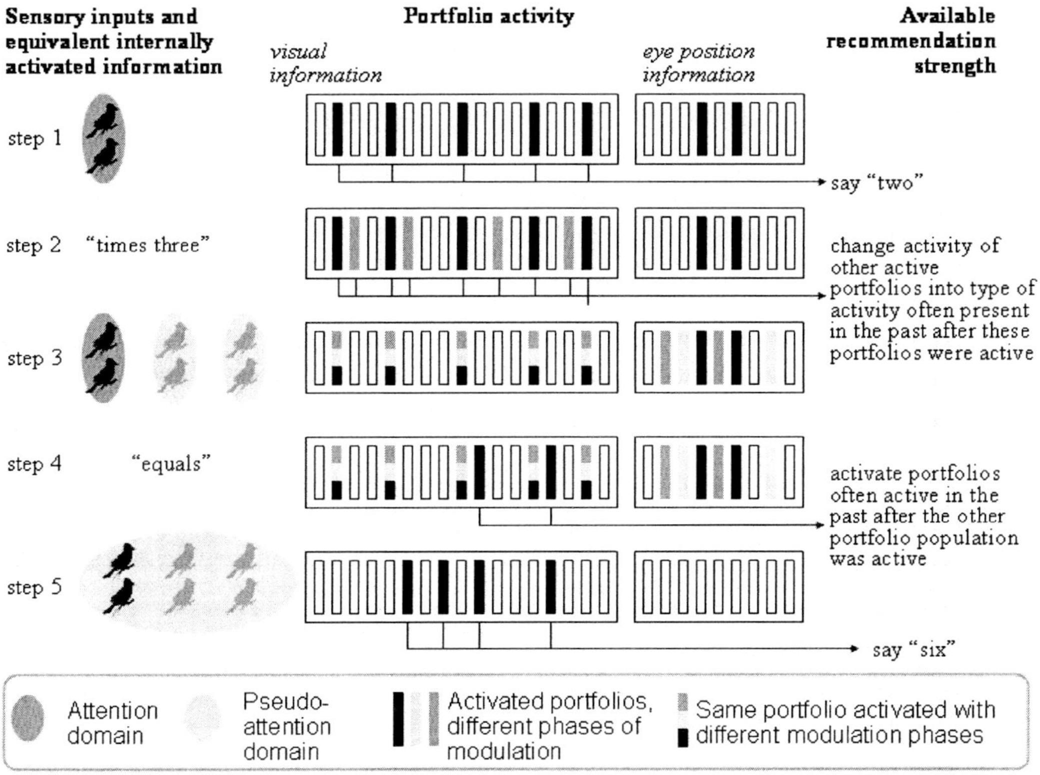

Figure 12.24. Process for learning multiplication using the already learned the attention behaviour sequence (figure 12.23). Learning this behaviour sequence has resulted in portfolios acquiring a number of indirect activation strengths. In step 1 the learner is encouraged to focus the attention domain on a group of objects and say the number in the group. In step 2 the learner responds to hearing "times three" by activating portfolios often active when the words were heard in the past. The various portfolios active at this point have recommendation strengths in favour of generating a pattern of activation similar to patterns which occurred in the past shortly after they were active. Because of the prior attention sequence learning, the result is the activation shown in step 3. The step 3 activation resembles a pseudovisual activation of three groups of two. In step 4 the word "equals" activates portfolios which when added to the populations already active result in acceptance of a further recommendation to indirectly activate portfolios often active in the past just after they were active. Because of the prior attention sequence learning the resultant activation is a pseudovisual image of the group of six objects. This pseudovisual portfolio population has recommendation strength in favour of saying "six". That behaviour is rewarded. The reward increases the recommendation strengths of recent behaviours including the various indirect activation behaviours.

Now consider the problem illustrated in figure 12.24. In this problem the learner sees just one group of two objects, and hears (and is encouraged to imitate) the words "times three equals". In step 1 of figure 12.24, the focus of the visual domain on the group of two objects generates portfolio activity which would have recommendation strength (as one of many recommended behaviours) in favour of saying "two". In the second step, the portfolios active in response to "two" have some recommendation strength in favour of changing their activation to multi phase modulation, because that activation occurred in the past after their current activation, and was followed by positive consequence feedback. In addition, the portfolios activated in response to "times three" have recommendation strength in favour of changing the activation of the portfolios active in response to "two" to three phase modulation, for the same reason. This recommendation is accepted, and the resultant pattern of activation illustrated as step 3 has some similarity at higher condition complexities to that which would be generated if three groups of two objects were actually perceived. At this point the word equals activates a set of portfolios. The pattern of activation is illustrated in step 4 of the figure. The portfolios activated in response to "equals" of course have recommendation strength in favour of saying "equals", but also have recommendation strength in favour of activating portfolios active in the past shortly after the other active portfolios were active. The other active portfolios are those corresponding with the three groups of three, and they have recommendation strength in favour of activating portfolios active shortly after their own past activity. The strongest recommendation strength therefore activates the portfolios activated in the past when six objects were present as illustrated in step 5, and have recommendation strength in favour of saying "six".

The ability to multiply using pseudo-visual images of the process has thus been bootstrapped using only a genetically defined capability to activate and change the activity of portfolios on the basis of past temporally correlated activity, a genetically defined tendency to imitate, and feedback from a teacher which is treated as a sensory input and does not need to access the internal processes of the learner.

Step 7: Learning Standard Tables

The pseudo-visual multiplication process can be used to guide the behavioural response to any multiplication problem, but can only generate an exact answer if the product has an accessible portfolio activation which recommends the number. As discussed in step 1, this limits the process to six or less. For higher numbers the visualization process can only generate an estimate. The learning of multiplication tables permits accurate answers for higher products. The process for learning is recitation of the entries in the table, such as eight times nine equals seventy-two. The process by which a response can be generated after such learning is illustrated in figure 12.25. Hearing the word eight activates portfolios directly in auditory array A2 and these activate portfolios in visual array V4 on the basis of frequent past simultaneous activity. Similarly with the other two words.

The presence of portfolios activated directly or indirectly by hearing the word "times" is interpreted as a recommendation to synchronize the outputs from the three populations in visual information array V3 and release to array V4. These recommendation strengths have developed because such a behaviour has been followed by positive consequences in the past. The resultant activity in array V4 has recommendation strength in favour of the behaviour of saying "seventy-two" has often been present in the past when those portfolios have also been active, and the behaviour has been rewarded.

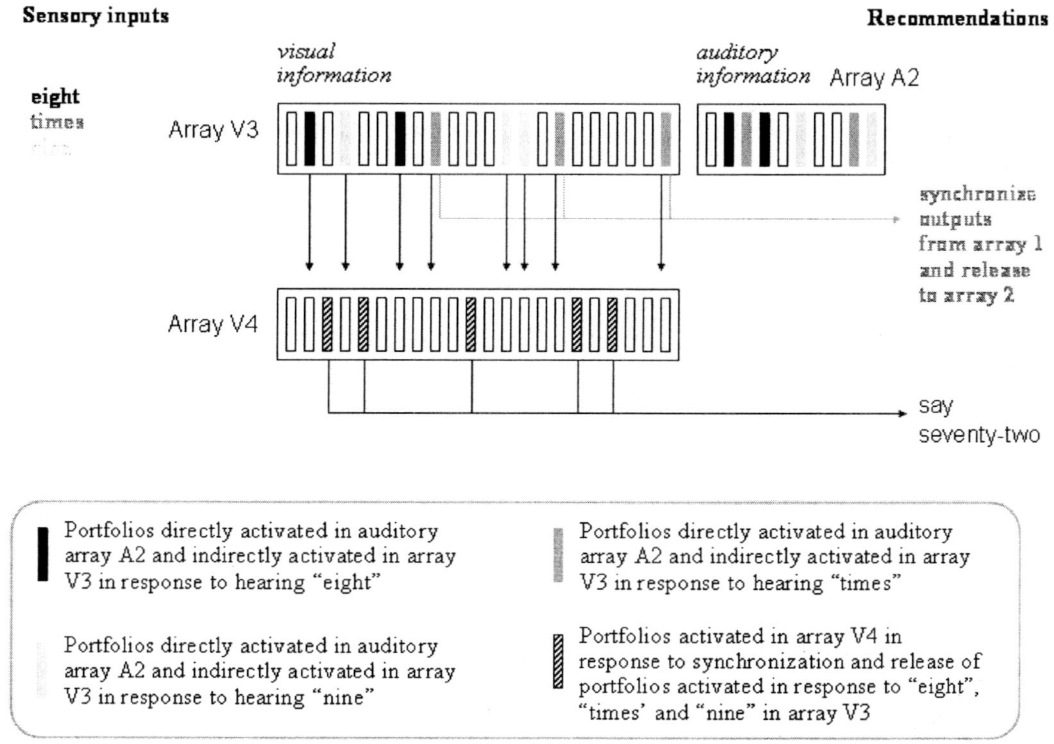

Figure 12.25. Learning sound driven multiplication behaviours. Three pseudovisual portfolio populations are indirectly activated in array V3 in response to hearing the three words "eight", "times" and "three". The outputs from these populations are synchronized and released to V4 where they activate a portfolio population. A teacher is heard to say "equals seventy-two", and rewards imitation behaviour. The reward results in the active portfolios in V4 acquiring recommendation strength in favour of saying "equals seventy-two".

Step 7: Acquisition of Recommendation Strengths by Input Numbers

Once a process for learning the solution to a problem is in place, various short cuts can develop. In particular, consider again the multiplication problem 8 x 9 illustrated in figure 12.25. Learning proceeded because each member of the active portfolio population in array V4 acquired recommendation strength in favour of speaking "seventy-two" because the word was imitated when the population was active, and the imitation was rewarded. However, the separate populations in array V3 were also active shortly before speaking "seventy-two" and therefore shortly before the reward. Portfolios in these populations can therefore also acquire recommendation strength in favour of saying "seventy-two". In other words, the portfolios activated in array V3 in response to hearing just "eight" will have some recommendation strength in favour of saying "seventy-two". If "eight" is heard alone, this recommendation strength will not be sufficient to generate the behaviour, but if populations are activated in array V3 in response to "eight", "times" and "nine" there may be enough recommendation strength to generate the spoken "seventy-two" with relatively little assistance from array V4 activations. As further discussed below, the activation in array V4 may be unnecessary if the portfolio activations in array V3 are sufficiently strong.

General Considerations

The objective of this section has been to demonstrate that arithmetic behaviour can be bootstrapped using portfolios which have been defined heuristically and do not correlate unambiguously with arithmetical concepts. The bootstrapping does not need any a priori information on such concepts, it only requires a genetically programmed tendency towards imitation and a genetically programmed way to interpret certain behaviours of an external supervisor as rewards for recent behaviours. The end point of the bootstrapping process is a set of portfolio arrays which can discriminate effectively between arithmetically different circumstances, a set of components corresponding with frequently used sequences of attention and portfolio activation behaviours, and recommendation weights from portfolios into these components which ensure that the appropriate behaviour is implemented at each point. The bootstrapping process has also resulted in a number of alternative ways in which a particular problem could be solved.

Arithmetical capabilities for subtraction and division can be acquired by analogous learning processes to define the attention sequence and internal activation behaviours in response to verbal and visual inputs.

The end point of the learning process would include a range of different types of specialized behavioural components in competition. These different types include components for "concept of number" behaviours which would be relevant for estimating relative magnitudes. Another type would be "concept of addition, multiplication etc." components which would be relevant for solving very simple problems and generating estimates for other problems. A third type would be components managing the detailed attention sequences for arithmetical problems, such as the equation handling process discussed in detail earlier. A fourth type would be components using recommendation weights of portfolios activated relatively directly in response to sensory inputs like numbers to directly drive behaviours of speaking or writing answers, generating faster responses by bypassing the detailed attention sequences.

Note that different components or different types could receive outputs from some of the same clustering arrays. For example, portfolios activated in response to perceptions of numbers could have outputs to both attention sequence and direct answer generation components.

A particular type of behaviour such as arithmetic could also develop somewhat specialized arrays on some levels of complexity. This situation could occur if the brain has a genetically defined tendency to specialize in this way. Such a tendency would be based on detection of a situation in which a relatively small group of behavioural components were often invoked by a relatively small group of portfolios. The arrays which were the locations of the relevant portfolios could be duplicated (approximately but to an operationally effective degree) by assignment of new array resources, provision of connectivity similar to that into the existing arrays, and forced condition recording when the original arrays were active and the behavioural group selected. The benefit of this capability is that detection of contradictory consequences for just behaviours of the limited type could be used to evolve the new arrays to discriminate with higher effectiveness between input state differences relevant to the behaviour. This higher effectiveness would occur because discrimination for other behavioural types was not needed. The disadvantage is the much larger use of resources. The brain would therefore have limits to its capability to organize in this fashion.

Note, however, that such a separation of a behaviour type is based only on relative frequency of conditions and behaviours, and will therefore not in general correspond exactly with some external cognitive definition of arithmetic. In other words, some of the assigned portfolios and components will contribute to behaviours which are not arithmetic as defined by an outside observer, and some arithmetic behaviours will require resources outside those assigned.

Accounting for Major Experimental Phenomena

There are two important types of empirical effects observed in human arithmetic processing [Ashcraft 1992]. These effects are the problem size/difficulty effect, and various error effects including relatedness effects.

Problem Size/Difficulty Effect

The problem size/difficulty effect is that reaction time to generate a solution to basic arithmetic facts (i.e. single digit addition problems from 0 + 0 to 9 + 9 and the equivalent multiplication problems) is longer for problems with larger numbers.

In the recommendation architecture cognitive model, response time is primarily the time to reach an active portfolio population with enough recommendation strength to generate a response. The key factor in determining response speed is the size of the activated portfolio populations generated in response to the sensory inputs and at subsequent processing stages. The relevant consideration here is that for smaller numbers, strong visual portfolios are defined which instantiate the concept of the number as discussed under step 1 of the learning process. Such portfolios exist for numbers up to about six. For a larger number, the population activated is simply a set of portfolios often active in the past when the number has been heard or seen. The process of generating an answer depends upon development of three adequate active populations corresponding with the two numbers and the symbol (for example in array V3 in figure 12.21 in the case of arithmetic). Because the populations directly activated by larger numbers are smaller, there may be some enhancement required, achieved by activating other portfolios often active at the same time in the past, to reach a total degree of activation in array V4 adequate to generate a response. If the result is large, the activation in array V3 may itself require additional activation to reach a recommendation level adequate to drive behaviour. The time for the additional activation steps makes response time longer.

As a result, the fastest responses will be if all the numbers including the answer are small, and response time will increase as the numbers making up the problem and result increase. It is also observed that tie problems generate slightly faster responses that typical for their size. Tie problems are those in which both numbers in the problem to be solved are the same (e.g. 6 x 6 = 36). For such tie problems, there will be overlap in the portfolio populations activated in array V3 in response to the two tie numbers. Because the frequency modulation mechanism is only a bias, not a synchronization, the presence of input activity into the same portfolios at two different phases will tend to result in larger activity in each phase. This larger activity will result in more rapid response.

Error Effects

Error effects refers to the type of errors which are most frequent in human arithmetic performance. Error effects include the observation that 90% of adult multiplication errors are "table errors" in which the answer is given to a different number combination rather than an answer which is not the product of any pair of numbers in the 1 x 1 to 9 x 9 table [Campbell and Graham 1985; Campbell 1987; Miller, Perlmutter and Keating 1984]. Furthermore, errors tend to be incorrect multiples of one of the numbers in the product rather than unrelated products (e.g. 32 as an answer to 4 x 6 rather than 35).

Another type of error is confusion of operation in which a result is generated which would be correct for a different operation [Ashcraft and Battaglia 1978; Winkelman and Schmidt, 1974]. An example would be 5 + 4 = 20, which would be correct for multiplication.

In the recommendation architecture cognitive model, errors arise because any activated portfolio has recommendation strengths into a number of different behaviours, and any population of activated portfolios recommends a wide range of different behaviours. Portfolios activated as a result of sensory processing or other internal activation processes taking place in parallel with arithmetic may result in activation of portfolios which have some arithmetic recommendation weights, and may therefore contribute "arithmetic noise" to the behavioural selection, and result in an error.

However, such errors can only result in a shift to an arithmetic behaviour which already has a strong recommendation strength in the populations activated in response to the problem. In other words, the noise will only be able to tilt the balance in favour of an already fairly strongly recommended behaviour. The error phenomena can be understood from this perspective.

For example, in the problem 4 x 6, the active portfolio populations in response to 4 and x will both have recommendation strength in favour of 24 and other multiple of 4 including 28, 32 and 36. Active portfolios in response to x will have recommendation strength in favour of any table product but very little recommendation strengths in favour of non table products. In the example 5 + 4 = 20, portfolios activated in response to 5 and 4.

Cognitive Deficits

As pointed out by McCloskey [1992], the impaired cognitive performance of a brain damaged patient reflects the functioning of a previously normal cognitive system with one or more components damaged. Subject to the assumption that the cognitive processes disrupted by brain damage do not undergo substantial reorganization following damage, the observation of a specific combination of performance impairments can lead to inferences about the major processing components and their interactions in the normal cognitive system. Such inferences can be made by asking: what must a normal system be like in order that damage could result in just this combination?

In the recommendation architecture model, arithmetic is handled by portfolio arrays in clustering communicating with behavioural components in competition. The arrays detect conditions on a number of different levels of complexity that discriminate between sensory input states in which different arithmetic behaviours are appropriate.

Some specialization of arrays for arithmetic will generally occur, but the specialization will only be partial. The specialization will not be for arithmetic exactly as commonly defined, rather it will be for a collection of similar behaviours which tend occur frequently under similar conditions, and specialized arrays will only be defined if the degree of similarity justifies the extra resources required. Furthermore, if the output of an array was often present when some different type of behaviour was often appropriate, that output could become an input to that behaviour.

Hence arithmetic behaviour as commonly defined may use array outputs from some specialized arrays, but also from some arrays relevant to a wide range of other behaviours. Most of the outputs from the specialized arrays will be used by behaviours with a similarity in an information sense to arithmetic, but some of the outputs may be used for quite different behaviours.

In competition, a number of specialized behavioural components are created. These components favour different sequences of attention and internal clustering activation behaviours. The order of behaviours in the sequence is determined by the component, but the timing of each behaviour being determined by the presence of appropriate clustering outputs. There may be multiple components corresponding with different ways of addressing any one problem, and there is competition between these components, on the basis of current clustering outputs and past consequence feedback, to determine which component will be selected for a particular problem.

In this model, local damage could have a number of effects. Firstly, damage to a clustering array targeting some arithmetic behaviours would remove the recommendation strengths of the affected outputs into all their targeted behavioural components including arithmetic. The guidance of those behaviours would become less effective. Subsequent consequence feedback could add and/or adjust the recommendation strengths of inputs from other arrays, but if the damaged array was specialized for arithmetic, complete recovery would be unlikely unless a long learning period to replace the specialized array occurred. The deficit could either be in the invocation of some component sequences, or in the performance of some particular types of behaviour used in many sequences.

Secondly, if all the connectivity of an array to a group of components managing similar behaviours goes by a similar physical pathway, local damage to the pathway would degrade the performance of only those behaviours. Note that if outputs from one array to a component develop to gain processing speed by bypassing an intermediate array which combines those outputs into higher complexity portfolios, it is plausible that both the original and bypass outputs will use a similar physical path to the component. For example, if portfolios activated in response to numbers in one array acquire connectivity directly to answer generation and can bypass the portfolios activated in response to groups of numbers and symbols, the two arrays will tend on average to be physically much closer to each other than to the competition component, and the simplest physical route for the new connectivity will substantially follow the route of the old connectivity.

Damage to a component or collocated group of components corresponding with similar behaviours would have relatively similar results. Such damage again could result in deficits to the invocation of behavioural sequences or in the performance of a particular type of behaviour.

Deficits of these types are indeed observed as a result of local damage. Thus patient DRC [Warrington, 1982] was much slower and less accurate than normal subjects in the performance of even simple problems like 5 + 7, but could read and write numbers, judge which number was larger, and give reasonable numerical estimates (e.g. how tall is the average English woman?). He reported that he often knew the approximate answer but no longer knew the exact answer, and used counting to work out answers. Damage affecting the connectivity into components managing the behaviour sequences for generating answers, but not the components managing counting, "concept of number" and "concept of addition etc." behaviours would result in this deficit.

The patient studied by Benson and Denckla [1969)] was a 58 year old man who could point to the correct answer from a multiple choice list of written Arabic numbers when presented with simple visual Arabic numeral problems like 4 + 5. He could also point to the correct Arabic numeral when he heard a verbal numeral. However, his production of written Arabic or spoken answers was severely impaired. For example, in response to the problem 4 + 5 he said eight and wrote 5 (but chose 9 from a multiple choice list). He also exhibited errors when writing to dictation, such as writing 215 in response to hearing two hundred and twenty one. Damage affecting the connectivity into the speech generation components but not into motor behaviours would have these effects.

Other Architectural Approaches

A widely discussed architecture is that proposed by McCloskey, Caramazza and Basili [1985] illustrated in figure 12.26. As can be seen from the figure, the model defines components corresponding with different cognitive functions. Models of this type have been criticized on a number of grounds. One limitation is that such models do not recognize that arithmetical problems can be solved through a variety of routes. A second limitation is that functional analyses of this type have no clear relationship with physiology [Smith 1998]. A further issue is that experimental evidence indicating strong interactions between number reading and number fact retrieval processes [Campbell 1994], processes which are independent in the McCloskey, Caramazza and Basili model.

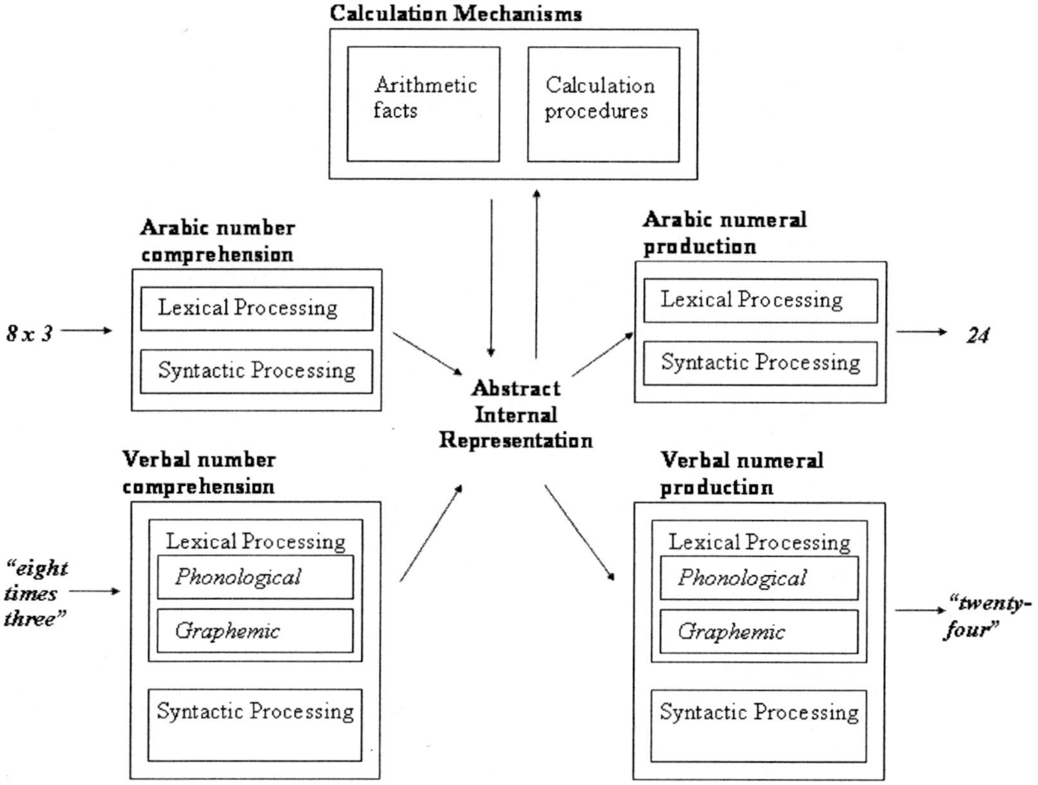

Figure 12.26. The arithmetic processing architecture proposed by McCloskey, Caramazza and Basili [1985]. Although this proposal may provide a description of how arithmetic processing appears to an outside observer, it does not address economical use of system resources and is therefore a "user manual" type description rather than a system architecture.

An alternative modeling approach has been proposed by Campbell and Clark (1988). In this approach there are multiple numerical codes which activate each other in the course of numerical processing. These activations are implemented by some kind of associative network. Sokol, Goodman-Schulman and McCloskey [1989] have argued that until both the codes and the processes by which the codes perform arithmetic processing have been specified, this approach is not a plausible model.

The recommendation architecture provides a model for arithmetic processing of the type that Campbell and Clark argued was necessary in which the codes and processes for interaction are specified. The McCloskey, Caramazza and Basili approach is essentially to construct a "user manual" type description of arithmetic processing. As discussed in chapter 3, although such an approach may be the easiest way to document how the features of a system take place, it is generally a poor guide to how system operations support the performance of those features. Furthermore, the recommendation architecture cognitive model supports multiple routes for solving problems and has a much more plausible relationship with physiology.

References

Ashcraft, M. H. (1992). Cognitive Arithmetic: A review of data and theory. 44, 75 – 106.

Benson, D. F. and Denckla, M. B. (1969). Verbal paraphasia as a source of calculation disturbance. 21, 96 – 102.

Campbell, J. I. D. and Graham, D. J. (1985). Mental multiplication skill: Structure, process, and acquisition. 39, 338 – 366.

Campbell, J. I. D. (1987). Production, verification and priming of multiplication facts. 15, 349 – 364.

Campbell, J. I. and Clark, J. M. (1988). An encoding complex view of cognitive number processing: Comment on McCloskey, Sokol and Goodman (1986). 117, 204 - 214.

Campbell, J. I. (1994). Architectures for numerical cognition. Cognition 53, 1 – 44.

Clark, J. M., & Campbell, J. I. D. (1991). Integrated versus modular theories of number skills and acalculia. 17, 204-239.

Coward, L. A. (2001). The Recommendation Architecture: lessons from the design of large scale electronic systems for cognitive science. 2(2), 111-156.

Dallaway, R. (1994). . CSRP 306, Cognitive Science Research Papers. University of Sussex.

Marr, D. (1982). . W. H. Freeman and Company, San Francisco, CA.

McCarthy, R. A. & Warrington, E. K. (1990). . San Diego: Academic Press.

McCloskey, M. (1992). Cognitive mechanisms in numerical processing: Evidence from acquired dyscalculia. 44, 107 – 157.

McCloskey, M., Caramazza, A. and Basili, A. G. (1985). Cognitive mechanisms in number processing and calculation: Evidence from dyscalculia. 4, 171 – 196.

Miller, K. F., Perlmutter, M. and Keating, D. (1984). Cognitive Arithmetic: Comparison of operations. 10, 46 – 60.

Rumelhart, D. E., Smolensky, P., McClelland, J. L. & Hinton, G. E. (1986). Schemata and sequential thought processes in PDP models. In Rumelhart, D. E. & McClelland, J. L., eds, , Vol. 2: Psychological and biological models. MIT Press, Cambridge, MA.

Sokol, S. M., Goodman-Schulman, R. A. and McCloskey, M. (1989). In defense of a modular architecture for the number processing system: Reply to Campbell and Clarke. 118, 105 - 110.

Smith, D. J. (1998). Applied Cognitive Psychology. Chapter 5. The Psychology of Numeracy. Cardiff: UWIC.

Warrington, E. K. (1982). The fractionation of arithmetical skills: A single case study. Quarterly Journal of Experimental Psychology 34A, 31 – 51.

Chapter XIII

The Nature of Human Consciousness

The nature of human consciousness has long been viewed as a central topic for human thinking and investigation. However, this long history of intellectual inquiry has not resulted in an agreed definition of consciousness. Modern psychology traces the origins of its investigation of consciousness to William James. It is interesting that a paper he published late in his career was entitled "Does 'Consciousness' Exist?" [James 1904]. The thesis of this paper was that although consciousness is clearly a function in experience which thoughts perform, it is not an entity or quality of being which can be contrasted with the material world. Some more recent investigators have departed from this position with the claim that some of the phenomena of consciousness are a "hard problem" for which no scientific explanation may be possible.

In chapter 2, the nature of scientific theories was discussed. The conclusion was reached that a genuine scientific theory describes a body of natural phenomena with a hierarchy of descriptions. Each description establishes causal relationships which are consistent with observation, but the degree of detail differs between levels. At the most detailed level, the degree of detail means that only very small segments of phenomena can be described if the description is to remain within human comprehension. However, because it is possible to map descriptions between different levels, understanding of an entire phenomenon is possible on the basis of understanding the elements that make up that phenomenon on a somewhat more detailed level, the elements that make up those elements on a yet more detailed level and so on down to the most detailed level. In this context, understanding means being able to map causal relationships at the highest level into equivalent causal relationships at the most detailed level.

In this sense of "understanding", an attempt to understand higher human consciousness in terms of physiology will be made in this chapter. The attempt will build on the description hierarchy established in earlier chapters. In particular, the level at which a description of conscious phenomena will be constructed will be portfolios in clustering and components in competition, with the more detailed descriptions of portfolios and components in terms of devices being the same as those described in earlier chapters. Thus, for example, the properties of portfolios used for the description of consciousness are provided by columns of devices in the cortex. The argument will be made that a wide range of phenomena commonly labeled conscious can be understood on this basis, including the physiological reasons why, for example, the experience of the colour red feels as it does and not like something else, a phenomenon often defined as a "hard problem".

The Definition of Consciousness

The ability to respond to a range of stimuli from the environment in ways that are appropriate for the continued existence of the organism is a definition of consciousness at the most simple level. Consciousness of this type is possessed by all living organisms at least down to the level of single cell bacteria. A somewhat higher level of consciousness is the ability to create complex, appropriate responses to complex combinations of stimuli. This type of consciousness does not necessarily involve extensive learning capability, and is of the type possessed by insects and reptiles. A yet higher level of consciousness is the ability to record information during one experience and use this information to create appropriate responses to later experiences. This is the type of consciousness possessed by mammals. Human consciousness includes this type, but also includes the ability to generate potential responses to experiences which have not yet occurred but may occur in the future, utilizing a constant stream of pseudosensory internal experiences or images.

The ability to experience and talk about this constant succession of mental images generated independently of sensory input is how human consciousness is classically defined. Mental images can include objects never actually perceived by the senses. Images can be of self, acting and experiencing. The acting and experiencing self can be viewed as "I" or "me", in other words from the inside or from the perspective of an outside observer. "We can look out from within the imagined self at the imagined vistas, or we can step back and see ourselves perhaps kneeling down for a drink of water at a particular brook" [Jaynes 1976]. This "me" perspective is generally not a part of our normal direct sensory experience.

Everyday human experience includes some experiences and activities regarded as being performed "unconsciously", meaning that these experiences and activities do not appear to be able to generate verbal reports. This inability can appear absolute. For example, although we can remember and talk about the time when we learned to ride a bicycle, the detailed process by which the appropriate combinations of muscle movements were acquired is apparently inaccessible to any verbal reporting. In other situations the inability to generate verbal reports can be controllable. For example, we can sometimes travel a familiar route and on arrival have no memory of the journey, but in this case "paying attention" while traveling results in memory which can be expressed verbally. Furthermore, an apparent verbal reporting capability can be inaccurate. An interesting example is the observation that when an expert in some skill is asked to describe the operation of the skill, there can be considerable mismatch between the description and the actual operation [Speelman 1998]. Descriptions tend to resemble the way the expert was instructed verbally during training, and the generation of a verbal description will sometimes result in the behaviour of the expert shifting towards the verbal description, with a deterioration in performance [Bainbridge 1977; Berry 1987].

Thus conscious behaviour when contrasted with unconscious is generally assumed to mean that accurate verbal reports of the behaviour can readily be generated both in parallel with the behaviour and subsequent to the behaviour. Such reports can include descriptions of sensations and feelings.

Phenomenal consciousness refers to the qualitative nature of experience, for example why there is sometimes an immediacy and vividness to the experience of the colour red and why it feels as it does and not like something else. A question which immediately follows is whether different people experience the colour red in the same way. A further example of phenomenal consciousness is that it is possible to unconsciously identify and act in response to the presence of a tree, for example walk around it, but it is also possible to become aware of the tree as a tree, and this awareness feels qualitatively different. Phenomenal consciousness is in some ways an exception to the general conscious capability to generate verbal reports, because although it is possible to say that the experience is vivid and immediate, it does not appear possible to generate more detailed descriptions of the experience. It has been argued that for phenomenal consciousness in particular it will be difficult to create a scientific explanation [Nagel 1974].

The Causal Phenomena of Consciousness

As a first step towards a theory of consciousness as defined, the causal relationships which constitute the phenomena of consciousness at the psychological level must be specified. The following relationships will be used in this chapter, based on the discussion of the definition of consciousness in the previous section.

Firstly, there are causal relationships that current sensory inputs are capable of initiating a stream of mental images where one image is associatively generated by a combination of sensory inputs and earlier images. In some but not apparently all cases this image stream determines or strongly influences future behaviour. Images can include objects never actually perceived, objects which do not exist and even objects which could not exist. There can be images of self including internal feelings and of self viewed from the outside. In general these images are not visual hallucinations (i.e. they are not perceived as directly visible to the eyes).

Secondly, there are some behaviours and experiences for which verbal reports are or could be generated, and others for which no such report is possible. There is a continuum of situations in which verbal reporting is not possible, ranging from situations in which a degree of effort could result in a report through situations in which such effort results in verbal reports that are inaccurate to situations in which no report at all can be generated in any circumstances. Some sensory experiences do not result in any memory which can be verbally expressed, although they appear to be presented to the senses and influence behaviour, while others result in memory which can subsequently be verbally expressed.

Thirdly, when a feature or object is perceived, it is possible to take some internal action which results in a qualitative change in the experience of the feature or object. This qualitatively different experience can generate general verbal reports but not detailed reports of its content.

Consciousness Models at the Psychological Level of Description

Lorentz (1977) proposed that the phenomena of human consciousness derive from capabilities which exist in other animals or have developed from such capabilities. In Lorenz' approach, the major change between other animals and human beings is in the way in which these capabilities interact. These capabilities are the abilities to extract constant objects from sensory input; to maintain mental models of relative positions; to treat every object as biologically relevant in order to find objects which are genuinely significant (i.e. curiosity); to compose complex actions from a vocabulary of muscle movements; to try past behaviors in new situations; and to imitate behavior. Humans have added the abilities to represent time with spatial models and to imitate behavior by indirect instruction.

Lorenz does not directly address the source of the constant stream of mental images, or discuss any architectural models or neurophysiological mechanisms. However, the capabilities which Lorenz identifies are all typical of a system with the recommendation architecture. For example, the capability to extract constant objects from sensory input reflects the search for portfolios which can be associated with appropriate behaviours.

The argument of this chapter is that the human consciousness indeed rests upon mechanisms which in many cases are present in animals, but the relevant capabilities are better expressed in terms of mechanisms by which information derived from the senses can be recorded and indirectly activated. In other words, Lorenz' model is more of a "user manual" type, and a system architecture approach is required to relate physiology to consciousness.

Jaynes (1976) offers a well developed psychological theory of consciousness along with some ideas on the evolutionary and historical origins of consciousness. In Jaynes' view, as we interact with an object we perceive the object in many different ways over a succession of instants in time. These different ways Jaynes labels percepts. An object is represented in consciousness by a reactivation of a percept. A conscious image is thus the reproduction of one of the instantaneous mental images which occurred during direct perception of the object in the past. New objects are assimilated into existing categories developed to define behavioral responses. Mental objects are arranged in a spatial framework called a mindspace. This mental arrangement is used to represent relationships in space, in time, and in emotion (e.g. containing or letting out anger). Central to Jaynes' theory are the concepts of analog I and metaphor Me, which are conscious representations of self acting and self experiencing. The ability to tell stories is the basis for arranging events in a sequence and thus to modeling causal relationships. Consciousness for Jaynes is a linguistic invention, developed in response to a need for behavioral responses to extremely complex social situations. Implicit in Jaynes' theory is the recommendation architecture concept of development of condition similarity clusters to define behavioral responses, and the behavioral role of consciousness in developing behavior in complex social situations developed in detail in this chapter is the functional role originally proposed by Jaynes. A modified version of Jaynes' model will be a large part of the highest level of description in the theory of consciousness developed in this chapter.

At a somewhat more detailed level, Baars (1988) proposed a phenomenological theory of consciousness around the concept of the global workspace, and attempted to define an architecture which could support the more general psychological descriptions. According to Baars, the contents of consciousness are the contents of a global workspace located in the primary sensory projection areas of the cortex (e.g. VI for the visual sense). A set of input processors generate visual images, inner speech etc. which compete through structures like the brainstem reticular formation and the nucleus reticularis for access to the global workspace. The contents of the global workspace are distributed widely to specialty unconscious processors including perceptual analyzers, output systems, action systems, syntax systems, and planning and control systems.

In Baars' model, "the overall function of consciousness is to provide very widespread access to unconscious brain knowledge including autobiographical memory ... ; the lexicon of natural language ..... ; automatic routines that control actions ..... ; and, by way of sensory feedback, even the detailed firing of neurons and neuronal populations". The fundamental difference from the recommendation architecture approach is that Baars' model is essentially a user manual type description which utilizes components that correspond with cognitive functions like perceptual analyzers, output systems, action systems, planning and control systems. As argued in chapter 3, such models may be helpful for organizing easier to understand descriptions of the phenomena of consciousness at the psychological level, but will have little relationship with how those phenomena are supported in a system architecture which is organized to make effective use of information handling resources.

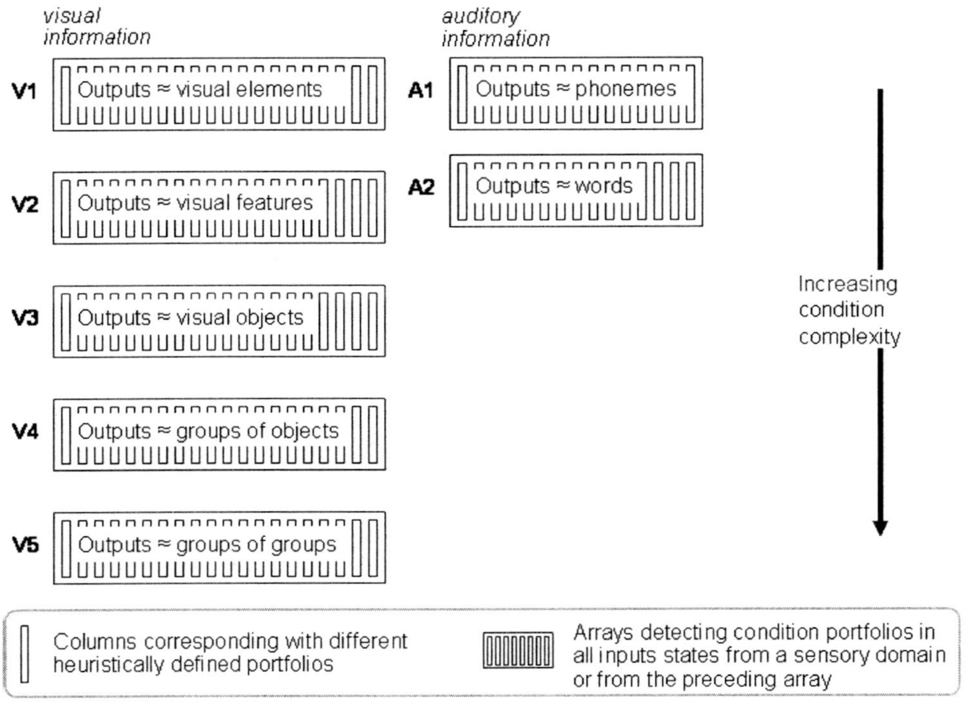

Figure 13.1. Organization of clustering for visual and auditory information in cognitive model.

The Cognitive Model

The recommendation architecture model for consciousness is the same model as described in chapter 9, used in chapters 10 and 11 to account for the phenomena of memory, and used in chapter 12 to account for the phenomena of arithmetic processing. The mapping of this model into human brain physiology is as described in chapter 8. The human cortex corresponds with clustering and columns in the cortex correspond with portfolios. Subcortical structures correspond with competition, with a general tendency for the thalamus to manage selection of the sensory information which will be allowed to influence behaviour, the basal ganglia to manage selection of general types of behaviour, and the cerebellum to manage selection of specific behaviours.

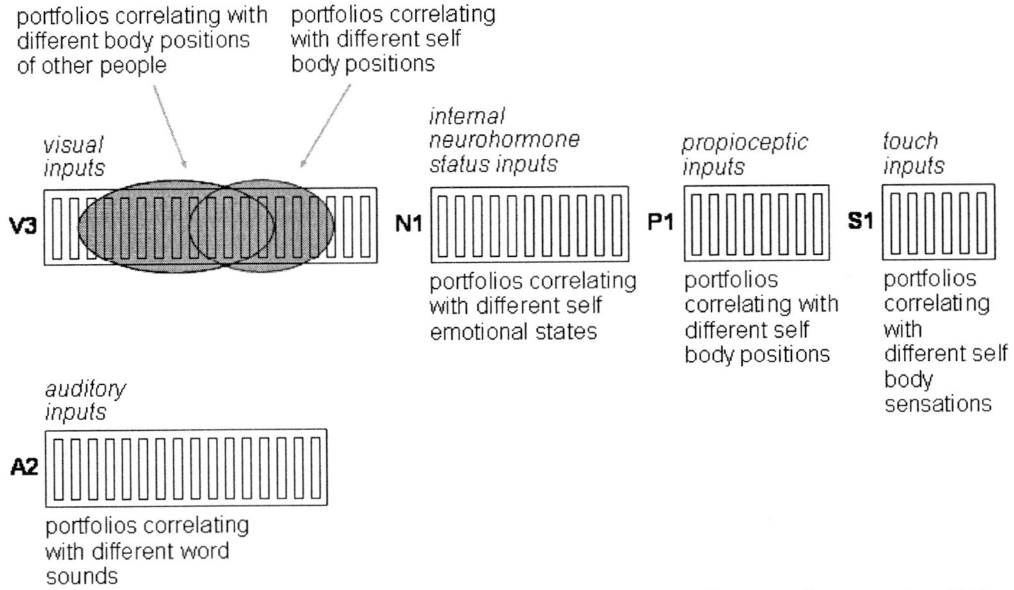

Figure 13.2. Some heuristically defined arrays important for understanding consciousness. In addition to visual and auditory information, inputs indicating concentrations of different neurohormones are used to heuristically define portfolios that can discriminate between different emotional states. Propioceptic inputs are used to heuristically define portfolios that can discriminate between different self body positions and touch inputs are used to heuristically define portfolios that can discriminate between self body sensations. As always, portfolios do not correlate exactly with specific behaviourally relevant emotional states, positions or sensations but are on levels of condition complexity which can discriminate between such states, positions and sensations. In practice the illustrated clustering capability may be instantiated in a number of arrays at different levels of complexity in each sensory input domain, but simplifying to one array does not change the essential description.

This model, with the aspects more relevant for consciousness emphasized, is illustrated in figures 13.1, 13.2 and 13.3 and table 13.1. Figure 13.1 illustrates two sequences of arrays in clustering, one processing visual sensory inputs, the other auditory inputs. Each array detects condition portfolios in a different range of condition complexity, with the ranges defined to give the best discrimination between cognitive categories at different levels. On the visual side these category levels are visual elements, visual features, visual objects, groups of objects and groups of groups. The inputs to an array are the outputs from the preceding array, or a sequence of such outputs derived from a sequence of different points in time as discussed in detail in chapter 12. An example of a visual element might be a boundary. An example of a feature might be a leg and an example of an object might be a dog. A group of objects might be a dog barking at a cat up a tree, and a group of groups might be the animal group and some people getting out of a car at the same time. Figure 13.2 illustrates some of the arrays which are heuristically defined in different information spaces including visual, auditory, propioceptic, touch, and internal detections of neurochemical levels in the brain.

As discussed in chapter 9, it is likely that discrimination between behaviourally relevant features will in practice require arrays within several different ranges of condition complexity. However, this consideration does not make any qualitative difference to process descriptions and for the purposes of simplicity will be omitted.

A. Activation on the Basis of Recent Activity

1. Reactivate portfolios which have recently been active
2. Reactivate portfolios which recently recorded conditions

B. Activation on the Basis of Frequent Past Activity

1. Activate portfolios which have often been active at the same time in the past as the currently active portfolio
2. Activate portfolios which have often been active just after past activity of the currently active portfolio
3. Activate portfolios which have often been active just before past activity of the currently active portfolio

C. Activation on the Basis of Frequent Past Condition Recording

1. Activate portfolios containing conditions recorded at the same time in the past as some conditions in the currently active portfolio
2. Activate portfolios containing conditions recorded just after some conditions in the currently active portfolio
3. Activate portfolios containing conditions recorded just before some conditions in the currently active portfolio

D. Managing Information Availability

1. Prolong the activity of currently active portfolios
2. Shorten the activity of recently activated portfolios
3. Synchronize the activity of several different populations of currently active portfolios in the same array
4. Release outputs of one array to the next array

E. Attention management

1. Perform a general sequence of attention and internal activation behaviours
2. Perform a specific sequences of attention and internal activation behaviours
3. Perform an individual attention behaviour

F. Speech Behaviour Management

1. Generate a sound
2. Speak a word
3. Say a phrase
4. Express meaning verbally by a sentence

G. Motor Behaviour Management

1. Perform a general sequence of motor behaviours
2. Perform a specific sequence of motor behaviours
3. Perform an individual motor behaviour

Figure 13.3 illustrates how the outputs from one clustering array are distributed to different competition components corresponding with different behaviours, and table 13.1 lists many of the possible behaviours relevant to consciousness. A competition component does not change clustering outputs, rather it selects all or a subset of those outputs to be released to a target array in clustering or to drive a behaviour.

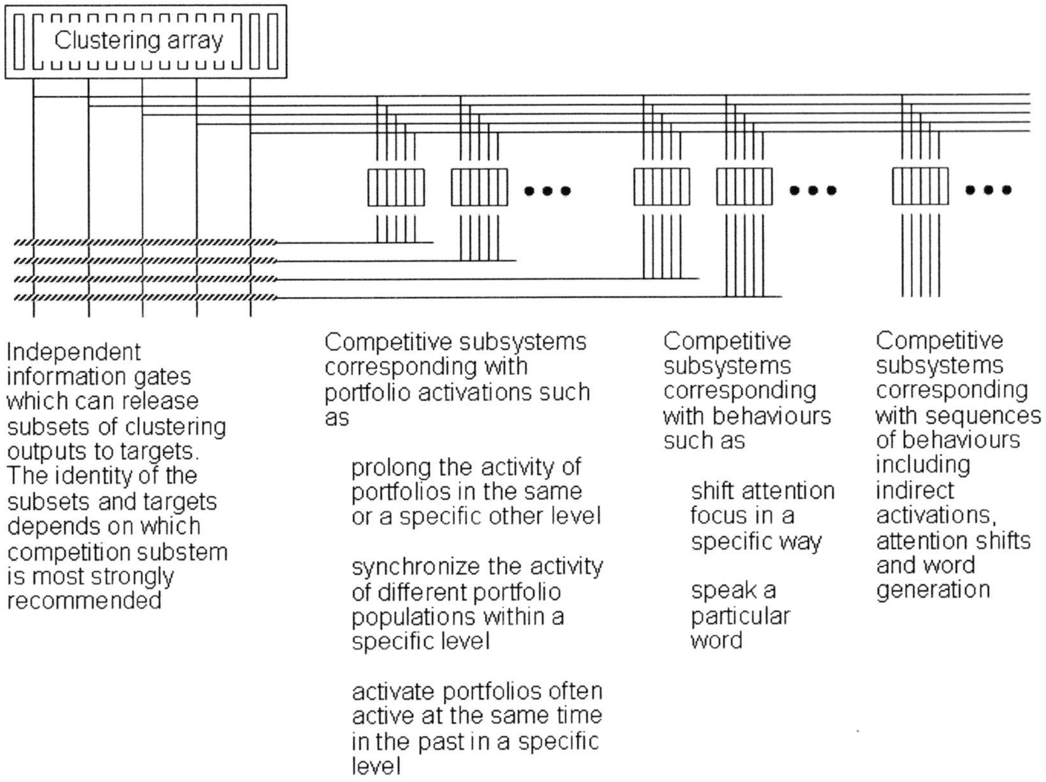

Figure 13.3. Organization of competition to manage the outputs of one clustering array, Each competition component may receive inputs from a number of different arrays.

Some Key Characteristics of the Model

In this section, some characteristics of the model discussed in earlier chapters are reviewed and the aspects relevant to consciousness emphasized.

Ambiguity of Portfolios

Because of the ambiguity of all information within clustering, the population of portfolios activated in response to an object belonging to one cognitive category will contain portfolios activated in the past in response to many other categories. These portfolios will therefore have recommendation strengths appropriate not only to the currently perceived category, but also recommendation strengths appropriate to all of those other categories.

Indirect Portfolio Activations

Some of the recommendation strengths of a currently active population will be in favour of indirect activations of other portfolios on the basis of temporally correlated past activity with the currently active portfolios. Some of these indirect activation behaviours do not interfere with the production of current (for example) motor behaviour, either because the relevant portfolio populations are in different arrays or because motor behaviour and indirect activation populations irrelevant to the motor behaviour can both be supported in the same array at different frequency modulation phases. Hence in some cases indirect activation behaviours and externally directed motor behaviours can proceed in parallel.

Indirect Activations and Mental Images

Indirect portfolio activations are generated on the basis of temporally correlated past activity. However, such correlations will be weaker for portfolios closer to the raw sensory level. To use a specific example, the portfolios in the visual information arrays in figure 13.1 which are activated on hearing the word "dog" are those which have often been active in the past when the auditory portfolios activated by the word were also active. When a dog is perceived visually, portfolios will be activated in the arrays discriminating between visual elements, visual features and visual objects in figure 13.1. However, the portfolios in the arrays close to sensory inputs will tend to be more specific to individual perceptions of dogs and will therefore be active much less consistently than portfolios in higher arrays when the word "dog" is heard. This is essentially restating that portfolios at the ~ level provide the best discrimination between visual objects.

Indirect activations will therefore be generally limited to higher arrays and not the arrays close to sensory inputs. As a result, an activation in response to hearing the word "dog" is similar to activations in response to seeing a dog within higher arrays, but does not include an activation in arrays closer to sensory inputs. The brain will therefore experience a mental image of a dog, but without the activity close to the senses, in other words there will not be a visual hallucination.

Given the lower consistency of indirect activations in arrays closer to sensory input, and therefore lower behavioural value, it is plausible that genetic information would in any case prevent connectivity etc. to support such activations. Furthermore, the absence of such activations will reduce the overall recommendation strengths in favour of behaviours only appropriate when the corresponding object is physically present. Hence the brain can "distinguish" behaviourally between a direct experience and a mental image.

Derivation of Consistent Portfolio Populations

As discussed in detail in chapter 10, there is a tendency for an activated population of portfolios to evolve towards a set which is internally consistent on some basis. For example, if at a particular point in time activation on the basis of a particular type of temporally correlated activity is favoured (for example by activation of the corresponding component type) then the activated population will tend to converge on a population in which a high proportion of the activated portfolios have had correlated activity of the type.

One example in chapter 10 was targeted recall episodic memory. In this phenomenon, a bias is created in favour of activation on the basis of simultaneous past condition recording, and a group of words are spoken which have been selected to correspond generally with a past experience. The portfolio population activated in response to hearing the words tends to evolve towards a population similar to that present during the earlier experience.

Internal Verbalizations

As discussed in chapter 9, if a visual object is the focus of attention at the same time as the word "dog" is being heard, portfolios activated in response to the visual input acquire recommendation strength in favour of saying "dog".

Suppose that a population is activated with recommendation strength in favour of speaking some word. Then suppose that this word is spoken by self, and an indirect population is activated in response to hearing the word spoken by self. This indirect population is also active shortly after the population which recommended speaking the word. That "word" population can therefore acquire recommendation strength in favour of activating portfolios in the "word" population.

This is not particularly significant for the dog example, but is more significant for words which correspond with different but related direct perceptions. In this case there may be separate populations directly activated by different direct perceptions but all with recommendation strength in favour of speaking the same word. These separate populations could acquire recommendation strength in favour of activating each other.

For example, as developed further in the next section, words like "anger" or "body" or "swimming" are present when attention is focused either on self or on some other person. As a result, a population activated in response to the perception of anger in some other person could result in activation of some portfolios which are directly activated only by a perception of self anger and vice versa. Another example is that a population activated in response to a perception of self swimming could result in activation of portfolios which are directly activated only by a perception of someone else swimming.

Note that such recommendation strengths may be assigned provisionally, but will only become permanent if used shortly before an occurrence of positive consequences.

Learning to Generate Appropriate Responses to Self Related Words

Learning the appropriate indirect activations in response to words occurs by the type of processes outlined in chapter 9. The indirect activations which occur in response to some self related words, and the processes by which these words acquire the appropriate activation capabilities are important for understanding consciousness.

Portfolios can be heuristically defined within any source of input information. Some of the sources and portfolios relevant to self images are illustrated in figure 13.2. Visual inputs derived from observations of the self body will result in some portfolios in array V3 which are activated in response to and have the capability to discriminate between different body parts and positions including arms, legs, hands and fingers. Visual inputs derived from observations of the bodies of other people will result in some portfolios with the capability to discriminate between different body parts and positions of these "external" bodies. Given the general similarity between human limbs, digits etc. there could be some overlap between these two groups of portfolios. Inputs derived from propioceptic senses and from the sense of touch will also result in portfolios which are activated by and can discriminate between different self body positions and sensations.

As discussed in chapter 9, emotions like anger, fear, sadness and shame and feelings like hunger, tiredness and sexual arousal correspond with higher concentrations of neurochemicals which modulate the relative probabilities of different types of behaviour being accepted. Pleasure corresponds with release of neurochemicals that increase the recommendation strengths of recently active portfolios into recently implemented behavioural components (including indirect portfolio activation components). The releases of these neurochemicals are behaviours which must be recommended by activated portfolios. In many cases these portfolios are located within special purpose clustering subsystems managing behaviour type priorities and rewards as illustrated in figure 9.1. However, it is possible for portfolios in the primary clustering system to acquire connectivity (and therefore recommendation strength) into these neurochemical release behaviours.

Furthermore, the concentrations of such neurochemicals can be detected and such detections are a source of information. Portfolios able to discriminate between different combinations of concentrations of the neurochemicals can therefore be defined as illustrated in figure 13.2.

All the portfolios in figure 13.2 can be directly activated by the presence of their conditions within the appropriate sensory inputs. They can also be indirectly activated by the various available mechanisms listed in table 13.1 A, B, C and D. Such indirect activations will be experienced by the brain as a partial re-experience of an earlier state, with the portfolios containing simple conditions close to raw inputs being absent.

However, another factor can increase the degree of such re-experiences. Suppose that in some direct experience, anger is present. The presence of anger means that the behaviour of increasing the concentration of the relevant neurochemical has been accepted and the portfolios correlating with the presence of the anger neurochemical will be activated. If the perception of positive consequences follows the anger, recommendation strengths of recently active portfolios into the behaviour of increasing the concentration of the relevant neurochemical will be increased. These recently active portfolios include the portfolios detecting the presence of the "anger" neurochemical. These portfolios can therefore acquire some recommendation strength in favour of creating the circumstance that they detect.

This recommendation strength has two effects. One is a tendency to amplify any emotion or feeling. The second is that if the portfolios are activated indirectly, for example by the activation of other portfolios often active in the past at the same time, the emotion or feeling can be indirectly activated. In other words, the actual feeling present during a past experience can be indirectly activated. It is likely that there is some genetic bias which tends to limit the possible recommendation strengths of such indirect activations.

Consider now the indirect activations learned in response to different words and phrases.

"Finger" and "My Finger"

The word "finger" directly activates portfolios in array A2. During an earlier learning period, these portfolios have been active when portfolios in V3 have also been directly activated both in response to attention directed to a self finger and in response to attention directed to other people's fingers. The similarity between fingers means that there will be some overlap between these different groups of "finger" portfolios in V3. The portfolios activated in A2 will therefore have recommendation strength in favour of activating the V3 portfolios, especially those activated directly by both self and other fingers. All the "finger" portfolios in V3 will have recommendation strengths in favour of activating each other on the basis of past simultaneous activity.

The portfolios activated in A2 in response to hearing self say "my finger" will have greater recommendation strength in favour of activating V3 portfolios previously active in response to perception of self fingers, but in general some portfolios which would be activated in response to other peoples fingers will also be activated.

"Body" and "My Body"

The word "body" directly activates a different set of portfolios in A2. During a learning period these portfolios will have been active when portfolios in V3 have also been directly activated both in response to attention directed to the self body as a whole (i.e. the attention domain covering the whole body) and in response to attention directed to other peoples bodies. There will again be some overlap between these different groups of "body" portfolios in V3. The similarity between bodies means that there will be some overlap between these different groups of "body" portfolios in V3. The portfolios activated in A2 will therefore have recommendation strength in favour of activating the V3 portfolios, especially those activated directly by both self and other bodies. All the "body" portfolios in V3 will have recommendation strengths in favour of activating each other on the basis of past simultaneous activity.

The portfolios activated in A2 in response to hearing self say "my body" will have greater recommendation strength in favour of activating V3 portfolios previously active in response to perception of the self body, but in general some portfolios which would be activated in response to other peoples bodies will also be activated. This can include portfolios in V3, but also portfolios in P1 and S1 correlating with particular body positions.

"Angry" and "You are Angry" Applied by Another Person to Self

The word "angry" is heard both when another angry person is observed and when self is angry. The portfolios activated in A2 in response to hearing the word therefore acquire recommendation strength in favour of activating portfolios in V3 directly activated by observations of others and portfolios in N1 directly activated by detections of self neurochemical levels. As discussed earlier, the portfolios activated in N1 have some recommendation strength in favour of releasing "anger" neurochemical, and hearing the word "angry" can therefore generate the feeling of anger to a weak degree.

"Swimming" and "I am Swimming"

As illustrated in figure 13.4, portfolios are activated in V3 and V4 in response to the perception of another person swimming. In response to the personal experience of swimming, different portfolios are activated in V3 and V4, although there may be a small degree of overlap with the other person portfolios. In addition, in response to personal swimming experiences, portfolios are activated in P1 and S1 in response to propioceptic and touch sensory inputs (and perhaps also in N1 of figure 13.2 in response to the perception of feelings). The word "swimming" activates portfolios in A2 which through use of the word during swimming acquire recommendation strength in favour of activating portfolios in both populations.

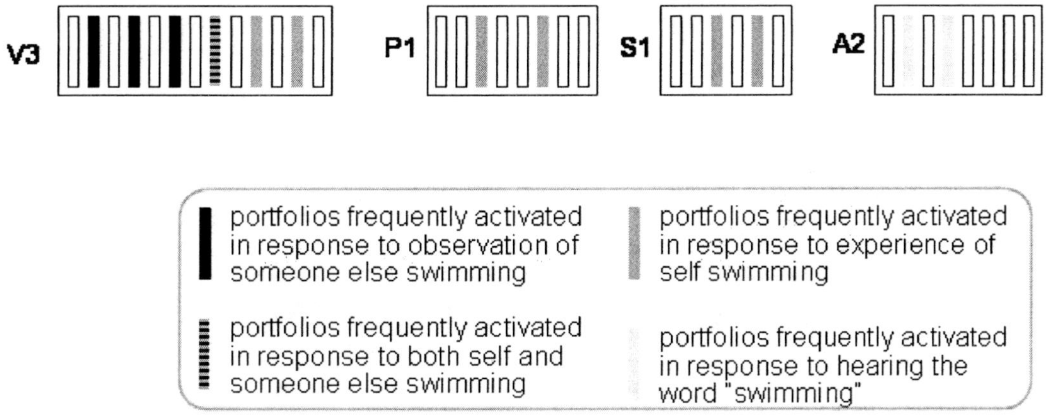

Figure 13.4. Portfolios frequently activated in response to the experiences of swimming, observing someone else swimming, and hearing the word swimming. Simultaneous activity means that the portfolios activated directly in response to hearing the word "swimming" acquire recommendation strength in favour of indirect activation of the portfolios activated directly by perceptions of both self and someone else swimming. As a result, portfolios activated directly by perceptions of both self and someone else swimming and sometimes active at the same time and can acquire recommendation strength in favour of indirectly activating each other. Thinking about self swimming can therefore activate portfolios which would only be directly activated by observation of someone else swimming.

Furthermore, by the process discussed in the previous section, portfolios in one population can acquire recommendation strength in favour of activating portfolios in the other population. The effect is that even when attention is focused on self swimming, portfolios activated directly only in response to observation of someone else swimming may be indirectly activated.

The portfolios activated in response to self hearing the words "I am swimming" may therefore include "someone else swimming" portfolios.

Note that in general "I am swimming" will be handled as a group of objects generating a combined population in V4. However, this V4 population could then indirectly activate a population in V3.

The Personal Pronouns "I", "Me" and "You"

Different groups of portfolios are directly activated in array A2 in response to hearing the words "I", "me", "you", "my" and "your" etc. These portfolios will have recommendation strengths in favour of activating different portfolio populations in V3 on the basis of frequent past simultaneous activity. Hearing someone else say a word will not have, for example, propioceptic information indicating self speaking. These type of differences in direct portfolio activations will result in hearing someone else say "you" tending to activate the same population in V3 as is activated by hearing self say "I".

The portfolios indirectly activated response to hearing self say the word "I" will include those active in the past in V3 in response to visual attention on the self body, but also any portfolios in N1, P1, or S1 which have often been active at the same time. The word "I" spoken by self will therefore tend to activate a pseudo-experience which will include portfolios on higher levels of condition complexity which would be directly activated by perceptions of visual, propioceptic, touch and emotional inputs. This pseudo-experience will be a weighted average of past experience. As noted earlier, an actual emotional state may be partially activated. As discussed earlier, the activation may also include some portfolios only directly activated by visual observation of other people.

Over time, the portfolios making up this total indirect population will tend to be active at the same time, and can acquire recommendation strength in favour of activating each other on this basis. Hence this "I" population forms a relatively consistent population as discussed earlier. The particular subset of this population activated at some point in a mental image sequence will depend upon the earlier image populations in the sequence.

Heuristically Defined Competition Components Corresponding with Specific Sequences of Behaviours

Any frequently occurring sequence of behaviours will tend to become instantiated in a higher level competition component, particularly if the sequence is often followed by positive consequences. This component when activated will activate the specific component corresponding with the first behaviour in the sequence, increasing the effective weights of clustering outputs into that component. Once the first behaviour has been performed, the higher level component will activate the component corresponding with the second behaviour in the sequence and so on.

Such components can be heuristically defined for sequences of motor behaviours, but also for sequences of internal activation behaviours. Examples of sequences including internal activation behaviours include the internal activation sequences in support of episodic memory discussed in chapter 10 and the sequences of attention and internal activation behaviours associated with performing different arithmetic processes discussed in chapter 11.

In the context of consciousness, an example might be a behaviour to add an "I" activated population to whatever other populations were active in V3, resulting in activation of portfolios in V4 corresponding with a pseudo-visual perception of self participating in the group of objects previously activated in V3.

Processes Supporting the Causal Relationships of Consciousness

Consider now how this architecture provides an account for the various types of causal relationship at the psychological level which constitute human consciousness as defined.

Generation of Mental Image Sequences

A portfolio population which happens to be activated at some particular point in time, perhaps in response to direct perception of an object, will have a wide range of recommendation strengths. Consider, for example, the population activated in response to seeing a dog as illustrated in figure 13.5. Portfolios activated in array V2 will correlate partially with features like legs, tails etc. and these portfolios will frequently have been active in response to other animals. The portfolios activated in array V3 will correlate somewhat better with dogs, but a significant proportion will have been activated in the past in response to different types of animals such as cats, wolves and badgers. As a result, portfolios will have recommendation strengths in favour of saying "that is a cat" etc. but the predominant speech related recommendation strength will be in favour of saying "that is a dog".

The portfolios activated in array V4 will correlate partially with the group of objects made up of the dog, its owner, and the table at which they are sitting, but will overlap with portfolios which would be activated in response to direct perception of different groups such as a dog sitting outside its kennel, or a dog chasing at a cat.

At every level, the predominant externally directed recommendation strengths will be in favour of behaviours appropriate to the presence of a dog. However, suppose that no externally directed behaviour has enough recommendation strength to be implemented under current circumstances. Recommendation strengths in favour of indirect activations could then be accepted (and could possibly even be accepted in parallel with a dog directed behaviour). What are the available recommendation strengths at this point in time?

Any one portfolio will have recommendation strengths of all the indirect activation types A through D listed in table 13.1, and each portfolio will have such strengths in favour of many other portfolios.

Suppose now that there is recommendation strength in favour of activating a component corresponding with activation of portfolios not currently active but often active in the past at similar times to currently active portfolios is activated. Such a component would combine activation of components corresponding with activation type B1 (to activate portfolios which have often been active at the same time in the past as the currently active portfolio) and type D2 (to shorten the activity of currently active portfolios).

A relatively high proportion of the portfolios often active in response to dogs in V3 are already active and will not remain active in the resultant secondary population. The set of currently inactive "dog" portfolios which are available to be activated is relatively small. The subset of the primary population overlapping with populations active when other types of animal are present have larger sets of portfolios available for indirect activation. The resultant secondary portfolio population in V3 will therefore tend to contain a higher proportion of portfolios active when, for example, a different type of animal with significant similarity to dogs was perceived in the past.

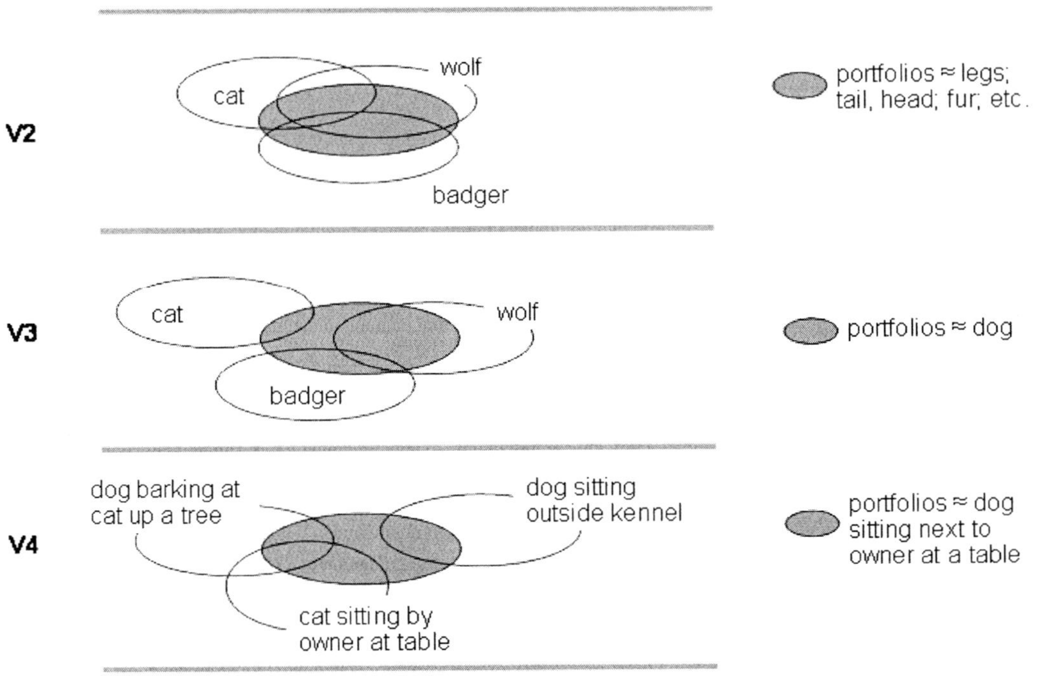

Figure 13.5. Examples of overlaps between portfolios directly activated in response to perception of a dog sitting next to its owner, and the portfolios which would be activated in response to direct perceptions of a number of other situations. The overlaps are a result of the partial behavioural ambiguity of portfolios, and there are overlaps with a very large number of possible direct perceptions. Because of these overlaps, mental images of another situation could be generated by indirect activation on the basis of simultaneous past activity. The situation which would actually be the mental image generated will be determined by the relative recommendation strengths of the different overlaps, and various other factors such as the identities of recently active portfolios for which there would be supplementary recommendation strengths in favour of their activation.

The effect of acceptance of this type of recommendation is therefore to generate a secondary population in which the largest subset of portfolios will correspond with a different animal from dog. Alternatively, it could correspond with a type of dog with a different physical appearance. If this secondary population then activates a tertiary population on the basis of frequent past simultaneous activity, the tertiary population will have substantial overlap with the population which would be activated in V3 in response to actual perception of the different type of animal or dog. In other words, a mental image of the different type of animal would be generated.

This relatively simple picture is made considerably more complex because of a number of factors. Firstly, an analogous process may operate in array V4. Secondly, the resultant population in V4 may contribute to the evolution of the populations in V3 on the basis of frequent past simultaneous activity. Thirdly, recommendation strengths in favour of indirect activations on the basis of recent activity or past or recent condition recording are present and may contribute to the derived populations. Fourthly, sensory inputs derived from other objects in the visual field may directly activate other portfolios.

There will be a tendency for populations to settle briefly into states in which most portfolios were active in the past on the same basis, but then to shift to other such states. The effect will be the experience of a constant succession of mental images in which the succession is determined by current visual inputs, recent visual inputs, and past and recent temporal correlations. Small differences in the initial portfolio population can result in large differences in the derived population, and the final population will be very sensitive to the past experience of the system.

Sequences of activation behaviours which often occur will tend to become instantiated in special purpose behavioural components. If the activation of such components is often followed by positive consequences, they will tend to by activated more often. Certain ways of manipulating image streams will therefore be learned, and may differ between different individuals. Episodic memory as discussed in chapter 10 is a special case of image stream manipulation and forms part of the general phenomenon of consciousness. Note that this mechanism for accessing episodic memories implies that it is a skill which can be improved by practice.

Because conditions could be recorded during image streams, an image stream could potentially be recalled later. A specific image stream could become self reinforcing if, for example, significant condition recording occurred and reactivation of the image stream on the basis of past temporally correlated condition recording was therefore possible.

In this model there is no qualitative difference in arrays V3 and higher between direct sensory experience and reminiscence and stream of consciousness, all are constructed from the activation of widely shared portfolios. However, direct experience activates repetitions close to sensory input in arrays V1 and V2 which are generally not activated in imagined images. Clusters indicating the absence of such repetitions could generate recommendations to behave accordingly, enabling a behavioral distinction between direct experience and reminiscent experience.

Images of Self

Consider now an example offered originally by Hebb [1969] and developed further by Jaynes [1976]. Suppose that you are asked to imagine when you last went swimming. The initial images are probably of a place: a beach or a pool. But the actual images of yourself swimming are of a person in the water, a perspective of self which is never experienced directly. However, it is also possible to imagine the feel of water on your face, swimming motions of arms and legs, the view of the shore or edge of the pool as they were experienced. These two perspectives, self viewed by an outside observer and viewed from the inside looking out are what Jaynes labels the "me" and "I" perspectives.

At a more detailed level, when asked to imagine when you last went swimming, words like "imagine when" result in indirect activations in V3 and V4 with general recommendation strengths in favour of activating portfolios on the basis of past simultaneous condition recording. The population indirectly activated by hearing the word "last" biases this recommendation strength in favour of the most recent condition recording.

The words "you" and "swimming" indirectly activate populations in V3 and V4 which have often been active in the past at the same time as the words. This population is of the type illustrated in figure 13.4, and will contain portfolios which are directly activated in response to both observing others swimming and personal experiences of swimming. Some of the portfolios will have recorded conditions during the most recent swimming experience, and will have recommendation strength in favour of activating other portfolios active during the same experience. This recommendation strength will be favoured by the recommendation strengths in favour of such behaviours. A secondary population will therefore be activated with a higher proportion of such portfolios, and perhaps a tertiary population with an even higher proportion.

This tertiary population will be an approximation to the population active during the swimming experience, and will be experienced by the brain as a partial re-experience of that experience. Portfolios in the population will have recommendation strength in favour of speaking about that experience.

If speaking about self experiences where self is viewed from the perspective of an outside observer is a behaviour which has been rewarded in the past, the activation of portfolios which have such recommendation strengths will be rewarded. The presence of such portfolios in self re-experience populations will therefore be encouraged.

Frequent occurrence of rewarded self re-experience populations will result in a competition component corresponding with the appropriate behaviour sequence. Such a behaviour sequence could acquire recommendation strength in favour of activating o particular population of portfolios corresponding with external and internal observation of self, where the external observation portfolios were originally defined by perceptions of others, but have been evolved by condition recording during self re-experiences.

Images of Non Existent Objects

Consider the portfolio populations which could be activated in response to hearing the phrase "pig with wings". The evolution of the portfolio populations is illustrated in figure 13.6.

The word "pigs" indirectly activates a population in V3 which is similar to the population which would be activated in response to seeing a pig. The word "wings" indirectly activates a population in V2 which is similar to the population which would be activated in response to seeing a wing feature. The population in V3 indirectly activates an additional population in V2 made up of portfolios which have often been active in the past at the same time. These V2 portfolios will tend to correlate with the visual features of pigs. The presence of the word "with" encourages the activation of "pig features" in V2 and the subsequent synchronization and release to V3 of outputs from the V2 populations corresponding with wings and pig features. The resultant population in V3 is similar to the population which would have been activated by a direct perception of a pig with wings.

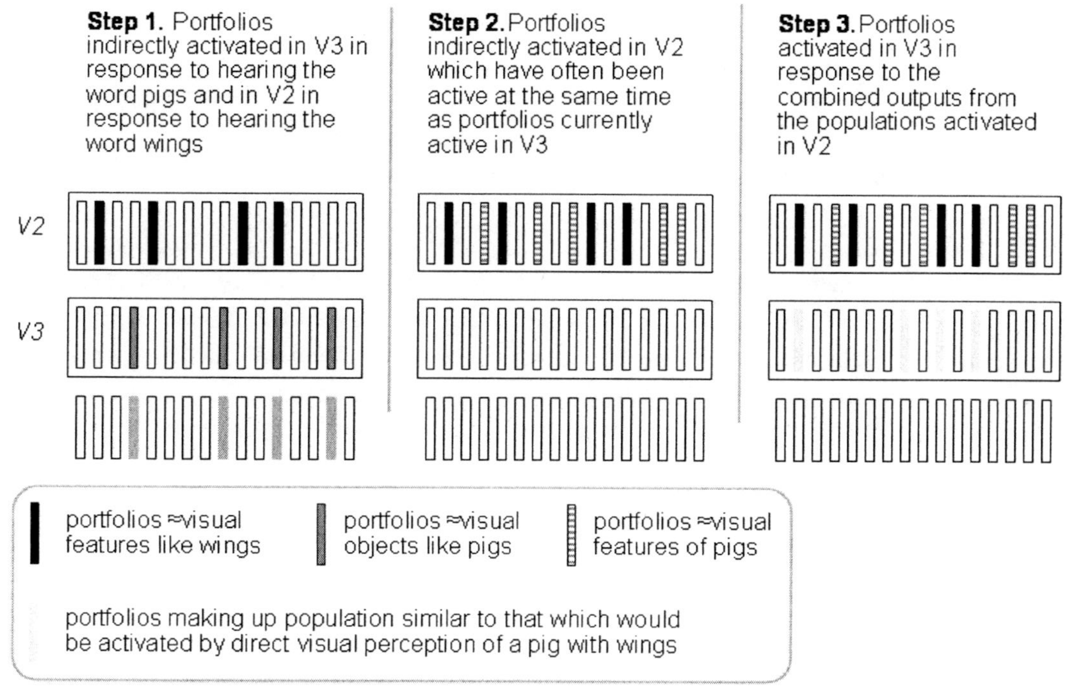

Figure 13.6. A process for generating a mental image of a pig with wings. The first step is that portfolios directly activated by the sound of the word "pig" indirectly activate portfolios in array V3 which would be directly activated by seeing a pig, and portfolios directly activated by the sound of the word "wing" indirectly activate portfolios in array V2 which would be directly activated by seeing a wing. The second step is that the active portfolios in array V3 corresponding with pigs indirectly activate portfolios in V2 which have often been active at the same time in the past, in other words corresponding with the features of pigs. The third step is that the activation in V3 is extinguished and the outputs of the pig feature and wing populations in V2 are synchronized and released to V3. The resultant portfolio population in V3 is the population which would be activated by direct perception of a pig with wings. Some condition recording will occur to establish this V3 population if a pig with wings have never been imagined before.

If this is the first exposure to the pig with wings concept, the population in V3 which can be activated without condition recording will be limited, and condition recording will be initiated until the population reaches the minimum level. Future exposures to the concept will tend to reactivate these portfolios.

Generation of Verbal Reports

To generate a verbal report, there must be portfolios activated with recommendation strength in favour of appropriate speech. The requirement is therefore that it must be possible to reach an activated population with the appropriate recommendation strengths from some population which can be activated by appropriate sensory inputs.

Consider first the phenomenon of inability to recall a (generally very familiar) recent experience such as driving from one point to another. Activation on the basis of temporally correlated condition recording is essential in order to evolve a population towards a specific past experience. If the experience is very familiar and very little condition recording occurs, the experience will not be accessible. However, if attention is paid during the experience, the combination of perceived objects into groups and groups of groups is forced, and some condition recording occurs at these higher levels which can be the basis for later recall.

The inability to describe the actual performance of an expert skill reflects the fact that although early learning by verbal instruction may well involve portfolios which also have recommendation strength in favour of speech generation, as the skill develops there may be special purpose portfolios defined which are optimized for the motor behaviours making up the skill and do not have speech recommendation strength. Speech generation recommendation strength could of course be acquired by such portfolios, but such acquisition would be a complex process of finding word combinations which could indirectly activate the relevant skill portfolios as they are directly activated by sensory inputs in the course of the skilled behaviour. These word combinations might not activate the appropriate portfolios in the mind of an inexpert learner because such portfolios do not exist.

The inability to describe the details of the process by which a skill such as riding a bicycle in learned partially results from the same specialized portfolio issue, but also from the fact that skills are largely instantiated in recommendation strengths. No record of past recommendation strengths is preserved, and even current recommendation strengths can only be accessed through the process of performing the behaviour, not indirectly on the basis of past temporally correlated usage.

Phenomenal Consciousness

Consider what happens when a person pays attention to a red object and focuses attention on the red. Portfolios are directly activated in V1, V2 and V3 and even V4 in response to the perceived object and other objects around it. All of these portfolios will have recommendation strengths in favour of different behaviours in response to the object or group of objects, including motor behaviours (such as pointing to the object) or speech behaviours (such as saying “that object is red”).

These portfolios will include many that have often been active in the past or recorded conditions in the past when different red objects have been present, or different groups of objects have been present which included a red object. There will therefore be recommendation strengths in favour of activating a range of other portfolios which will include many portfolios that would be directly activated in response to the direct perception of other red objects.

Suppose that the components corresponding with indirect activation strengths of these types are activated. The effect would be activation of a large population of portfolios, with the only common link being that they had been active or recorded conditions in the past when some red object had been present. Within such a population there would be many different groups of portfolios which would be activated when different other objects were directly perceived, but these groups will in general be small subsets of the full group which would be directly activated by such an object.

Because the subsets are small, there will not be enough recommendation strength in favour of behaviours appropriate to these objects for such a behaviour to be accepted. For example, it would not be possible to name such objects. The experience would therefore be of a rich experience associated with the colour red, but it would not be possible to describe the nature of the experience in detail.

Note that such a population will tend to be activated to some degree around any current population, even if that population was itself indirectly activated. Thus each mental image in the stream of images discussed earlier will be associated with an extended population of more loosely associated portfolios.

The experience of phenomenal conscious can thus be understood as the "halo" of information which can be activated around any direct sensory input or mental image. The existence of the "halo" is of course a significant factor influencing the evolution of the sequence of mental images. The detailed content of this "halo" will be very sensitive to the past experience of the brain, and particularly the recent experience which will have resulted in some of the strongest recommendation strengths. The information content of a "red" conscious experience will be unique to each individual, and each experience of one individual will be somewhat different.

What is the behavioural value of such an activation ? The effect is to activate a large population of portfolios with a possible connection to current sensory inputs. This population expands the range of recommendations available in response to the experience. Such an expansion may not be required for a simple response to a single object, but will b essential for developing an appropriate response when many factors including the past experience of many other objects must be taken into account. Thus, as suggested originally by Jaynes [1976], a critical role for consciousness is in developing appropriate behaviours in very complex social situations in which the experience on many different unique individuals must be taken into account. This point is further discussed in chapter 14.

References

Baars, B. (1988). . Cambridge University Press.

Bainbridge, L. (1977). Verbal reports as evidence of the process operator's knowledge. International , 11, 411-436.

Berry, D.C. (1987). The problem of implicit knowledge. , 4, 144-151.

Hebb, D. O. (1969). The Mind's Eye. Psychology Today, May 1969.

James, W. (1904). Does 'Consciousness' Exist? 1, 477 - 491.

Jaynes, J (1976), . Boston: Harvard.

Nagel, T. (1974). What is it like to be a bat? . 4, 435-450.

Speelman, C. (1998). Implicit Expertise: Do We Expect Too Much from Our Experts. In Kirsner, K., Speelman, C., Maybery, M., O'Brien-Malone, A., Andersen, M., and MacLeod, C. (Eds.), . New Jersey: Erlbaum.

Chapter XIV

Future Directions

There are many directions in which a deeper understanding can be developed by application of the recommendation architecture approach. In this chapter some of the possible future areas for such a deeper understanding are considered. These areas include the evolution of higher cognitive processes including consciousness, the development of self awareness, speech processing, the understanding of the structure and processes of cities and other complex learning systems quite different from the brain, and the design and implementation of machines with higher cognitive capabilities.

The Evolution of Higher Cognition

A critical issue for any theory addressing the evolution of the higher cognitive capabilities observed in human beings is the relative shortness of the time available for that evolution. Only 6 million years of evolution separates homo sapiens from the great apes, and for all but the last 2 million years, there is no evidence in the archeological record of anything but great ape cognitive skills. The first indisputable signs of distinctive cognitive skills appear only in the last 250,000 years with the emergence of modern homo sapiens [Tomasello 1999].

To put the same issue another way, there appears to be about 98% overlap in genetic material between homo sapiens and the great apes [Sagan and Druyan 1993], a similar overlap as for lions and tigers, horses and zebras, or rats and mice. The genetic changes involved in the emergence of higher cognition must therefore be limited.

Given the available timescale, any plausible explanation for the development of higher cognition can only include at most one qualitative change plus some changes which are merely quantitative modifications to existing mechanisms rather than introduction of a new mechanism. The qualitative change must also be limited enough that there is time for it to develop gradually from its initial state.

The key mechanisms which support higher cognition in the human brain include all the indirect portfolio activation capabilities listed in table 13.1. These mechanisms include activation on the basis of past temporally correlated activity or condition recording, management of information availability, and attention management. Given the evolutionary timescale, addition of one of these types of mechanism is implausible, and furthermore, given observed cognitive capabilities, all the mechanisms would appear to be present in the great apes and many mammal species much less closely related to homo sapiens. Hence the distinction between humans and the great apes can only depend on the degree to which these mechanisms occur.

Furthermore, at each point in the evolutionary process, incremental changes must have survival advantages in their own right which outweigh any costs at that point. In other words, if a change has a cost but only a future benefit, natural selection will tend to eliminate it.

The Critical Change

This section describes a model for the evolution of higher cognition in which there is one qualitative change and some limited quantitative changes. The qualitative change is the addition of a "voice box" making it possible to generate a wide range of different sounds which can be discriminated by a hearer. This voice box would of course be expected to have evolved from a relatively primitive initial mutation. It is clear that this evolution has occurred on the available timescale, given the absence of voice box capabilities in the great apes. The limited qualitative changes are gradual increases in the tendency to indirectly activate portfolios on the basis of simultaneous past activity. In this model, all of the mechanisms in table 14.1 would be present in the great apes, the only change in the emergence of homo sapiens being a gradual increase in the tendency for indirect activations to occur.

Note that there is a biological cost to such changes. Addition of a "voice box" requires physical resources and might at least in its early stages affect other capabilities such as swallowing food. An increase in the strength of indirect activation behaviours increases the risk that under some circumstances the response to a critical environmental situation will be delayed or inappropriate. If there are large populations of indirectly activated portfolios in addition to or instead of portfolios directly activated by, for example, some sounds from the environment, the ability to detect and respond to such sounds will be reduced. The value of the increased level of indirect activation behaviours must exceed the cost of this risk at each point in the evolutionary process.

A plausible evolutionary scenario must therefore indicate how a slight increase in the tendency towards indirect activation has immediate behavioural benefits (i.e. survival advantages) which outweigh the immediate costs. Then a further increase must have additional immediate survival advantages which again outweigh the further immediate costs and so on.

The Evolutionary Scenario

There are two areas in which behavioural enhancements could have provided survival advantages in favour of a gradually increasing tendency for indirect portfolio activation behaviours. Firstly, enhancements to tool making through imagining better ways in which tools could be used leading to better tools and more skillful use. Secondly, enhancements to signaling and communication between individuals which initially improve group hunting capabilities, then lead to increases in possible group sizes, and then to the capability to generate appropriate social behaviours within large groups made up of unique individuals. The proposal is that it was the value of enhanced signaling (once mutation had resulted in an initial voice box) which provided the primary selection pressure, reinforced at certain points by the value of better tool making. These two sources provide selection pressure to support a steady increase in the tendency for indirect portfolio activation behaviours. This increased tendency for indirect portfolio activation led to human cognition.

Tool Making

For tool making to be possible, a population of portfolios must be activated with recommendation strengths in favour of the sequence of motor behaviours required to make the tool. The problems are firstly how an appropriate population of such portfolios could be activated from a plausible starting point (i.e. one which could be reached starting from sensory experience), and secondly how portfolios could acquire such recommendation strengths.

There are a number of different degrees of sophistication to toolmaking. A simple level is the use without modification of an object which happens to be available in the circumstances that require a tool. A second level is acquiring such an object for later use. A third level is modification of an object to make a better tool, in the circumstances that require the tool. A fourth level is modification of an object to make a better tool in advance of the circumstances that require it.

All of these levels have been observed in, for example, chimpanzees. Some chimpanzees have the ability to use a grass stalk or reed to lure termites out of a termite mound. The reed may be broken off an hour or more in advance of reaching the mound and hundreds of metres away. Leaves and twigs are removed and the reed is shaped and shortened for the task. The tool must be used with skill, inserting it to follow the termite tunnels and removing it in a way that does not shake off too many termites [Sagan and Druyan 1993].

Imitation is frequently the way in which the discovery of how to make and use a tool is transmitted between the original discoverer and other animals. Imitation is a very widespread capability in mammals and birds, and processes analogous with those described in chapter 9 would be applicable. However, there are a number of different degrees of sophistication to making such discoveries and improving on existing tools. The first level is physical experimentation in the actual situation requiring the tool. Behaviours available for relatively unrelated situations are utilized in a different order and/or applied to different objects to make and use a tool. A second level is mental experimentation, again in the actual situation requiring the tool. Behaviours and objects are manipulated mentally to find a tool and behaviour to solve a problem. A third level is mental experimentation in the absence of the situation requiring the tool.

An example of the first level is the discovery by a crow that a wire can be bent and used to extract a food object from a container within which it is otherwise inaccessible [Weir,

Chappell and Kacelnik 2002]. In this example, the birds in the wild create hooked probes from twigs, but the wire bending is a different material applied to a different problem.

An apparent example of the second level is the chimpanzee Sultan as observed by Wolfgand Kohler [Sagan and Druyan 1993]. The chimpanzee was placed in a room with a banana suspended out of reach in one corner. In the middle of the room was a wooden box with its open side up. Sultan first tried to reach the banana by jumping but this was not possible. He paced for a while, then suddenly stood still in front of the box, seized it, tipped it, positioned it under the banana, climbed on it and sprang upwards with all his force to successfully reach the banana. Later Sultan was put in a room with a higher ceiling and a suspended banana, the wooden box and a stick. After failing to reach the banana with the stick alone, Sultan sat, gazed around and scratched his head for a while. He then suddenly stared at the box, leaped up, seized the box and the stick, pushed the box under the banana and knocked the fruit down with the stick after first climbing on the box. Kohler emphasized that there was an apparently "thoughtful" period preceding sudden directed action.

There appear to be no examples of tool invention at the third level other than in human beings.

Many animals can use tools to achieve their purposes. The distinctiveness of human tool making is thus more quantitative than qualitative. Tools can be made by humans a long time in advance of the need by a long manufacturing process, in both cases much longer than observed in chimpanzees. Chimpanzees even appear capable of mental manipulations to design tools, but only in the circumstances in which the tool is required.

Mechanisms for Tool Making and Designing in Non-humans

As pointed out earlier, making a tool requires activation of a population of portfolios with recommendation strengths in favour of the necessary motor movements. If the tool making has been learned by imitation as described in chapter 9, the populations with the appropriate strengths are those directly activated by the visual and propioceptic circumstances in which the tool making was learned. Thus in the chimpanzee reed tool example, the portfolios activated in response to seeing reeds have recommendation strengths in favour of breaking off a reed. The portfolios activated in response to a broken off reed being held have recommendation strengths in favour of trimming behaviours and so on. Portfolios activated in response to hunger and in response to visual and other cues indicating the termite eating season may also contribute supplementary recommendation strengths generally required to reach a total strength at which the behaviour sequence will be initiated. Thus indirect portfolio activation capabilities are not essential to tool making of this type.

Design of a tool as demonstrated by the crow example occurs when the behaviours required for tool making are available for other purposes. Some portfolios in the bird brain will have recommendation strengths in favour of activating these behaviours. These portfolios will be those activated in response to the object on which the behaviour was performed successfully in the past. If a different object in the local environment activates a population of portfolios containing some with recommendation strength in favour of such a behaviour, the behaviour may be attempted on the different object. Thus if the perception of the wire activates a portfolio population which overlaps with that activated by a twig, twig related behaviours may be activated. Success will result in the portfolios of the somewhat different population acquiring greater recommendation weights in favour of the behaviours. Random experimentation could then lead to a useful behaviour sequence. Again, indirect portfolio activation capabilities are not essential to tool design of this type.

The tool design capabilities exhibited by the chimpanzee Sultan appear to require a degree of indirect portfolio activation capabilities. Portfolios are activated in response to seeing the group of objects made up of the suspended banana, the floor and the ceiling. Portfolios are also activated in response to the box and the stick. By activation on the basis of temporally correlated past activity, a pseudosensory activation is eventually created similar to the one which would have been activated by pushing the box under the banana, climbing on it, and using the stick.

In the case of the chimpanzee, the immediate presence of all the relevant objects means that the degree of indirect activation required is limited. The ability to learn to create a tool by a long process carried out in the absence of the circumstances in which the tool will be needed would require much more extensive indirect activation capabilities. A tool making process which has already been discovered by a process which took place in the presence of the problem could later be applied in the absence of the problem for later use of the tool without indirect activation capabilities. Hence chimpanzee tool making capabilities can be accounted for by a definite but limited indirect portfolio activation tendency. The apparent absence of the ability to design tools in the absence of the specific problem situation suggests that such improvements to tool making will alone not generate sufficient advantage to compensate for the disadvantages of increasing the degree of indirect portfolio activation.

Signaling

Many animals have the ability to signal other animals. Warning cries are a typical example. In the recommendation architecture model, the physiological basis for such signaling behaviour is that visual perception of a behaviourally relevant object (such as a threatening predator) by some individual animal activates portfolios in the brain of the animal with recommendation strengths in favour of both evasive behaviour and signal generation (a sound or some visual signal). The perception of the signal by other animals activates portfolios in their brains with recommendation strengths in favour of the same evasive behaviour.

Consider how past experience influences the response of the other individuals perceiving the signal. If the other individuals have often heard the same signal in the past at the same time as direct perception of a predator, and the perception was followed by evasive behaviour which was then followed by perception of a reward, the signal was also followed by evasive behaviour and reward. Portfolios activated directly in response to the perception of the signal can therefore acquire recommendation strength is favour of evasive behaviour.

In addition, the portfolios directly activated in response to perception of the signal were active in the past at the same time as the portfolios directly activated in response to perception of the threat. The portfolios directly activated by the threat could therefore indirectly activate the portfolios which would be activated by direct perception of the threat.

This indirect activation will bring into play the recommendation strengths of the direct threat perception portfolios.

The disadvantages of the indirect activation mechanism are that it requires somewhat longer response times, and may activate portfolios which are irrelevant to current environmental circumstances. Its advantage is that two or more different signals have a much greater ability to interact and generate a composite behaviour, resulting in support for much more complex behaviours. For example, a sequence of signals can activate portfolios which would be directly activated by a group of objects, and hence recommendation strengths appropriate to the group rather than just the individual objects.

Such composite behaviours might be useful, for example, in coordinating the hunting behaviour of a group. A mutation resulting in the capability to produce a wider range of signals could therefore have an initial survival benefit of making it possible to trigger a wider range of behaviours, and a gradual increase in an existing tendency towards indirect activations would increase that survival benefit.

As emphasized by Jaynes [1976] in the context of the development of consciousness, communication is an important limit on the possible size of a group. For example, above about twenty individuals, it is hard for a leader to dominate his followers if dependent only on daily face-to-face encounters without speech. The ability to issue signals which can trigger behaviours (i.e. commands), and perhaps to pass such commands through intermediaries, greatly expands the possible group size. As the tendency towards indirect portfolio activation increases, it would be possible to increase group sizes.

The Evolutionary Sequence

The first step in the proposed sequence is a mutation which increases the number of different sounds which the hominid can generate. This greater number expands the range of signals which can be provided to other individuals, with advantages in cooperative behaviour like hunting. This advantage favours later mutations to further increase the number of sounds.

The second step is mutation which slightly increases the tendency towards indirect portfolio activation behaviours. The effect of this mutation is that the portfolios directly activated by different sounds have a greater tendency to activate visual portfolios often active at the same time in the past. As a result, portfolios corresponding with groups of objects can be activated as a result of signals corresponding with members of the group. Behaviours learned in response to the presence of groups can then be initiated by groups of signals. These groups of signals are effectively commands for appropriate behaviours The coordination within groups is further enhanced by this ability.

Further increases in the tendency towards indirect activations results in activation of portfolio populations corresponding with more and more complex groups and groups of groups of objects. Condition recording will occur during such indirect activations, since they are not internally distinguishable from direct activations. The resultant changes to portfolios are therefore available to guide subsequent behaviours. As a result, sequences of mental images can be used to experiment with possible future behaviours, and the behaviour with the best imagined result applied in practice. The planning of behaviours in extremely complex social situations therefore becomes possible.

This scenario is the same as that described by Jaynes [1976] at the psychological level. The recommendation architecture approach thus provides a more detailed physiological model for that scenario.

In the course of this evolutionary process, the ability to effectively manage the direction in which populations develop by indirect activation must be acquired. A population activated in response to an object or group of objects will have recommendation strength in favour of speaking the signal corresponding with that object or group of objects. Speaking the signal will have the effect of reinforcing the corresponding population. At an early stage this reinforcement may be important in activating an adequate population for the next stage of processing. For example, a weak population corresponding with a group of objects may be created, and the behaviour of speaking the signals corresponding with the group would have the effect of strengthening the population. This strengthened population could then indirectly activate other populations. Speaking to oneself would then be important in guiding population development.

At some point tool making would begin to benefit from increased indirect activation capabilities. For example, speaking the name of the tool to oneself would activate a population with recommendation strengths in favour of tool making behaviours, and tool making could be sustained for long periods in the absence of the situation in which the tool would be required. At a later stage of indirect activation capability, the mental image of a tool could be used to create mental images of different parts of the tool which have no individual names, enabling more complex tool making. At a yet later stage, imagining the use of new or different tools for some problem would be supported, with such imagining proceeding for long periods in the absence of the problem situation. More sophisticated tools could then be designed.

The cost of each increase in the tendency towards indirect portfolio activation is that the recommendation strengths of portfolios activated directly by current sensory inputs may be outweighed by those of indirectly activated portfolios in circumstances in which an immediate response to current sensory inputs has high survival value. Human beings are much less sensitive to environmental cues than many other mammals, a typical example is that dogs detect sounds which cannot be detected by their owners. Such differences may be the result of the increased volume of indirect portfolio activation in the human brain, perhaps coupled with a loss of sensory sensitivity as the indirect activation tendency has increased.

The overall process is thus a gradual increase in the tendency towards indirect portfolio activation, each increase providing survival advantages to outweigh the costs of such activations.

The Development of Self Awareness

The recommendation architecture makes it possible to develop an understanding of how self awareness develops in human children in physiological terms. As an example, the process by which a two-year old could achieve the first-person perspective needed to make the statement: "Billy help Mommy. Billy good boy." Here, the child identifies himself as an actor in a way that carries both evaluative and probably affective aspects. The problem is to describe at the physiological level both the cognitive processes which could support such statements and the learning processes by which such cognitive processes could be learned. I am indebted to Jack Martin at Simon Fraser University for suggesting this example.

In the recommendation architecture model, the process for generating any statement is activation of portfolio populations in the cortex with recommendation strength in favour of the words.

As outlined in chapter 9 and discussed in somewhat more detail later in this chapter, the capability to generate appropriate word behaviours is acquired if three things happen at the same time. Firstly, a word heard and imitated. Secondly, visual portfolios are activated by the presence of the object corresponding with the word. Thirdly, speaking the word is rewarded. In these circumstances the portfolios directly activated in response to the visual perception of the object acquire recommendation strengths in favour of saying the word. In addition the portfolios directly activated in response to hearing the word acquire recommendation strengths in favour of indirect activation of the portfolios directly activated in response to the visual perception of the object.

Consider first the word "Mummy". Large numbers of perceptions of a similar object (i.e. mother) in various circumstances are seen. These perceptions result in creation of a population of portfolios which are sometimes activated in response to perceptions of mother. Many of the portfolios are of course also activated in response to other perceptions as well. Within the population a core subset is activated very frequently whenever attention is directed to mother. The whole of the subset is not activated for all perceptions of mother, but portfolios in the subset are on average activated fairly frequently for mother perceptions. These portfolios will acquire recommendation strengths in favour of many behaviours appropriate to the presence of mother.

The word "mummy" is used by adults when Billy's attention is directed towards mother. Portfolios directly activated be hearing the word therefore acquire recommendation strength to activate some of the visual "mother" population, especially members of the core subset. The "mother" population portfolios therefore also acquire recommendation strength in favour of saying "Mommy".

Consider next the word "Billy". Large numbers of perceptions of a similar object (i.e. self) in various circumstances are seen when Billy's attention is directed towards himself. Some of the activated portfolios may also be activated when others are observed. For example, portfolios activated when Billy looks at his own hands may overlap with those activated when he looks at other hands; portfolios activated when looks at himself in a mirror may overlap with those activated when he looks at others either directly or in a mirror). These perceptions result in creation of a population of portfolios within which a significant core subset tend to be activated very frequently when attention is directed at self. Some of these conditions may include information derived from perception of feelings (sad, happy, angry, pain etc.).

The word "Billy" is then used when Billy's attention is directed towards himself, and word usage acquired. Note that the use by adults of phrases like "Billy sad", "Billy happy" etc. will mean that the word qualifiers will preferentially activate portfolios which contain information which is often present when Billy is sad etc. Hence the word may create a subset of the experience of the feeling. Some of the conditions containing feeling information may be originally recorded when observing others and being told they are sad etc.

Consider next the word "good". Portfolios within the auditory perception of the word similarly become associated with other portfolios active when the word "good" was spoken. Those other portfolios include portfolios directly activated in response to self observation, action observation etc.. Furthermore, these other portfolios may have been active when past behaviours were rewarded, and may even acquire recommendation strength on to generation of positive consequences. Hence portfolios within the word may acquire (direct or indirect) recommendation strength on to the generation of positive consequences.

Consider next the word "boy". Portfolios which occur in the perception of other children become attached to the sound of the word. These conditions may include portfolios often present in self observations. The speech behaviour is learned that if there is significant overlap in the population activated by self and by boy, that overlap is a condition which recommends saying "I am boy". Any other overlaps would have the same effect.

Consider next the word "help". As a preliminary step, consider first the word "do". Billy has many perceptions of people acting on objects. Portfolios are shared across different people perceptions and across different object populations. Portfolios incorporating information from perception of people and objects and a doing relationship acquire an association with the word "do", especially the core population members. So the word "do" activates significant subsets of this core population. This process is analogous with the process by which the word "chase" activated portfolios as described in chapter 9. The portfolios indirectly activated in response to hearing the word "help" are portfolios directly activated in response to perceptions of situations in which one person helped another, including when Billy helped someone else and was rewarded for the help.

Note also that speech has in some circumstances been a strongly rewarded behaviour in the past. Certain circumstances will therefore directly activate portfolios which have general recommendation strengths in favour of speech behaviours. Such recommendation strengths would tend to activate all the components in competition corresponding with speech behaviours.

Given all this learning, the scenario for generating the speech is firstly that Billy is doing something with his mother. Portfolios are activated directly by his perception of his mother, himself, and the activity. Some of these portfolios have acquired general recommendation strengths in favour of speech behaviours.

The portfolios activated in response to perception of his mother have recommendation strengths in favour of saying "mother". The portfolios activated in response to perception of himself have recommendation strengths in favour of saying "Billy". The portfolios activated in response to perception of the group of objects made up of his mother, himself and the objects with which the activity is occurring have recommendation strengths in favour of saying words like "do" or "help" or "good". The combination of the different portfolios thus recommends the type of sentence spoken as discussed in somewhat more detail in the next section.

Note that on this model self awareness is a learned behaviour. In particular, given the capability for substantial indirect portfolio activation, such behaviour will be learned if appropriate teaching is provided, but not necessarily otherwise. The model is therefore consistent with Jaynes theory [Jaynes 1976] that self awareness only developed over a period of a few hundred years around 2500 years ago, since no physiological evolution would be required over the few hundred year period.

Speech Processing

In the book up to this point, there has been discussion (in chapter 9) of how auditory inputs corresponding with nouns or adjectives can generate pseudovisual activations, and some discussion of how such activations can be generated in response to verbs. The objective of this section is to summarize the earlier discussion, and outline the directions in which a more complete theory of speech processing in terms of portfolio activation management could be achieved.

In the case of nouns, hearing the word at the same time as the corresponding object is seen results in auditory portfolios directly activated by the sounds acquiring recommendation strengths in favour of activation of the visual portfolios often directly activated when the visual object was present. These portfolios will tend to be at the ~objects level of complexity (i.e. level V3 in figure 14.1). Imitating the sound when the object is present results in the visual portfolios acquiring recommendation strengths in favour of speaking the word.

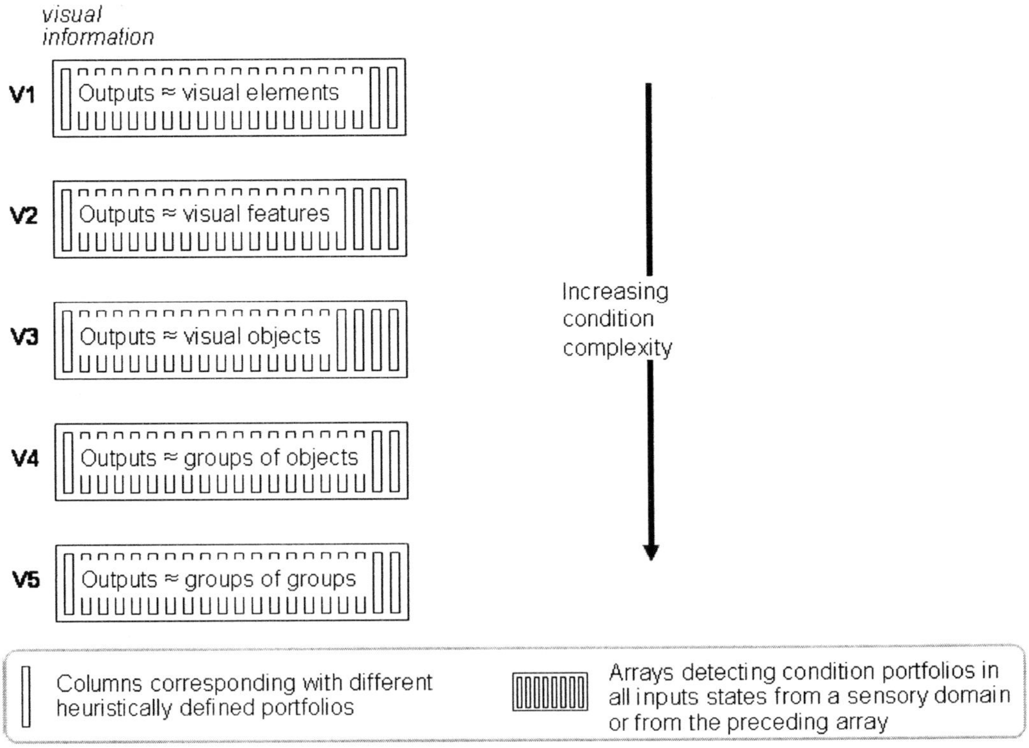

Figure 14.1 Arrays of portfolios detecting conditions at different levels of complexity in visual sensory inputs.

In the case of adjectives, the presence of a similar environmental condition such as red on a number of occasions when the word "red" is heard results in the portfolios directly activated in response to the word acquiring recommendation strength in favour of activating the portfolios most often active when the colour red is present. Such portfolios will tend to be at the at the ~features level of complexity (i.e. level V2 in figure 14.1). Again, the visual portfolios also acquire recommendation strength in favour of speaking the word.

In the case of verbs like "chase", the presence of the word at the same time as a number of different visual chase circumstances (dog chasing cat, cat chasing mouse, policeman chasing suspect etc.) results in portfolios directly activated in response to the sounds acquiring recommendation strengths in favour of activating the visual portfolios at the ~group-of-objects level of complexity which were often active during the different chase scenes (i.e. level V4 in figure 14.1). The direct activation of that group of portfolios and no others could only occur in response to the visual experience of some vague, amorphous object chasing some other amorphous object.

As pointed out in chapter 9 and elsewhere, the organization of portfolios into layers of different complexity as illustrated in figure 14.1 is simplified for the purposes of explanation. A somewhat more realistic view is illustrated in figure 14.2. There are rather more different levels of portfolio condition complexity in this figure. A number of different levels may correspond with, for example, the approximate object level of complexity. The portfolios corresponding with different objects may be distributed differently across the levels. Furthermore, the levels corresponding with the approximate feature level may overlap with the levels corresponding with the approximate object level and so on.

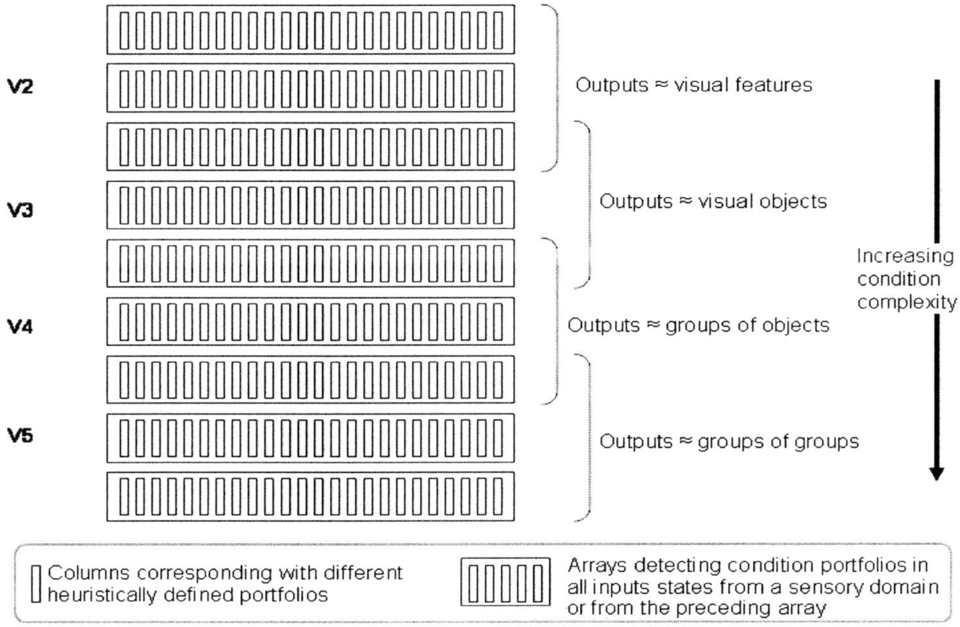

Figure 14.2. More realistic version of clustering arrays handling visual inputs. Several sequential arrays all contribute to discrimination between cognitive categories on different levels of complexity such as features, objects and groups of objects. Furthermore, the arrays providing discrimination between categories on one level may overlap with those providing discrimination on another level.

Pronouns will be handled in a similar fashion to nouns, but where the name of a person will indirectly activate portfolios often present when that specific person was present, the second and third person pronouns will activate only portfolios present when a range of people were present. The first person pronouns will also tend to indirectly activate portfolios of conditions containing information derived from propioceptic senses and internal neurohormone concentrations as discussed in chapter 13.

The articles "the" and "a" will activate portfolios with recommendation strengths in favour of different subsets of portfolios often active when the associated noun word was present in the past.

Components will be defined in competition to reflect frequently used and rewarded sequences of activation behaviours, and these components will instantiate grammatical rules.

For example, the English order of object, verb then subject will be instantiated in a component. This component will result in the first word activating a population in levels V3 corresponding with a subject. Next the middle word will activate a population of portfolios in arrays V4 which correspond with something performing an action. Outputs from the population in V3 combined with the activity in V4 will result in an activation which would have occurred if the subject had been observed performing the action. The third word will generate an activity in V3 which will be combined with the activity in V4 to generate the type of activity which would be observed if the subject had been observed performing the action on the object.

The behaviours in the sequence managed by the component will be indirect activations on the basis of past simultaneous activity which have been rewarded in the past and releases of information between arrays at appropriate point in the sequence. As described in chapter 9, the component will generate bias in favour of the next behaviour in the sequence, but the initiation of that behaviour will also depend on the recommendation strengths of the portfolios active at that point in the sequence.

Other components will instantiate the processes for handling other types of word sequences. Important components will be those managing the handling of phrases and sentences as described in chapter 9. These components will often activate more detailed components such as subject-verb-object, sometimes putting together the activations derived from multiple words as subject, verb, and object.

On this model, grammar is the solution to the problem of managing and integrating the generation of sequences of indirect activations. There could be many different solutions to this problem within the constraints that the mechanisms available are limited to those in table 14.1.

Design of Intelligent Machines

As described in chapters 5 and 6, electronic systems with the recommendation architecture have been designed and build. There is no issue in principle with the construction of such systems with much greater capabilities.

However, the critical element of a recommendation architecture system is the heuristic organization of a rich and varied environmental experience into a hierarchy of information repetition. Intelligence lies in organizing the experience of a complex environment in such a way that the organization can be used to guide behaviour. The construction of an intelligent machine therefore requires that the machine has access to such an environment, and the resources and experience to organize it effectively.

However, the implementation of a system to perform a complex combination of features which are well understood will always be much more resource efficient if it is a von Neumann machine designed under external intellectual control. The reason is that a system with the recommendation architecture must record and subsequently detect very large numbers of information conditions, many of which will have relatively little behavioural value. To put the same issue a different way, in the recommendation architecture a high integrity behaviour for a particular situation will need to be supported by the detection of a large number of conditions which together will have adequate recommendation weight in favour of that behaviour. A von Neumann system in which each condition detection is a command will need to detect far fewer conditions.

Hence a recommendation architecture system will only be cost effective if the problem to be solved is very complex and not accessible to intellectual understanding. If the problem is less complex and intellectual understanding limited, the full constraints of the recommendation architecture will not apply, and implementations based on perceptron algorithms will be practical.

The most interesting application of electronic recommendation architecture systems would therefore be in the exploration of the possibilities and alternatives available in the general area of cognition, consciousness and self awareness.

Some Other Considerations

There are many areas in which the recommendation architecture perspective opens up an alternative way of understanding biological systems. As an example, human cities are systems which must perform a complex combination of tasks concerned with the coordination of the activities of many different human beings. These tasks include education, production, distribution of resources, research, setting priorities on the use of resources, providing shelter, maintaining order, defense etc. At a detailed level there is considerable interaction between these different tasks. In general resources are limited and the way in which the tasks are performed must be changed heuristically. The recommendation architecture therefore provides an approach to understanding the structure of cities [Coward and Salingaros 2004]

Another area is understanding the capability differences between human beings and designing optimized educational approaches. For example, one human might have a greater than average (genetically defined) capability to define new portfolios at various levels of complexity. Another might draw provisional condition defining inputs from a greater than average (genetically defined) range of sources, in other words to make such connections with a somewhat lower criterion for frequent past simultaneous activity. The first human might have greater than average capability to create a novel organization of some domain, and the second might have a greater than average capability to establish the interactions within a very large domain. In a general sense, this corresponds with the difference between skill in mathematics or physics and skill in a subject like history.

A further area follows understanding of the way in which arrays of portfolios are defined and the ways in which sequences of attention and internal activation behaviours must be defined to support cognitive processing. Such understanding has implications for improving education.

Conclusions

The recommendation architecture is a set of constraints which exist for any system which must learn a complex combination of features with limited resources. The theory that the architecture of the human brain has been forced within these constraints by natural selection pressures leads to understanding of why the brain has certain physiological structures, and an understanding of how these structures support higher cognitive processes up to and including consciousness and self awareness. A number of experimental prediction can be made on the basis that the human brain has this architecture.

Implemented electronic systems have demonstrated the types of processes required to support cognition. More complex systems provided with a rich environment for learning can be expected to demonstrate sophisticated cognitive capabilities, not excluding consciousness.

References

Coward, L. A. and Salingaros, N. A. (2004). An Information Architecture Approach to Understanding Cities. 30(2), 107 - 118.

Jaynes, J (1976), . Boston: Harvard University Press.

Sagan, C. and Druyan, A. (1993). . Ballantine Books.

Tomasello, M. (1999). . Boston: Harvard University Press.

Weir, A. A. S., Chappell, J. and Kacelnik, A. (2002). Shaping of Hooks in New Caledonian Crows. 297, 981.

Index

A

D

G

H

I

J

K

L

M

Q

R

S